5G Mobile and Wireless Communications Technology

Written by leading experts in 5G research, this book is a comprehensive overview of the current state of the 5G landscape. Covering everything from the most likely use cases, to a wide range of technology options and potential 5G system architectures, to spectrum issues, it is an essential reference for academics and professionals involved in wireless and mobile communications.

- Describes and explains key technology options, including 5G air interfaces, device-to-device communication, mm-wave communications, massive MIMO, coordinated multi-point, wireless network coding, interference and mobility management, and spectrum issues.
- Summarizes the findings of key global 5G research collaborations such as METIS and outlines key scenarios, network requirements, and system architectures.
- Demystifies the relation between IoT, machine-type communications, and cyber physical systems, and describes the impact of 5G on sectors such as automotive, building, and energy.
- Equips readers with a solid insight into the impact and opportunities of 5G.

Afif Osseiran is Director of Radio Communications at the Ericsson Chief Technology Officer (CTO) department. He previously managed the EU 5G flagship project, METIS, and was Technical Manager of the Eureka Celtic project WINNER+. He has co-edited two books on IMT-Advanced (aka 4G) and is a senior member of the IEEE.

Jose F. Monserrat is an associate professor in the Communications Department of the Universitat Politècnica de València. He is senior member of IEEE and has been involved in several European projects, including NEWCOM, PROSIMOS, WINNER+, METIS and METIS-II.

Patrick Marsch is a manager at Nokia Bell Labs, where he leads a wireless system research department and is the Technical Manager of the 5G-PPP project METIS-II. He was the Technical Project Coordinator of the project EASY-C, where the world's largest research test beds for LTE-Advanced were established. He is co-editor of *Coordinated Multi-Point in Mobile Communications* (Cambridge, 2011).

5G Mobile and Wireless Communications Technology

EDITED BY

AFIF OSSEIRAN
Ericsson

JOSE F. MONSERRAT
Universitat Politècnica de València

PATRICK MARSCH
Nokia

CAMBRIDGE
UNIVERSITY PRESS

University Printing House, Cambridge CB2 8BS, United Kingdom

Cambridge University Press is part of the University of Cambridge.

It furthers the University's mission by disseminating knowledge in the pursuit of
education, learning and research at the highest international levels of excellence.

www.cambridge.org
Information on this title: www.cambridge.org/9781107130098

© Cambridge University Press 2016

First published 2016

Printed in the United Kingdom by TJ International Ltd. Padstow Cornwall

A catalogue record for this publication is available from the British Library

Library of Congress Cataloguing in Publication data
Osseiran, Afif, editor.
5G mobile and wireless communications technology / [edited by] Afif Osseiran, Ericsson,
Jose F. Monserrat, Polytechnic University of Valencia, Patrick Marsch, Nokia Networks.
New York : Cambridge University Press, 2016.
LCCN 2015045732 | ISBN 9781107130098 (hardback)
LCSH: Global system for mobile communications. | Mobile communication systems – Standards.
LCC TK5103.483 .A15 2016 | DDC 621.3845/6–dc23
LC record available at http://lccn.loc.gov/2015045732

ISBN 978-1-107-13009-8 Hardback

To my new born son S., my twin sons H. & N., my wife L. S-Y for her unwavering encouragement, and in the memory of a great lady, my aunt K. E.

A. Osseiran

To my son, the proud fifth generation of the name Jose Monserrat. And with the warmest love to my daughter and wife, for being always there.

J. F. Monserrat

To my two small sons for their continuous energetic entertainment, and my dear wife for her amazing patience and support.

P. Marsch

Contents

Contributors

Danish Aziz, Alcatel-Lucent (now Nokia)

Kumar Balachandran, Ericsson

Robert Baldemair, Ericsson

Paolo Baracca, Alcatel-Lucent (now Nokia)

Slimane Ben Slimane, KTH – Royal Institute of Technology

Mats Bengtsson, KTH – Royal Institute of Technology

Carsten Bockelmann, University of Bremen

Mauro Renato Boldi, Telecom Italia

Ömer Bulakci, Huawei

Luis Miguel Campoy, Telefonica

Icaro Leonardo da Silva, Ericsson

Jose Mairton B. da Silva Jr., KTH – Royal Institute of Technology

Elisabeth De Carvalho, Aalborg University

Heinz Droste, Deutsche Telekom

Mikael Fallgren, Ericsson

Roberto Fantini, Telecom Italia

Peter Fertl, BMW

Gabor Fodor, Ericsson

David Gozalvez-Serrano, BMW

Katsuyuki Haneda, Aalto University

Jesper Hemming Sorensen, Aalborg University

Andreas Höglund, Ericsson

Dennis Hui, Ericsson

Tommi Jämsä, was with Anite Telecoms, now Huawei

Andreas Klein, University of Kaiserslautern

Konstantinos Koufos, Aalto University

Katsutoshi Kusume, NTT DOCOMO

Pekka Kyösti, Anite Telecoms

Eeva Lähetkangas, Nokia

Florian Lenkeit, University of Bremen

Zexian Li, Nokia

Ji Lianghai, University of Kaiserslautern

David Martin-Sacristan, Universitat Politècnica de València

Patrick Marsch, Nokia
Michał Maternia, Nokia
Jonas Medbo, Ericsson
Sanchez Moya Jose F. Monserrat, Universitat Politècnica de València
Afif Osseiran, Ericsson
Olav Queseth, Ericsson
Petar Popovski, Aalborg University
Nandana Rajatheva, University of Oulu
Leszek Raschkowski, Fraunhofer Heinrich Hertz Institute
Peter Rost, Nokia
Joachim Sachs, Ericsson
Fernando Sanchez Moya, Nokia
Malte Schellmann, Huawei
Hans Schotten, University of Kaiserslautern
Erik G. Ström, Chalmers University of Technology
Lars Sundström, Ericsson
Ki Won Sung, KTH – Royal Institute of Technology
Satoshi Suyama, NTT DOCOMO
Tommy Svensson, Chalmers University of Technology
Emmanuel Ternon, NTT DOCOMO
Lars Thiele, Fraunhofer Heinrich Hertz Institute
Olav Tirkkonen, Aalto University
Antti Tölli, University of Oulu
Hugo Tullberg, Ericsson
Mikko Uusitalo, Nokia
Petra Weitkemper, NTT DOCOMO
Wolfgang Zirwas, Nokia

Foreword

The ICT industry has settled into the fourth round of the game, where everyone is guaranteed to win; the successes of 2G and 3G in the past and the promise of 4G in the current decade are leading to consensus on the new fifth generation (5G) of mobile systems. These successes started off as a movement of telephony to the mobile environment, and have, by 2015, already brought the Internet into the end user's hand. This new generation of mobile systems feels different. The global scale of enthusiasm and motivation is unprecedented. Even marketing has not been shy in proclaiming the advent of 5G on the roadmap, quite in contrast to the resistance in applying the name "4G" to LTE until Release 10 of the 3GPP standards.

We are still painting the empty canvas of that system which will appear as a small icon one day on our smartphones (or equivalent) as "5G". Can history help us predict what this system will all be about? Indeed, 2G was about global voice; 3G was about voice and data; 4G was about voice, data and applications. What about 5G?

We have witnessed mobile systems becoming an essential social infrastructure, mobilizing our daily life and facilitating digital economy. This trend will expand for 5G, boosting user experience and empowering industries with ICT, and the Internet of Things (IoT) will emerge as a new paradigm.

Credible details on the technology roadmap have started to emerge, which are largely articulated in this excellent book. 5G – so it seems – will require scale mainly in three dimensions.

First, rather traditionally, we need a massive scale in rate beyond the 4G capabilities of LTE Release 10. Spectrum is scarce in traditional cellular bands below 6 GHz, and improvement of spectrum efficiency is increasingly challenging. The only ways out seem to be through fresh approaches in system design, such as massive MIMO, mm-wave communications, relaying, network coding, advanced techniques in interference and mobility management, among others. Early prototypes and studies indicate that much of that is indeed feasible!

The world is starting to consume media such as video programming in more interactive ways, and the prospect for more immersive experiences in the form of Virtual Reality (VR) and Augmented Reality (AR) shows great challenge and promise. This places incredible requirements on mobile systems; large amounts of data have to be delivered to the user on demand, and end users can become the producers of copious amounts of information. These requirements do not merely affect the capacity of air interfaces and will cause re-architecture of transport networks and cloud systems to form a more distributed topology that extends to the converged mobile core, with storage and computing being spread all the way to the wireless edge.

Second, quite unsurprisingly, we need massive scale in the number of devices within the IoT that we want to connect. 5G will play an instrumental role in ensuring universal connectivity for myriad devices of very different characteristics. Indeed, prior system designs have not delivered the required IoT capabilities – an opportunity which 5G may want to capitalize on.

Third, rather excitingly, mobile technologies must attend to criticality, articulated in terms of much quicker round-trip times and higher system reliability. This will underpin the emerging Tactile Internet, manufacturing and industrial process control, utilities, intelligent transportation systems and all the fascinating derivative applications that these areas will engender. Some dramatic changes to system design, however, are needed to make this reality. Notably, ultra-low end-to-end delays are not possible unless we witness a major overhaul of the wireless air interface and system architecture. As with media delivery, designers will have to bring computation and storage closer to the end user.

All these approaches will undergo rigorous standardization activities that will commence leading up to and beyond an agreed agenda item for IMT-2020 during the WRC-19 meetings. This will ensure global harmonization in the form of common frequency bands, common global standards and a common framework for requirements, capability and performance. Various 5G initiatives have absorbed diverse ideas on what 5G may be and have shaped a common conceptual understanding of 5G. Although 3GPP has been and will continue capturing the requirements of the machine-type communications, differences in requirements for various market segments of the IoT remain and will have to be dealt with in future standards.

We don't completely know every use that 5G will be put to, but we are not worried about this. As one CEO observed recently: "We started developing 3G before the Internet was really operational and we started with 4G before the iPhone came around"[1]. It is hence a perfect time to commence with 5G.

Now, will that 5G be something we have not witnessed to date? You will find out in this fascinating book written by some of the most prominent experts in mobile system design, people who always live 10 years into the future.

We hope you enjoy the read, as much as we did!

Prof. Mischa Dohler	Takehiro Nakamura
Head, Centre for Telecom Research	VP and Managing Director
Chair Professor, King's College London	5G Laboratory
Fellow and Distinguished Lecturer, IEEE	NTT DOCOMO INC. R&D Center
Board of Directors, Worldsensing	Yokosuka, Japan
Editor-in-Chief, ETT and IoT	
London, UK	

---

[1] Statement by Hans Vestberg, CEO of Ericsson, 2015.

Acknowledgments

This book would never exist without the EU project Mobile and wireless communication Enablers for the Twenty-twenty Information Society (METIS), which was funded under the Seventh Framework Program between 2012 and 2015.

The journey began in April 2011 when a small group of engineers from Ericsson, Alcatel-Lucent[1], Huawei Europe, Nokia Corporation[1] and Nokia Siemens Networks[1] started to reflect on what may lay the foundation for a 5G project with a global impact. Their collaboration materialized into an EU project proposal that was later accepted by the EU commission (under the Seventh Framework Program). METIS included the following 25 companies and institutions that deserve our gratitude for their support in developing the basis for this book and helping to finalize it: Ericsson, Aalborg University, Aalto University, Alcatel-Lucent, Anite, BMW Group Research and Technology, Chalmers University of Technology, Deutsche Telekom, NTT DOCOMO, France Telecom-Orange, Fraunhofer-HHI, Huawei Technologies European Research Center, KTH – Royal Institute of Technology, National and Kapodistrian University of Athens, Nokia Corporation, Nokia Siemens Networks, University of Oulu, Poznan University of Technology, RWTH Aachen, Institut Mines-Télécom, Telecom Italia, Telefónica, University of Bremen, University of Kaiserslautern and Universitat Politècnica de València. It should be mentioned that the views expressed are those of the authors and do not necessarily represent METIS.

The EU commission has been unwavering in their support all through the project. Luis Rodriguez-Rosello, now retired, had been an encouraging influence from the beginning. The support and encouragement from the Commission continued over the lifetime of METIS from many other persons as well, a few key names being Bernard Barani, Mario Campolargo, Pertti Jauhiainen and Philippe Lefebvre. Barani and Lefebvre had been supportive when it came to strengthening METIS external exposure on 5G. Pertti Jauhiainen, the METIS project officer, must be acknowledged for his very pertinent advice throughout the project. At the highest level of the EU commission, especially the digital Single Market, EU commissioners have provided strong support in raising awareness about future wireless communication technologies across the world.

The bulk of the material in this book has been extracted from or based on several of the public deliverables of METIS. However, to provide the comprehensive picture on

---

[1] Now Nokia.

current 5G considerations, this was complemented by substantial additional material from authors and entities from outside of the METIS project (e.g. iJoin and 5GNow projects). We would therefore like to thank all our colleagues involved in the book for the support and cooperation that made the book possible.

The authors of this manuscript have shown great commitment and dedication during the writing process. Many worked during their free time, in the evenings and over weekends. They have demonstrated an exemplary spirit of collaboration, always being available when interrupted in the midst of their professional and private lives.

We wish to also thank those who reviewed the various chapters in this book, many drawn from the pool of authors for other chapters of this book. We are particularly indebted to Dr. Kumar Balachandran for his scrutiny and review of several parts of the book, including his significant edits to the introductory chapter. We are likewise thankful to our external reviewers: Dr. Jesus Alonso-Zarate, Prof. Mischa Dohler, Dr. Klaus Doppler, Salah-Eddine Elayoubi, Dr. Eleftherios Karipidis, Per Skillermark, Stefano Sorrentino, Dr. Rapeepat Ratasuk, Dr. Stefan Valentin, Dr. Fred Vook, Dr. Gerhard Wunder and Prof. Jens Zander.

Dr. Osseiran would also like to acknowledge the generosity of Dr. Magnus Frodigh and Mikael Höök of Ericsson. They were helpful in making the resources available that made this book possible.

We would like to thank Cambridge University Press for their help in finalizing this book.

Finally, some specific thanks in

- Chapter 1: to Hugo Tullberg for his careful review of the text and input on security. Mikael Fallgren and Katsutoshi Kusume are also thanked for their input on economic sectors.
- Chapter 2: to the colleagues in METIS who contributed to the 5G scenarios, use cases and system concept.
- Chapter 3: to the colleagues in work package 6 in METIS and work package 5 in iJOIN. Special thanks go to Joachim Sachs for his careful review and helpful comments.
- Chapter 4: to Erik Ström for his input regarding the representation of the reliability/ latency targets.
- Chapter 5: to Byungjin Cho, Riku Jäntti and Mikko A. Uusitalo for their contributions related to multi-operator D2D operation.
- Chapter 6: to Johan Axnäs for his contributions related to mobility and beam finding.
- Chapter 7: to Frank Schaich, Hao Lin, Zhao Zhao, Anass Benjebbour, Kelvin Au, Yejian Chen, Ning He, Jaakko Vihriälä, Nuno Pratas, Cedomir Stefanovic, Petar Popovski, Yalei Ji, Armin Dekorsy, Mikhail Ivanov, Fredrik Brännström and Alexandre Graell i Amat.
- Chapter 8: to Paolo Baracca and Lars S. Sundström for their thorough review of the chapter.

- Chapter 9: to Antti Tölli, Tero Ihalainen, Martin Kurras and Mikael Sternad for their contributions. The authors would also like to thank Dennis Hui for his careful review and precious comments.
- Chapter 10: to Henning Thomsen for his contribution related to multi-flow wireless backhauling, and Sumin Kim and Themistoklis Charalambous for their contributions related to buffer-aided relaying.
- Chapter 11: to Patrick Agyapong, Daniel Calabuig, Armin Dekorsky, Josef Eichinger, Peter Fertl, Ismail Guvenc, Petteri Lundén, Zhe Ren, Paweł Sroka, Sławomir Stańczak, Yutao Sui, Venkatkumar Venkatasubramanian, Osman N. C. Yilmaz and Chan Zhou.
- Chapter 12: to their colleagues in work package 5 in METIS.
- Chapter 13: to David Martín-Sacristán for his thorough review of the chapter. The authors would also like to thank all the people who contributed to the METIS channel modeling.
- Chapter 14: to the colleagues in METIS who worked so intensively in the simulation activities.

Afif Osseiran
Stockholm, Sweden

Jose F. Monserrat
Valencia, Spain

Patrick Marsch
Wrocław, Poland

Acronyms

Acronym	Definition
3GPP	Third Generation Partnership Project
4G	Fourth Generation
5G	Fifth Generation
5G-PPP	5G Public Private Partnership
ABS	Almost Blank Subframe
ACK	Acknowledged Message
A/D	Analogue-to-Digital
ADC	Analogue-to-Digital Converter
ADWICS	Advanced Wireless Communications Study Committee
AEI	Availability Estimation and Indication
AF	Amplify-and-Forward
AI	Availability Indicator
AMC	Adaptive Modulation and Coding
AMPS	Advanced Mobile Phone System
AN	Access Node
AoA	Angle of Arrival
AoD	Angle of Departure
AP	Access Point
API	Application Programming Interface
AR	Availability request
ARQ	Automatic Repeat Request
ASA	Azimuth Spread of Arrival
A-SAN	Assistant Serving Access Node
ASD	Azimuth Spread of Departure
AWGN	Additive White Gaussian Noise
BB	Baseband
BER	Bit Error Rate
BF	Beamforming
BH	Backhaul
BLER	Block Error Rate
BP	Break Point
BS	Base Station

BW	Bandwidth
CA	Carrier Aggregation
CapEx	Capital Expenditure
CB	Coordinated Beamforming
CC	Channel Component
CDD	Cyclic Delay Diversity
CDF	Cumulative Distribution Function
CDMA	Code Division Multiple Access
CDPD	Cellular Digital Packet Data
CDR	Coordinated Direct and Relay Transmission
CEPT	European Conference of Postal and Telecommunications Administrations
CH	Cluster Head
Cloud-RAN	Cloud Radio Access Network
CMOS	Complementary Metal Oxide Semiconductor
cmW	centimeter Wave
CN	Core Network
CNE	Core Network Element
CoMP	Coordinated Multi-Point
CP	Cyclic Prefix
CPE	Common Phase Error
C-Plane	Control Plane
CPRI	Common Public Radio Interface
CPS	Cyber-Physical Systems
C-RAN	Centralized Radio Access Network
CRS	Common Reference Signal
CS	Coordinated Scheduler
CSI	Channel State Information
CSIT	Channel State Information at Transmitter
CSMA/CA	Carrier Sense Multiple Access/Collision Avoidance
CS-MUD	Compressed Sensing Based Multi-User Detection
CTS	Clear to Send
CU	Central Unit
CWIC	CodeWord level Interference Cancellation
D2D	Device-to-Device
DAC	Digital to Analog Conversion
dB	Decibel
DBSCAN	Density-Based Spatial Clustering of Applications with Noise
DCS	Dynamic Channel Selection
DEC	Decoder
Demod.	Demodulation
DER	Distributed Energy Resources
DET	Detection
DF	Decode-and-Forward

DFS	Dynamic Frequency Selection
DFT	Discrete Fourier Transform
DFTS-OFDM	Discrete Fourier Transform Spread OFDM
DID	Device-Infrastructure-Device
Div	Diversity
DL	Downlink
DMRS	Demodulation Reference Signal
DoA	Direction of Arrival
DoD	Direction of Departure
DoF	Degrees of Freedom
DPB	Dynamic Point Blanking
DPS	Dynamic Point Selection
DR	Decode-and-Reencode
D-RAN	Distributed Radio Access Network
DRX	Discontinuous reception
DyRAN	Dynamic Radio Access Network
E2E	End-to-End
EC	European Commission
EDGE	Enhanced Data rates for GSM Evolution
EGF	Enhanced Gaussian Function
eICIC	enhanced Inter Cell Interference Cancellation
EM	Eigenmode
EMF	Electromagnetic Field
eNB	enhanced NodeB
ENOB	Effective Number of Bits
EPA	Extended Pedestrian A
EPC	Evolved Packet Core
E-PDCCH	Enhanced PDCCH
ESA	Elevation Spread of Arrival
ESD	Elevation Spread of Departure
ESE	Elementary Signal Estimator
ETSI	European Telecommunications Standards Institute
ETU	Extended Typical Urban
EVA	Extended Vehicular A
EVM	Error Vector Magnitude
FBC	First bounce cluster
FBCP	Fixed BF and CSI-Based Precoding
FBMC	Filter-Bank Multi-Carrier
FCC	Federal Communications Commission
FD	Full duplex
FDD	Frequency Division Duplexing
FDM	Frequency Division Multiplex
FDMA	Frequency Division Multiple Access
FEC	Forward Error Correction

FFT	Fast Fourier Transform
FinFET	Fin-Shaped Field Effect Transistor
FoM	Figure-of-Merit
FP7	Seventh Framework Programme
FRN	Fixed Relay Node
FWR	Four-Way Relaying
GaAs	Gallium Arsenide
GaN	Gallium Nitride
GHz	Giga Hertz
GLDB	Geolocation Database
GoB	Grid of Beams
GP	Guard Period
GPRS	General Packet Radio Service
GSCM	Geometry-Based Stochastic Channel Model
GSM	Global System for Mobile communications
HARQ	Hybrid Automatic Repeat Request
HBF	Hybrid Beamforming
HD	Half Duplex
HetNet	Heterogeneous networks
HO	Handover
HPBW	Half Power Beam Width
HSCSD	High Speed Circuit Switched Data
HSDPA	High Speed Downlink Packet Access
HSM	Horizontal Spectrum Manager
HSPA	High Speed Packet Access
HSUPA	High Speed Uplink Packet Access
HTC	Human-Type Communication
i.i.d. or iid	independently and identically distributed
I2I	Indoor to Indoor
IA	Interference Alignment
IBC	Interfering Broadcast Channel
IC	Interference Cancellation
ICI	Inter-Cell Interference
ICIC	Inter-Cell Interference Coordination
ICNIRP	International Commission on Non-Ionizing Radiation Protection
ICT	Information and Communications Technologies
IDFT	Inverse Discrete Fourier Transform
IDMA	Interleave Division Multiple Access
IEEE	Institute of Electrical and Electronics Engineers
IFFT	Inverse Fast Fourier Transform
IMF-A	Interference Management Framework from Artist4G
IMT	International Mobile Telecommunications
IMT-2000	International Mobile Telecommunications 2000
IMT-A	International Mobile Telecommunications-Advanced

InH	Indoor Hotspot
InP	Indium Phosphide
IoT	Internet of Things
IR	Impulse Response
IRC	Interference Rejection Combining
IS	Interference Suppression
ISA	International Society for Automation
ISD	Inter-Site Distance
IT	Information Technology
ITS	Intelligent Transport Systems
ITU	International Telecommunication Union
ITU-R	International Telecommunications Union – Radiocommunication Sector
ITU-T	International Telecommunications Union – Telecommunication Standardization Sector
JSDM	Joint Spatial Division Multiplexing
JT	Joint Transmission
KPI	Key Performance Indicator
LA	Link Adaptation
LAA	Licensed-Assisted Access
LBC	Last-Bounce Cluster
LBS	Last-Bounce Scatterer
LDPC	Low Density Parity Check
LO	Local Oscillator
LOS	Line of Sight
LR-WPAN	Low-Rate Wireless Personal Area Networks
LaS	Large Scale
LS	Least Square
LSA	Licenced Shared Access
LSCP	Lean System Control Plane
LSP	Large Scale Parameters
LTE	Long Term Evolution
LTE-A	Long Term Evolution-Advanced
LTE-U	Long Term Evolution-Unlicensed
M2M	Machine to Machine
MAC	Medium Access Control
MAP	Maximum A Posteriori
MBB	Mobile Broadband
MCS	Modulation and Coding Scheme
MET	Multiuser Eigenmode Transmission
METIS	Mobile and wireless communications Enablers for Twenty-twenty (2020) Information Society
MF	Matched Filter
MH	Multi-Hop

MHz	Mega Hertz
MIIT	Ministry of Industry and Information Technology
MIMO	Multiple Input Multiple Output
ML	Maximum Likelihood
MME	Mobility Management Entity
MMSE	Minimum Mean Square Error
mMTC	massive Machine-Type Communication
mmW	millimeter Wave
MN	Moving Networks
MNO	Mobile Network Operator
MODS	Multi-Operator D2D Server
MOST	Ministry of Science and Technology
MPA	Massage Passing Algorithm
MPC	Multipath Components
MPLS	Multiprotocol Label Switching
MRC	Maximal Ratio Combining
MRN	Moving Relay Node
MRT	Maximum Ratio Transmission
MoS	Mode Selection
MS	Mobile Station
MTC	Machine-Type Communication
MU	Multi User
MU MIMO	Multi User MIMO
MUI	Multi User Interference
MUICIA	Multi User Inter Cell Interference Alignment
MU-MIMO	Multi User MIMO
MU-SCMA	Multi User SCMA
MUX	MUltipleXing
n.a.	not applicable
NA	Network Assistance
NAIC	Network Assisted Interference Cancellation
NA-TDMA	North American TDMA
NDRC	National Development and Reform Commission
NE	Network Element
NF	Network Function
NFV	Network Function Virtualization
NFVI	Network Function Virtualization Infrastructure
NGMN	Next Generation Mobile Networks
NLOS	Non-Line of Sight
NMSE	Normalized Mean Square Error
NMT	Nordic Mobile Telephone
NN	Nomadic Nodes
NOMA	Non-Orthogonal Multiple Access
NRA	National Regulatory Authorities

NSPS	National Security and Public Safety
O2I	Outdoor-to-Indoor
O2O	Outdoor-to-Outdoor
Ofcom	Office of communications
OFDM	Orthogonal Frequency Division Multiplexing
OFDMA	Orthogonal Frequency Division Multiple Access
OL	Open Loop
OLOS	Obstructed Line of Sight
OLPC	Open Loop Path Loss Compensating
OMD	OFDM Modulation/Demodulation
OP CoMP	OPportunistic CoMP
OPEX	Operational Expenditures
OPI	Overall Performance Indicator
OQAM	Offset QAM
ORI	Open Radio Equipment Interface
P2P	Peer to Peer
PAPC	Per Antenna Power Constraint
PAPR	Peak to Average Power Ratio
PAS	Power Angular Spectrum
PC	Power Control
PCC	Phantom Cell Concept
PDC	Personal Digital Cellular
PDCCH	Physical Downlink Control Channel
PDCP	Packet Data Convergence Protocol
PDSCH	Physical Downlink Shared Channel
PER	Packet Error Rate
P-GW	Packet data network Gateway
PHY	PHYsical layer
PiC	Pilot Contamination
PLC	Programmable Logic Controller
PLL	Phase Locked Loop
PMU	Phasor Measurement Unit
PN	Phase Noise
PNL	Power Normalization Loss
PPC	Pilot Power Control
PPDR	Public Protection and Disaster Relief
PRACH	Physical Random Access Channel
PRB	Physical Resource Block
ProSe	Proximity Service
P/S	Parallel to Serial
P-SAN	Principal Serving Access Node
PSD	Power Spectral Density
PSM	Power Saving Mode
PUSCH	Physical Uplink Shared Channel

QAM	Quadrature Amplitude Modulation
QoE	Quality of Experience
QoS	Quality of Service
QPSK	Quadrature Phase Shift Keying
RA	Random Access
RACH	Random Access Channel
RAN	Radio Access Network
RAT	Radio Access Technology
RB	Resource Block
Rel	Release
ReA	Resource Allocation
RF	Radio Frequency
RLC	Radio Link Control
RLS	Recursive Least Squares
RMT	Random Matrix Theory
RN	Relay Node
RNE	Radio Network Element
RRC	Radio Resource Control
RRM	Radio Resource Management
RS	Relay Station
RSRP	Reference Signal Received Power
RTL	Reliable Transmission Link
RTS	Request to Send
RTT	Round Trip Time
Rx	Receiver
SA	Service and System Aspects
SBC	Single Bounce Cluster
SC	Single Carrier
SCM	Spatial Channel Model
SCMA	Sparse Code Multiple Access
SCME	Spatial Channel Model Extended
SDF	Spatial Degrees of Freedom
SDN	Software Defined Networking
SE	Switching Element
SFBC	Space Frequency Block Coding
S-GW	Serving Gateway
SIC	Successive Interference Cancellation
SiGe	Silicon Germanium
SIMO	Single Input Multiple Output
SINR	Signal to Interference plus Noise Ratio
SIR	Signal to Interference Ratio
SLIC	Symbol Level Interference Cancellation
SLNR	Signal to Leakage Interference plus Noise Ratio
SM	Spatial Multiplexing

SMEs	Small and Medium-sized Enterprises
SMS	Short Message Service
SNR	Signal-to-Noise Ratio
SoA	State of the Art
SOCP	Second Order Cone Programming
S/P	Serial to Parallel
SS	Small Scale
SU-MIMO	Single User MIMO
SUS	Semi-orthogonal User Selection
SvC	Serving Cluster
SVD	Singular Value Decomposition
TACS	Total Access Communications System
TAU	Tracking Area Update
TCP	Transmission Control Protocol
TD-CDMA	Time Division CDMA
TDD	Time Division Duplexing
TDM	Time Division Multiplex
TDMA	Time Division Multiple Access
TeC	Technology Component
TTI	Transmission Time Interval
TV	Television
TVWS	TV White Space
TWR	Two-Way Relaying
Tx	Transmitter
UDN	Ultra-Dense Network
UE	User Equipment
UFMC	Universal Filtered Multi-Carrier
UF-OFDM	Universal Filtered OFDM
UL	Uplink
ULA	Uniform Linear Array
UM	Utility Maximizing
UMa	Urban Macro
UMi	Urban Micro
uMTC	ultra-reliable Machine-Type Communication
UMTS	Universal Mobile Telecommunication System
UPA	Uniform Planar Array
U-Plane	User Plane
UTD	Uniform Theory of Diffraction
V2D	Vehicle-to-Device
V2I	Vehicle-to-Infrastructure
V2P	Vehicle-to-Pedestrian
V2V	Vehicle-to-Vehicle
V2X	Vehicle-to-Anything
VCO	Voltage Controlled Oscillator

VM	Virtual Machine
VNF	Virtual Network Function
VPL	Vehicular Penetration Loss
VU	Vehicular User
w/	with
w/o	without
WCDMA	Wideband Code Division Multiple Access
WEW	Wireless-Emulated Wired
Wi-Fi	Wireless Fidelity
WMMSE	Weighted MMSE
WNC	Wireless Network Coding
WRC	World Radio Conference
WS	Workshop
WSR	Weighted Sum Rate
WUS	Without User Selection
xMBB	extreme Mobile BroadBand
XO	Crystal Oscillator
XPR	Cross-Polarization Ratio
ZF	Zero Forcing

1 Introduction

Afif Osseiran, Jose F. Monserrat, Patrick Marsch, and Olav Queseth

1.1 Historical background

The Information and Communications Technology (ICT) sector was born in the twenty-first century out of a consolidation of two major industry sectors of the last century, the telecommunications industry and the computing industry. This book is designated to harness the momentum of the mobile telecommunications industry to a fifth generation of technologies. These technologies will allow completing the consolidation of services, content distribution, communications and computing into a complex distributed environment for connectivity, processing, storage, knowledge and intelligence. This consolidation is responsible for a blurring of roles across the board, with computing and storage being embedded in communication infrastructure, process control being distributed across the Internet and communication functions moving into centralized cloud environments.

1.1.1 Industrial and technological revolution: from steam engines to the Internet

The ICT sector arose out of a natural marriage of telecommunications with the Internet, and is presiding over a tremendous change in the way information and communications services are provisioned and distributed. The massive and widespread adoption of mobile connected devices is further driving deep societal changes with tremendous economic, cultural and technological impact to a society that is becoming more networked and connected. Humanity is going through a phase of a technological revolution that originated with the development of semiconductors and the integrated circuit and continued with the maturing of Information Technology (IT) sector and the development of modern electronic communication in the 1970s and 1980s, respectively. The next frontier in the maturation of the ICT sector is to create an indistinguishable framework for service delivery across a variety of scenarios that span huge variations in demand, including the delivery of personalized media to and from the Internet, incorporating the Internet of Things (IoT) or the Internet of everything into the connected paradigm, and the introduction of security and mobility

5G Mobile and Wireless Communications Technology, ed. A. Osseiran, J. F. Monserrat, and P. Marsch. Published by Cambridge University Press. © Cambridge University Press 2016.

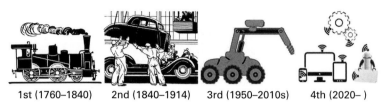

1st (1760–1840)   2nd (1840–1914)   3rd (1950–2010s)   4th (2020–)

Figure 1.1   The four stages of the Industrial Revolution.

functions as configurable features for any communication scenario. Some would call it the fourth stage of the Industrial Revolution [1].

The four stages of the Industrial Revolution are illustrated in Figure 1.1. The first stage of the Industrial Revolution (approximately 1760–1840) started in England with the introduction of the power loom and the steam engine. As a consequence, the agrarian economy of the eighteenth century underwent rapid transformation within decades to an industrial one, dominated by machinery for manufacturing goods.

The second stage of the Industrial Revolution (approximately 1840–1914) began with the introduction of the Bessemer steel process and culminated in early factory electrification, mass production and the production line. Electrification enabled mass production by dividing the labor into specialized activities on the production line, where a common example is the Ford production model in the car industry.

The third stage of the Industrial Revolution (approximately 1950–2010s) occurred thanks to electronics and IT, and in particular the introduction of Programmable Logic Controllers (PLCs). This allowed further automation of the production process and an increase in productivity.

The fourth stage of the Industrial Revolution may now be seen as the era where a new generation of wireless communications enables pervasive connectivity between machines and objects, which itself enables another leap in industrial automation.

It is expected that the 5th generation of mobile communications (5G) will provide the means to move into exactly this fourth stage of the Industrial Revolution, as it allows the currently human-dominated wireless communications to be extended to an all-connected world of humans and objects. In particular, 5G will have:

- connectivity as a standard for people and things,
- critical and massive machine connectivity,
- new spectrum bands and regulatory regimes,
- mobility and security as network functions,
- integration of content distribution via the Internet,
- processing and storage at the network edge and
- software defined networking and network function virtualization.

1.1.2   Mobile communications generations: from 1G to 4G

Figure 1.2 illustrates a short chronological history of the cellular radio systems from their infancy in the 1970s (i.e. 1G, the first generation) till the 2020s (i.e. 5G, the fifth

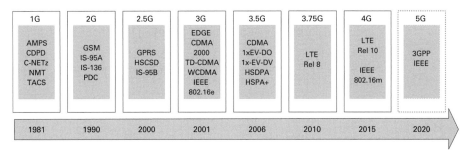

1G	2G	2.5G	3G	3.5G	3.75G	4G	5G
AMPS CDPD C-NETz NMT TACS	GSM IS-95A IS-136 PDC	GPRS HSCSD IS-95B	EDGE CDMA 2000 TD-CDMA WCDMA IEEE 802.16e	CDMA 1xEV-DO 1x-EV-DV HSDPA HSPA+	LTE Rel 8	LTE Rel 10 IEEE 802.16m	3GPP IEEE
1981	1990	2000	2001	2006	2010	2015	2020

Figure 1.2   Evolution of cellular standards.

generation). The major steps in the evolution of the cellular mobile systems are shown in Figure 1.2 and will be described hereafter.

The first commercial analog mobile communication systems were deployed in the 1950s and 1960s [2], although with low penetration. The year 1981 witnessed the birth of the first commercial deployments of the First Generation (1G) mobile cellular standards such as Nordic Mobile Telephone (NMT) in Nordic countries; C-Netz in Germany, Portugal and South Africa; Total Access Communications System (TACS) in the United Kingdom; and Advanced Mobile Phone System (AMPS) in the Americas. The 1G standards are called the *analog standards* since they utilize analog technology, typically frequency modulated radio signals with a digital signaling channel. The European Conference of Postal and Telecommunications Administrations (CEPT) decided in 1982 to develop a pan-European 2G mobile communication system. This was the starting point of the Global System for Mobile communications (GSM), the dominant 2G standard, which was deployed internationally from 1991. The introduction of 2G was characterized by the adoption of digital transmission and switching technology. Digital communication allowed considerable improvements in voice quality and network capacity, and offered growth in the form of supplementary services and advanced applications such as the Short Message Service (SMS) for storage and forwarding of textual information.

The primary purpose of GSM (i.e. 2G) was to create a common digital voice telephony network that allowed international roaming across Europe. GSM is based on a hybrid Time Division Multiple Access (TDMA)/Frequency Division Multiple Access (FDMA) method, in contrast with 1G systems based only on FDMA [3]. In parallel with GSM, other digital 2G systems were developed around the globe and competed with each other. These other main 2G standards include (1) TIA/EIA-136, also known as the North American TDMA (NA-TDMA) standard, (2) TIA/EIA IS-95A, also known as CDMAOne [4] and (3) Personal Digital Cellular (PDC), used exclusively in Japan. The evolution of 2G, called 2.5G, introduced packet-switched data services in addition to voice and circuit-switched data. The main 2.5G standards, General Packet Radio Service (GPRS) and TIA/EIA-95[1], were extensions of GSM and TIA/EIA IS-95A, respectively. Soon afterwards, GSM was evolved further into the Enhanced

---

[1] TIA/EIA-95 was a combination of versions TIA/EIA-IS95A and IS-95B.

Data Rates for Global Evolution (EDGE) and its associated packet data component Enhanced General Packet Radio Service (EGPRS), mainly by addition of higher order modulation and coding schemes. GSM/EDGE has continued to evolve and the latest release of the 3GPP standard supports wider bandwidths and carrier aggregation for the air interface.

Shortly after 2G became operational, industrial players were already preparing and discussing the next wireless generation standards. In parallel, the International Telecommunications Union, Radio Communications (ITU-R) developed the requirements for systems that would qualify for the International Mobile Telecommunications 2000 (IMT-2000) classification. In January 1998, CDMA in two variants – Wideband Code Division Multiple Access (WCDMA) and Time Division CDMA (TD-CDMA) – was adopted by the European Telecommunications Standards Institute (ETSI) as a Universal Mobile Telecommunication System (UMTS). UMTS was the major 3G mobile communication system and was one of the first cellular systems that qualified for IMT-2000. Six radio interfaces have been qualified to meet IMT-2000 requirements including three technologies based on CDMA, a version of GSM/EDGE known as UWC-136[2], and two technologies based on OFDMA [5]. Within the framework of the 3rd Generation Partnership Project (3GPP), new specifications were developed, together known as 3G Evolution and illustrated in Figure 1.2 as 3.5G. For this evolution, two Radio Access Network (RAN) approaches and an evolution of the Core Network were suggested.

The first RAN approach was based on the evolution steps in CDMA 2000 within 3GPP2: 1xEV-DO and 1xEV-DV.

The second RAN approach was High Speed Packet Access (HSPA). HSPA was a combination of High Speed Downlink Packet Access (HSDPA), added in 3GPP Release 5, and High Speed Uplink Packet Access (HSUPA), added in 3GPP Release 6 [6]. Both initially enhanced the packet data rate, to 14.6 Mbps in the downlink and to 5.76 Mbps in the uplink, and quickly evolved to handle higher data rates with the introduction of MIMO. HSPA was based on WCDMA and is completely backward compatible with WCDMA. While CDMA 1xEV-DO started deployment in 2003, HSPA and CDMA 1xEV-DV entered into service in 2006.

All 3GPP standards follow the philosophy of adding new features while still maintaining backward compatibility. This has been further applied in an evolution of HSPA known as HSPA+, which supports carrier aggregation for higher peak data rates without affecting existing terminals in the market.

The second UMTS evolution, commercially accepted as 4G, is called Long Term Evolution (LTE) [7][8], and is composed of a new air interface based on Orthogonal Frequency Division Multiple Access (OFDMA) and a new architecture and Core Network (CN) called the System Architecture Evolution/Evolved Packet Core (SAE/EPC). LTE is not backward compatible with UMTS and was developed in anticipation of

---

[2] Universal Wireless Communications-136 was an evolution of NA-TDMA to integrate GSM-EDGE. It was never deployed as specified by the ITU-R and was abandoned in favor of the 3GPP specification of GSM/EDGE.

higher spectrum block allocations than UMTS during World Radio Conference (WRC) 2007. The standard was also designed to operate with component frequency carriers that are very flexible in arrangement, and supports carriers from 1.4 MHz in width to 20 MHz.

The LTE standard offered significant improvements in capacity and was designed to transition cellular networks away from circuit-switched functionality, which provided a major cost reduction from previous generations. At the end of 2007, the first LTE specifications were approved in 3GPP as LTE Release 8. The LTE Release 8 system has peak data rates of approximately 326 Mbps, increased spectral efficiency and significantly shorter latency (down to 20 ms) than previous systems. Simultaneously, the ITU-R was developing the requirements for IMT-Advanced, a successor to IMT-2000, and nominally the definition of the fourth generation. LTE Release 8 did not comply with IMT-Advanced requirements and was initially considered a precursor to 4G technology. Although this statement was subsequently relaxed in common parlance and LTE is uniformly accepted as 4G, 3GPP LTE Release 10 and IEEE 802.16 m (deployed as WiMAX) were technically the first air interfaces developed to fulfill IMT-Advanced requirements. Despite being an approved 4G technology, WiMAX has had difficulties in gaining widespread acceptance and is being supplanted by LTE. LTE Release 10 added several technical features, such as higher order MIMO and carrier aggregation that improved capacity and throughput of Release 8. Carrier aggregation up to 100 MHz of total bandwidth allows an increase of the peak data rate to a maximum of 3 Gbps in downlink and 1.5 Gbps in uplink. Higher order MIMO configurations up to 8×8 in downlink and 4×4 in the uplink are also involved in the performance improvement.

3GPP standardization of LTE (i.e. Release 11 to Release 13) continues and is expected to proceed to Release 13 and beyond. LTE Release 11 refined some of the LTE Release 10 capabilities, by enhancing carrier aggregation, relaying and interference cancellation. New frequency bands were added, and the use of coordinated multipoint transmission and reception (CoMP) was defined. LTE Release 12, which was concluded in March 2015, added several features to improve the support of heterogeneous networks, even higher order MIMO, and aggregation between Frequency Division Duplexing (FDD) and Time Division Duplexing (TDD) carriers. Several features for the offloading of the backhaul and core networks were also defined. Further, in LTE Releases 12 and 13, new solutions (known as LTE-M and Narrow-Band IoT (NB-IoT)) were introduced in order to support massive Machine Type Communication (MTC) devices such as sensors and actuators [9] [10]. These solutions provided improvements in terms of extended coverage, longer battery life, and reduced cost. Release 13 also targets extreme broadband data rates using carrier aggregation of up to 32 carriers.

The cellular global mobile market was about 7.49 billion subscribers [11] by mid-2015, where the GSM/EDGE family including EGPRS for data connectivity is the dominant Radio Access Network (RAN) in use. GSM has a global market share of more than 57% (corresponding to 4.26 billion subscribers), is well beyond peak use and is currently in decline. On the other hand, the number of 3G subscribers including HSPA has risen since 2010 to 1.94 billion subscribers, which represents 26% of the market share. The Ericsson Mobility Report projects that WCDMA/

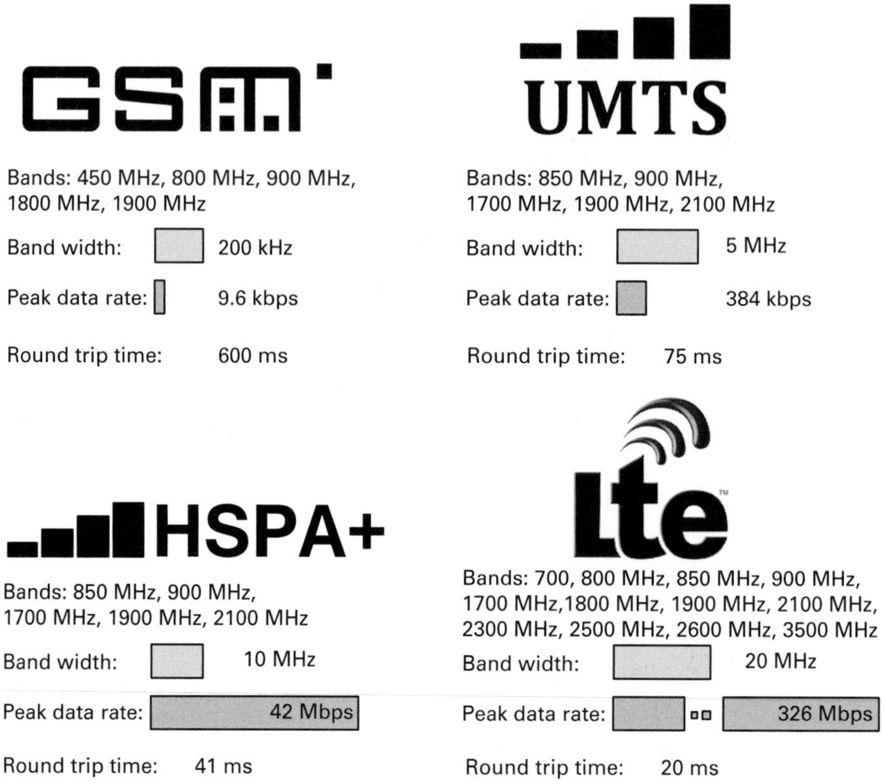

GSM

Bands: 450 MHz, 800 MHz, 900 MHz,
1800 MHz, 1900 MHz

Band width: 200 kHz

Peak data rate: 9.6 kbps

Round trip time: 600 ms

UMTS

Bands: 850 MHz, 900 MHz,
1700 MHz, 1900 MHz, 2100 MHz

Band width: 5 MHz

Peak data rate: 384 kbps

Round trip time: 75 ms

HSPA+

Bands: 850 MHz, 900 MHz,
1700 MHz, 1900 MHz, 2100 MHz

Band width: 10 MHz

Peak data rate: 42 Mbps

Round trip time: 41 ms

Lte

Bands: 700, 800 MHz, 850 MHz, 900 MHz,
1700 MHz,1800 MHz, 1900 MHz, 2100 MHz,
2300 MHz, 2500 MHz, 2600 MHz, 3500 MHz

Band width: 20 MHz

Peak data rate: 326 Mbps

Round trip time: 20 ms

Figure 1.3   Main characteristics of 3GPP/ETSI standards.

HSPA subscriptions will peak by 2020, and will decrease past that point [12]. The dominant 4G standard, LTE, captured around 910 million subscribers (or 12% of the total market) by the end of 2015 and is expected to reach 4.1 billion subscriptions by 2021 [12], hence making it the largest mobile technology. Figure 1.3 illustrates the main features of the 3GPP standards now in the market, highlighting the trend toward widespread use of spectrum, higher bandwidths, higher spectral efficiency and lower latency.

1.1.3   From mobile broadband (MBB) to extreme MBB

Extreme Mobile Broadband (xMBB) services will allow 5G to meet the continuing demand for high data rates and high traffic demands in the years beyond 2020.

The widespread increase in video traffic and the interest in virtual reality and ultra-high definition video streaming will create demand for data rates of the order of many Gbps. The introduction of 5G will allow wireless networks to match data rates and use cases that are currently handled by fiber access.

The Tactile Internet will additionally require support of very low delays through the network. This requirement along with the high user data rate requirements places even greater demands on the peak data rates supported by the system.

1.1.4 IoT: relation to 5G

Over the last few years, several terms such as the IoT, Cyber-Physical-Systems (CPS) and Machine-to-Machine (M2M) have been used to describe a key focus area for the ICT sector. These terms are each used with a specific emphasis:

1. IoT, also referred to as the "Internet of Everything", emphasizes the aspect of the Internet in which all objects (i.e. humans and machines) are uniquely addressable and communicate via a wire or wirelessly via a network [13],
2. CPS refers to the integration of computation and physical processes (such as e.g. sensors, people and physical environments) via a communication network. In particular, the physical processes can then be observed, monitored, controlled and automated in the digital (i.e. cyber) domain. Embedded computing and communication are the two key technical components that enable CPSs. A modern power grid can be seen as an example of CPSs [14].
3. M2M has been used to represent the way in which machines can communicate between themselves.

Digital processors have been embedded at all levels of industrial systems for many years. However, new communication capabilities (e.g. the ones offered by 4G and 5G) will enable the interconnection of many distributed processors and the possibility to move the digital observation and control from a local level to a system-wide and global level. Moreover, when objects are wirelessly connected via the Internet, and computing and storage are distributed in the network, the distinction among CPSs and IoT terms disappears. Hence, mobile and wireless communications are key enablers for the IoT. 5G in particular will enable IoT for new use cases (e.g. requiring low latency and high reliability) and economical sectors where so far mobile communication has been inexistent.

1.2 From ICT to the whole economy

In contrast to previous cellular generations, one of the major objectives of 5G is to meet projected mobile traffic demand and to holistically address the communications needs of most sectors of the economy, including verticals such as those represented by industries. In some of these economic sectors (such as consumer, finance and media) wireless communication has gradually been making inroads since the onset of the century. The years to follow are expected to push mobility and wireless adoption beyond the tipping point, and 5G will create the conditions where wireless connectivity changes from being an interesting feature to a necessity for a huge number of products in these sectors. The necessity of wireless arises due to the potential for data to build up knowledge, for knowledge to become useful information, and for information to enable higher orders of intelligence in various sectors of the society. At the very least, the data generated from various connected devices will lower the cost of delivering services, and at the very most, it will help accelerate all of humanity to degrees of efficient and productive activity that were impossible during the 255 years since the dawn of the modern Industrial Revolution. Improved wireless broadband connectivity will bring a cascade of secondary benefits to the economy and is capable of improving and bettering the lives of people in untold ways. Some of these economy sectors where wireless communication is expected to play a major role are as follows:

- **Agriculture**: Sensors and actuators are becoming more widely used, e.g. in order to measure and communicate soil quality, rainfall, temperature and wind, to monitor how the crops are growing and livestock movements.
- **Automobile[3]**: Wireless communication is interesting for a multitude of applications associated with intelligent transportation, e.g. to enable greater automation of moving vehicles, to provide Vehicle-to-Vehicle and Vehicle-to-Infrastructure communication for information, sensing and safety to prevent collisions, avoid road traffic congestion etc., as well as commercial applications such as media delivery to the vehicle.
- **Construction/Building**: Buildings are being constructed with sensors, actuators, integrated antennas and monitoring devices for energy efficiency, security, occupancy monitoring, asset tracking, etc.
- **Energy/Utilities**: The Smart Grid is affecting all parts of the value chain including exploration, generation and production, trading, monitoring, load control, fault tolerance and consumption of energy. Future systems where consumers also become producers of energy, appliances are connected and perhaps controlled by utilities, and the increase in the numbers of electric cars pose opportunities and challenges for power companies.
- **Finance (including banking)**: Financial activities, such as trading, banking and shopping are performed more and more over wireless links. Consequently, security, fraud detection and analytics are very important components of financial transactions that are improved due to the use of wireless connectivity.
- **Health**: Wireless communication can be used in a variety of ways ranging from the mundane to the complex; these include exercise monitoring, continuous consumer health sensing, medical alerts and health monitoring by health services, wireless connectivity within hospitals and for remote patient monitoring, remote health service delivery, remote surgery, etc.
- **Manufacturing**: Various engineering tasks and process control can be made more efficient, reliable and accurate with wireless communications; the use of 5G for ultra-reliable operation and extreme requirements on latency is interesting for factory cell automation, while massive machine connectivity will increase in the use of wireless communications in manufacturing for robots, autonomous operation of machines, RFIDs and low-power wireless communications for asset management, etc.
- **Media**: Video is a key driver of high bandwidth consumption, and it is expected that 5G will allow excellent user experience for viewing 3D and 4K formats on a mass scale. Today, the user experience for enjoying rich content like high-resolution video is limited to fixed networks and short-range wireless, while access to high-quality

---

[3] As it is defined herein, the automobile sector overlaps with the transport sector.

music is stressed in crowded areas where users might simultaneously consume unique content. New use cases such as Virtual Reality (VR) or Augmented Reality (AR) are also expected to become popular in mobile or nomadic situations.

- **Public safety**: Police, fire, rescue, ambulance and medical emergency services covered by this category require a high degree of reliability and availability. Just as 4G is being adopted for public safety, 5G radio access will be a very important component of the tools available for security services, law enforcement and emergency personnel to use. The use of SDN and NFV can help the network play a more direct role in public safety functions, such as fighting fires and assisting in earthquake or tsunami disasters, by efficiently managing local service connectivity between responders and from hazards toward the network. The network can also support rescue missions using location services.
- **Retail and consumer**: Wireless communication will continue to play an important role in areas such as retail, travel and leisure, including hospitality.
- **Transport (including logistics)**: Wireless communication is already playing an important role in this respect. This role will even further increase in the future with the advent of 5G. In fact, 5G will improve the infrastructure and communication functionalities in areas such as railway, public transport and transport of goods by terrestrial or maritime means.
- **Additional industries**: Aerospace and defense, basic resources, chemicals, industrial goods and support services will employ wireless communications increasingly in the coming years.

1.3   Rationale of 5G: high data volume, twenty-five billion connected devices and wide requirements

The necessity of wireless connectivity in society is primarily driven by an increased usage of mobile multimedia services, and has led to an exponential increase in mobile and wireless traffic demand and volume. Mobile traffic was first predicted to increase a thousand-fold over the decade 2010–2020 [15]. The figure was later revised to be in the order of 250 times [16]. It is important to note that in the already highly-developed communication societies, e.g. Western Europe and North America, traffic in cellular systems will increase by approximately a factor of 84 over the years 2010–2020 as shown in Figure 1.4.

Further, machine-type applications are becoming important in addition to the human-centric communications that have been dominating the cellular scene so far. In fact, the number of communicating machines was at some point forecasted to be trending toward the number 50 billion by 2020 [17]. That figure has been revised down to 25 billion connected devices based on more recent considerations [12].

The expected uptake of machine-type and human-type wireless communications in many economic sectors and vertical industries will lead to a large and wide diversity of communication characteristics imposing different requirements on mobile and wireless communication systems, e.g. in terms of cost, complexity, energy dissipation, data rate,

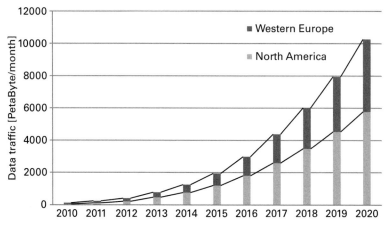

Figure 1.4   Data traffic volume for Western Europe and North America (Copyright 2015 Ericsson).

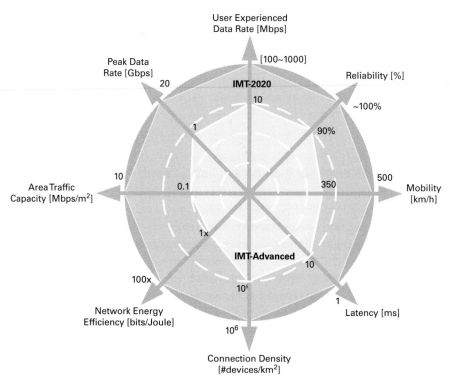

Figure 1.5   Spider diagram for IMT-2020 (i.e. 5G) and IMT-A requirements.

mobility, latency and reliability. For example, the so-called Tactile Internet will require radio latency down to 1 ms [18]. The spider diagram shown in Figure 1.5 is the best way to illustrate the wide range and expansion of the 5G requirements in comparison to prior cellular generations such as IMT-Advanced. In the diagram, the following most relevant key requirements are considered [19]:

- **Peak data rate [Gbps]**: The maximum achievable user/device data rate.
- **User experienced data rate [Mbps or Gbps]**: The achievable user/device data rate across the coverage area.
- **Radio Latency [ms]**: The time needed (on the MAC layer of the radio interface) for a data packet to travel from the source to the destination. Note that this refers to one-way latency.
- **Mobility [km/h]**: The maximum supported vehicular speed at which a nominal QoS can be achieved.
- **Connection density [#devices/km^2]**: The total number of connected devices per area.
- **Energy efficiency [bits/Joule]**: On the network side, the bits transmitted to/received from users, per unit of energy consumption of the RAN; on the device side, bits per unit of energy consumed by the communication module.
- **Reliability [%]**: The percentage of successful transmissions completed within a certain time period.
- **Area traffic capacity [Mbps/m^2]**: The total traffic throughput served per geographic area.

In addition to the above listed requirements, security is a prerequisite for any successful future wireless system.

1.3.1   Security

Security is one of the most important value offerings of the past four generations of wireless systems, and will continue to be a key requirement for any new generation technology. It is worth recognizing that wireless networks have in the past faced security vulnerabilities and have undergone revision over time. As computation becomes cheaper, systems are more susceptible to compromise, and newer generations of wireless systems must adapt to improving end-to-end security without compromising on legal intercept requirements from legitimate authorities. Mobile broadband will be increasingly used for Internet access and cloud services, and the vulnerability to attacks and the cost of damage caused by e.g. denial-of-service attacks [20] will increase. When relays are used or networks operate using mesh topologies, trust must be established between the network and the relay to avoid man-in-the-middle attacks [21], and also toward the device accessing the network through the relay to better shield the system against identity theft, e.g. through user identity caching.

5G networks will transport large amounts of IoT data. Here, each individual message may not be very sensitive, but sensitive knowledge may be extracted through data- or information-fusion. Hence, unauthorized access to even seemingly trivial data must be prevented. When the wireless device is engaging in actuation, encryption and validation of information and authentication of the controlling entity is even more important. Thus, connectivity to an Internet-connected household entry must be carefully audited at all times to prevent unlawful entry.

For ultra-reliable services it is important to guarantee the integrity and authenticity of the transmitted information. False emergency-brake messages or incorrect traffic light indication must be repudiated in vehicles before presentation to the occupants.

In industrial applications, physical processes are controlled and the transmitted information may carry sensitive information about the workflow that needs to be protected against eavesdropping or tampering [22]. Hence, appropriate security functions must be in place to guarantee the integrity and authenticity of the messages.

Additional security challenges arise since 5G networks will include new types of access nodes and support new kinds of services. For applications, e.g. banking, already existing end-to-end solutions will be applicable also in 5G. For other applications, e.g. IoT, new solutions are necessary.

Security research is an active field in itself. In this book, the advances on 5G communication technologies are reported, but the security aspect is beyond the scope of this book.

1.4   Global initiatives

There are a handful of 5G fora, research activities and projects across the globe. Europe took the lead in 2011 [23], and not long after, China, Korea and Japan followed suit with their own activities. These activities (and the corresponding 5G timeplan) are illustrated in Figure 1.6.

1.4.1   METIS and the 5G-PPP

METIS [24] is the first EU holistic 5G access project that has had world-wide impact on the 5G development. METIS falls under the umbrella of the European 7th framework program (FP7). The preparation for the project started in April 2011 and kick-off officially occurred on November 1, 2012. The project was finalized on April 30, 2015.

The METIS project has established itself as the global reference project on 5G, by determining key 5G scenarios, test cases and KPIs, which are now routinely referenced in commercial and academic literature. The project's key achievements included identifying and structuring the 5G key technology components. This has so far given Europe a clear leading position in 5G.

The 5G Public Private Partnership (5G-PPP) [25] succeeds the 7th framework program. The European ICT industry and the European Commission (EC) signed a commercial agreement in December 2013 to form a 5G Infrastructure Public-Private Partnership (5G-PPP). It is mainly a research program with a budget of 1.4 billion Euro for the 2014–2020 timeframe where both the EC and the ICT industry contribute equally (i.e. EUR 700 million each) to the budget. The 5G-PPP brings together industry manufacturers, telecoms operators, service providers, Small and Medium-sized Enterprises (SMEs) and researchers.

METIS-II, started in July 2015 within the framework of the 5G-PPP, builds on the momentum of METIS. METIS-II seeks to develop the overall 5G radio access network design, to a level of detail that is most suitable to support the expedient start of 5G standardization, within 3GPP Release 14 and beyond. METIS-II will provide the technical enablers needed for an efficient integration and use of the various 5G

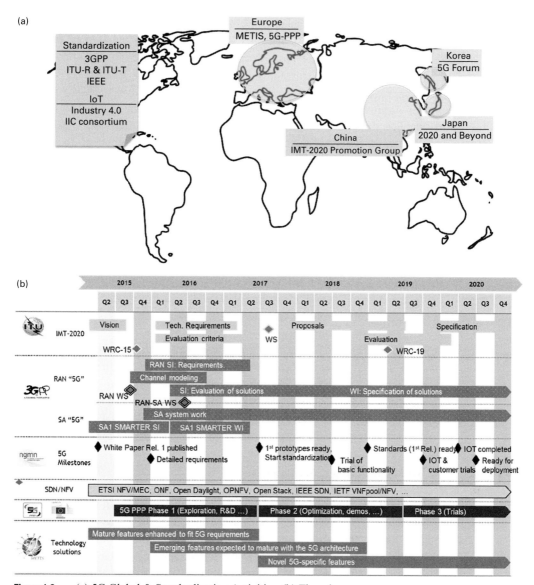

Figure 1.6 (a) 5G Global & Standardization Activities, (b) Timeplan.

technologies and components currently developed, and also the integration of these with evolved legacy technology such as LTE-A. To reach this goal, METIS-II puts a strong emphasis on the collaboration and discussion with other 5G-PPP or global projects, and will facilitate the discussion on 5G scenarios and requirements, key 5G technical components, spectrum aspects, and RAN performance evaluation within 5G-PPP.

The main objective of the 5G-PPP program is to secure Europe's leadership in the particular areas where Europe is strong or where there is potential for creating new markets such as smart cities, e-health, intelligent transport, education or entertainment and media [25]. The ultimate objective of 5G-PPP is to create the fifth generation of communication networks and services. The first projects under the 5G-PPP umbrella started in July 2015.

1.4.2   China: 5G promotion group

China established the IMT-2020 (5G promotion group) in February 2013 [26]. Three Chinese ministries initiated it: the Ministry of Industry and Information Technology (MIIT), Ministry of Science and Technology (MOST) and the National Development and Reform Commission (NDRC).

The group's main objective is to promote the development of 5G technologies in China and to facilitate cooperation with foreign companies and organizations.

While IMT-2020 (5G promotion group) deals with strategic decisions (e.g. through white papers on vision and requirements [27] as well as technologies), the research activities are carried under the MOST's 863-5G Program, which is a government sponsored research activity on 5G wireless access.

1.4.3   Korea: 5G Forum

South Korea's 5G Forum [28] is also a public–private partnership program formed in May 2013. The forum's main objectives are to develop and propose a national strategy for 5G, and to plan a strategy for technology innovation.

Its members include ETRI, SK Telecom, KT, LG-Ericsson and Samsung. The forum is also open to SMEs. The forum aims to ensure that a pre-commercial trial of some 5G technology subset is partly deployed for the 2018 Winter Olympics in Pyeongchang.

1.4.4   Japan: ARIB 2020 and Beyond Ad Hoc

The ARIB 2020 and Beyond Ad Hoc group was established in Japan in 2013 to study terrestrial mobile communications systems in 2020 and beyond. It was created as a sub-committee under the Advanced Wireless Communications Study Committee (ADWICS) in September 2013. ADWICS was established by ARIB in 2006.

The group's objectives are to study system concepts, basic functions and distribution architecture of mobile communications in 2020 and beyond. The expected deliverables will consist of white papers, contributions to ITU and other bodies relevant for 5G. In 2014, the ARIB 2020 and Beyond Ad Hoc group released a first white paper entitled "Mobile Communications Systems for 2020 and beyond" [29] that describes the group vision of the 5G system.

1.4.5   Other 5G initiatives

These 5G activities are smaller in scale and influence compared to those listed previously. To name a few of the organizations leading these initiatives: 4G Americas, 5G Innovation Centre at University of Surrey, and New York University Wireless Research.

1.4.6 IoT activities

There are a large number of IoT global initiatives tackling various aspects. Industrial Internet Consortium (IIC) [30] and Industrie 4.0 [31] are the two most relevant initiatives related to 5G.

The IIC was founded in March 2014 to bring together the organizations and technologies necessary to accelerate growth of the Industrial Internet. The main goals of the IIC are [30]:

- creating new industry use cases and test-beds for real-world applications;
- influencing the global development standards process for Internet and industrial systems.

Industrie 4.0 is a German initiative created in 2013 in order to keep the competitive edge of the German industrial production and to maintain its global market leadership. The objective is to integrate the IoT and services in production and in particular to create networks that incorporate the entire manufacturing process, hence converting factories into a smart environment [31].

1.5 Standardization activities

In the following, the 5G standardization activities in ITU, 3GPP and IEEE are briefly described.

1.5.1 ITU-R

In 2012, ITU's Radiocommunication Sector (ITU-R), led by the Working Party 5D, started a program to develop "IMT for 2020 and beyond". The objective is to develop the requirements for a 5G mobile communications air interface. The Working Party 5D has developed a work plan, timeline, process and deliverables for "IMT-2020". Note that WP 5D is currently using "IMT-2020" as an interim terminology to refer to 5G. According to the time plan, the "IMT-2020 Specifications" should be ready by the year 2020. As of September 2015, the following three reports have been completed:

- **Future technology trends of terrestrial IMT systems** [32]: This report provides information on the technology trends of terrestrial IMT systems considering the time-frame 2015–2020 and beyond. Technologies described in this report are collections of possible technology enablers which may be applied in the future.
- **Recommendation Vision of IMT beyond 2020** [19]: The report addresses the longer term vision for 2020 and beyond and will provide a framework and overall objectives of the future developments of IMT.
- **IMT feasibility above 6 GHz** [33]: This report provides information on the study of technical feasibility of IMT in the bands above 6 GHz.

These reports were instrumental to WRC 2015, where approximately 400 MHz additional spectrum were allocated for IMT, see Chapter 12 for more details.

1.5.2   3GPP

The 3GPP has endorsed a timeline for the 5G standard, a process that is scheduled to extend through to the year 2020 [34]. The 5G RAN Study Item and scope on key radio requirements started in December 2015. The corresponding Study Item on the 5G new radio access started in the 3GPP working groups in March 2016.

Furthermore, 3GPP has been addressing, in LTE [9][10][35] and GSM [36][37], the massive machine-type communications needs e.g. extended coverage, low-power, and low-cost devices. While in LTE the MTC tracks are called LTE-M and NB-IoT, in GSM it is called Extended Coverage GSM for IoT (EC-GSM-IoT).

1.5.3   IEEE

The primary organization in the Institute of Electrical and Electronic Engineers (IEEE) dealing with local and metropolitan networks is the IEEE 802 standards committee. Prominent among these is the IEEE 802.15 project dealing with Wireless Personal Area Networks (WPAN) [38] and the IEEE 802.11 project specifying Wireless Local Area Networks (WLAN) [39]. IEEE 802.11 technologies were initially designed to operate at the 2.4 GHz frequency band. Later, IEEE 802.11 developed a Gigabit standard in the amendments IEEE 802.11ac (at higher frequencies i.e. 5 GHz band) and IEEE 802.11ad (in the 60 GHz millimeter wave band). These systems have been commercially available since 2013, and will be followed around the year 2019 by systems supporting multi-gigabit operation at bands below 6 GHz (e.g. IEEE 802.11ax) and mmW band (e.g. IEEE 802.11ay). It is possible that IEEE will propose a candidate from one of their high throughput technology components for IMT-2020. The IEEE 802.11p amendment targets vehicular applications and will see widespread adoption for V2V communication after 2017. The IEEE is also active in the IoT arena, with the amendment IEEE 802.11ah supporting Wi-Fi operation over extended ranges in sub-GHz bands. The IEEE 802.15.4 standard has led to specification of Low-Rate Wireless Personal Area Networks (LR-WPAN) that are further specified for ad-hoc mesh connectivity by the Zigbee Alliance, and are also used for coordinated and synchronized operation by the International Society for Automation (ISA) with the ISA100.11a specification.

The 5G system is expected to utilize adjunct air interfaces such as those specified by the IEEE. The interfaces between such air interfaces and the 5G network will be specified with careful attention to addressing, identity management, mobility, security and services.

1.6   Scope of the book

The main 5G building blocks are shown in Figure 1.7: Radio Access, Fixed and Legacy RATs (e.g. LTE), Core Network, Cloud, Data Analytics and Security. The covered

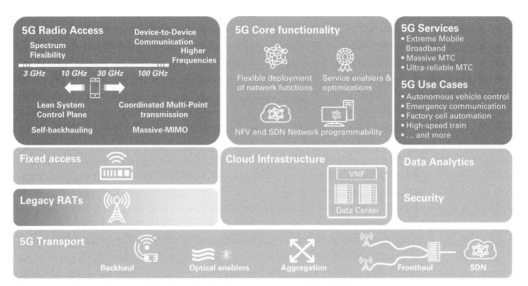

Figure 1.7   5G main areas. The icons in the figures are copyright of Ericsson.

blocks in the book are highlighted in dark grey in Figure 1.7, in particular the 5G Radio Access and use cases as dealt with in Chapter 2. The role of legacy systems such as LTE is briefly tackled in various chapters. Core Network and Cloud (including network function virtualization) are briefly covered in Chapter 3. Data analytics, fixed access and security are outside the scope of the book.

The book is organized as follows:

- Chapter 2 summarizes the defined key **5G use cases and requirements**, and the overall **5G system concept**.
- Chapter 3 provides key considerations on **the 5G architecture**.
- Chapter 4 elaborates on one of the key new use case fields in 5G mentioned before, namely **machine-type communications**.
- Chapter 5 provides more details on **device-to-device** communications in 5G.
- Chapter 6 ventures into centimeter and **millimeter Wave communications** that have the potential to unlock a huge amount of spectrum from 10 GHz to 100 GHz carrier frequency for both wireless backhaul and access.
- Chapter 7 overviews the most likely **5G radio-access technologies**.
- Chapter 8 presents a key technology for 5G, **massive MIMO**.
- Chapter 9 discusses **Coordinated Multi-Point transmission**, in particular with regard to how joint transmission may be better facilitated in 5G.
- Chapter 10 focuses on **relaying** and **wireless network coding** in 5G.
- Chapter 11 looks into technologies related to **interference and mobility management in 5G**.
- Chapter 12 treats the 5G **spectrum**, in particular investigating the expected 5G spectrum landscape and requirements, and the envisioned spectrum access modes.

- Chapter 13 elaborates on the main challenges of **5G channel modeling** and describes corresponding new channel models.
- Chapter 14 provides **guidelines for simulation** to align 5G assumptions, methodology and simulation reference cases.

References

[1] BITKOM e.V., VDMA e.V. and ZVEI e.V, "Umsetzungsstrategie Industrie 4.0," April 2015, www.bmwi.de/BMWi/Redaktion/PDF/I/industrie-40-verbaendeplatt form-bericht,property=pdf

[2] T. Farley, "Mobile telephone history," *Telektronikk Journal*, vol. 3, no. 4, pp. 22–34, 2005, www.telenor.com/wp-content/uploads/2012/05/T05_3–4.pdf

[3] F. Hillebrand, *GSM and UMTS: The Creation of Global Mobile Communication*. Chichester: John Wiley, 2002.

[4] A.J. Viterbi, *CDMA: Principles of Spread Spectrum Communication*. Redwood City: Addison Wesley Longman, 1995.

[5] International Telecommunications Union Radio (ITU-R), "Detailed specifications of the terrestrial radio interfaces of International Mobile Telecommunications-2000," Recommendation ITU-R M.1457-12, February 2015.

[6] H. Holma and A. Toskala, *HSDPA/HSUPA for UMTS*. Chichester: John Wiley, 2007.

[7] E. Dahlman, S. Parkvall, J. Sköld, and P. Beming, *3G Evolution: HSPA and LTE for Mobile Broadband*, 2nd ed. New York: Academic Press, 2008.

[8] S. Sesia, M. Baker, and I. Toufik, *LTE – The UMTS Long Term Evolution: From Theory to Practice*, 2nd ed. Chichester: John Wiley & Sons, 2011.

[9] Ericsson, Nokia Networks, "Further LTE physical layer enhancements for MTC," Work Item RP-141660, 3GPP TSG RAN Meeting #65, September 2014.

[10] Qualcomm Incorporated, "New work item: Narrowband IOT (NB-IOT)," Work Item RP-151621, 3GPP TSG RAN Meeting #69, September 2015.

[11] GSMA, Definitive data and analysis for the mobile industry [Online] https://gsm aintelligence.com/

[12] Ericsson, Ericsson Mobility Report, Report No. EAB-15:037849, November 2015, www.ericsson.com/res/docs/2015/mobility-report/ericsson-mobility-report -nov-2015.pdf

[13] Oxford Dictionaries. Oxford University Press. 2016. [Online] www .oxforddictionaries.com/definition/english/Internet-of-things

[14] K S. Khaitan and J.D. McCalley, "Design techniques and applications of cyber-physical systems: A survey," *IEEE Systems Journal*, vol. 9, no. 2, pp. 350–365, June 2015.

[15] Cisco, "Global mobile data traffic forecast update," 2010–2015 White Paper, February 2011.

[16] Ericsson, Ericsson Mobility Report, no. EAB-15:010920, February 2015, www .ericsson.com/res/docs/2015/ericsson-mobility-report-feb-2015-interim.pdf

[17] Ericsson, More than 50 billion connected devices, White Paper, February 2011, www.ericsson.com/res/docs/whitepapers/wp-50-billions.pdf

[18] International Telecommunications Union Telecomm (ITU-T), "The Tactile Internet," Technology Watch Report, August 2014, www.itu.int/dms_pub/itu-t/oth/23/01/T23 010000230001PDFE.pdf

[19] International Telecommunications Union Radio (ITU-R), "Framework and overall objectives of the future development of IMT for 2020 and beyond," Recommendation ITU-R M.2083, September 2015, www.itu.int/rec/R-REC-M .2083

[20] D. Yu and W. Wen, "Non-access-stratum request attack in E-UTRAN," in Computing, Communications and Applications Conference, Hong Kong, January 2012, pp. 48–53.

[21] N. Ferguson, B. Schneier, and T. Kohno, *Cryptography Engineering: Design Principles and Practical Applications*. John Wiley, Indianapolis, Indiana 2010.

[22] W. Ikram and N. F. Thornhill, "Wireless communication in process automation: A survey of opportunities, requirements, concerns and challenges," in UKACC International Conference on Control 2010, Coventry, September 2010, pp. 1–6.

[23] European Commission, "Mobile communications: Fresh €50 million EU research grants in 2013 to develop '5G' technology," Press Release, February 2013, http://europa.eu/rapid/press-release_IP-13-159_en.htm

[24] METIS, Mobile and wireless communications Enablers for the Twenty-twenty Information Society, EU 7th Framework Programme project, www.metis2020.com

[25] 5G-PPP. 2016. [Online] https://5g-ppp.eu/

[26] IMT-2020 (5G) Promotion group. 2016. [Online] www.imt-2020.cn/en

[27] IMT-2020 (5G) Promotion group, 5G Vision and Requirements, white paper, May 2014.

[28] 5G Forum. 2016. [Online] www.5gforum.org/eng/main/

[29] ARIB, 2020 and Beyond Ad Hoc (20B AH), "Mobile Communications Systems for 2020 and beyond." October 2014, www.arib.or.jp/english/20bah-wp-100.pdf

[30] IIC Consortium. 2016. [Online] www.iiconsortium.org/

[31] Industrie 4.0 Working Group, "Recommendations for Implementing the Strategic Initiative Industrie 4.0. Final Report of the Industrie 4.0 Working Group." 2013 www.acatech.de/fileadmin/user_upload/Baumstruktur_nach_Website/Acatech/ro ot/de/Material_fuer_Sonderseiten/Industrie_4.0/Final_report__Industrie_4.0_acc essible.pdf

[32] International Telecommunications Union Radio (ITU-R), "Future technology trends of terrestrial IMT systems," Report ITU-R M.2320, November 2014, www .itu.int/pub/R-REP-M.2320

[33] International Telecommunications Union Radio (ITU-R), "Technical feasibility of IMT in bands above 6 GHz," Report ITU-R M.2376, July 2015, www.itu.int/pub/ R-REP-M.2376

[34] 3GPP, "Tentative 3GPP timeline for 5G," March 2015, http://www.3gpp.org/ne ws-events/3gpp-news/1674-timeline_5g

[35] 3GPP TR 36.888, "Study on provision of low-cost Machine-Type Communications (MTC) User Equipments (UEs) based on LTE," Technical Report TR 36.888 V12.0.0, Technical Specification Group Radio Access Network, June, 2013.

[36] 3GPP TR 45.820, "Cellular system support for ultra-low complexity and low throughput internet of things (CIoT) (Release 13)," Technical Report, TR 45.820

V13.0.0, Technical Specification Group GSM/EDGE Radio Access Network, August 2015.

[37] Ericsson et al., "New Work Item on Extended DRX (eDRX) for GSM," Work Item GPC150624, 3GPP TSG GERAN, July 2015.

[38] IEEE 802.15 Working Group for WPAN. [Online] www.ieee802.org/15/

[39] IEEE 802.11 Wireless Local Area Networks. [Online] www.ieee802.org/11/

2 5G use cases and system concept

Hugo Tullberg, Mikael Fallgren, Katsutoshi Kusume, and Andreas Höglund

In the 5G vision, access to information and sharing of data are possible anywhere and anytime to anyone and anything. 5G expands the usage of human-centric communications to include both human-centric and machine-centric communications. Mobile and wireless communication will increasingly become the primary way for humans and machines to access information and services. This will lead to socio-economic changes not yet imaginable, including improvements in productivity, sustainability, entertainment and well-being.

To make this vision a reality, the capabilities of 5G systems must extend far beyond those of previous generations. 5G systems must exhibit greater flexibility than previous generations, and involve farther-reaching integration including not only the traditional radio access networks, but also core network, transport and application layers. Altogether, this requires a new way of thinking in 5G wireless access, network architecture and applications.

In this chapter, first, the needs of the end users are described in terms of use cases and requirements, and then an overview of the 5G system concept meeting these user needs is given.

2.1 Use cases and requirements

This section provides the vision based on the expected societal development toward the year 2020 and beyond from the end-user perspective described in Chapter 1. Concrete use cases that have specific goals and challenges are provided. To achieve the goals and to overcome the challenges, there are certain specific requirements for 5G systems to meet. A collection of diverse use cases gives a set of challenging requirements that have to be fulfilled by 5G systems. The material below is largely based on [1]–[8]. The technical solutions to address these requirements are then discussed in the later chapters of this book.

2.1.1 Use cases

In this section, the most relevant 5G use cases are presented. Further, the challenges and requirements for each of these are named. As mentioned in Chapter 1, 5G will become a cornerstone in many of the economic sectors. Table 2.1 shows as an example how the

5G Mobile and Wireless Communications Technology, ed. A. Osseiran, J. F. Monserrat, and P. Marsch. Published by Cambridge University Press. © Cambridge University Press 2016.

Table 2.1 Economic sectors versus use cases.

Use case	Agriculture	Automotive	Construction	Energy	Finance	Health	Manufacturing	Media	Public safety	Retail and consumer	Transport
Virtual and augmented reality							X			X	
Traffic jam		X							X		
Teleprotection in smart grid network				X							
Stadium			X							X	
Smart city			X	X					X		
Shopping mall							X				
Remote surgery and examination							X				
Media on demand	X								X		
Massive amount of geographically spread devices	X						X				
Large outdoor event							X				X
High-speed train		X									X
Factory cell automation								X			
Emergency communication		X							X		
Autonomous vehicle control		X								X	

addressed use cases map onto the major economic sectors. It should be noted that the list of use cases is far from being exhaustive. Only the most relevant ones from technical and business perspective are given. Finally, some of the use cases can be considered as a set of use cases (e.g. smart city or public safety).

2.1.1.1   Autonomous vehicle control

Autonomous vehicle control enables the autonomous driving of vehicles; see Figure 2.1(a). This is an emerging trend having various potential impacts on society. Autonomous driving may, for instance, assist humans and bring a number of benefits such as better traffic safety by avoiding accidents, lower stress and the possibility for drivers to concentrate on other productive activities (e.g. working in the vehicle).

Autonomous vehicle control requires not only vehicle-to-infrastructure communication, but also vehicle-to-vehicle, vehicle-to-people and perhaps vehicle-to-sensors, which may be installed on the roadside. These connections need to provide a very low latency and high reliability for vehicle control signaling, which is critical for safe operation. Although such signaling would typically not require high bandwidth, higher data rates will be necessary if an application requires exchanging video information among vehicles, for example, to enable controlled fleet driving for a group of cars to

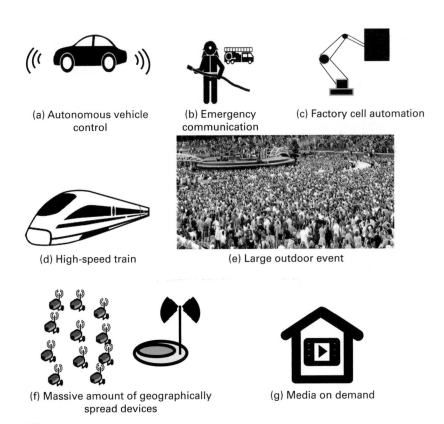

(a) Autonomous vehicle control   (b) Emergency communication   (c) Factory cell automation

(d) High-speed train   (e) Large outdoor event

(f) Massive amount of geographically spread devices   (g) Media on demand

Figure 2.1   5G use cases.

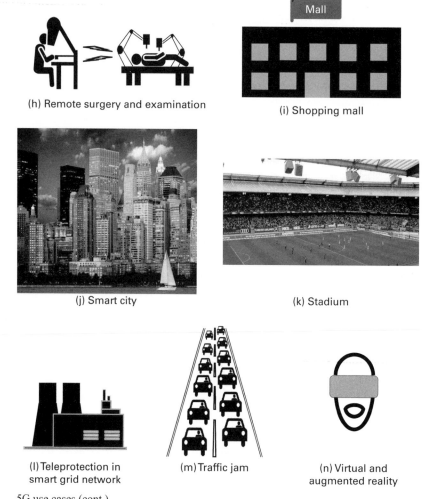

(h) Remote surgery and examination

(i) Shopping mall

(j) Smart city

(k) Stadium

(l) Teleprotection in smart grid network

(m) Traffic jam

(n) Virtual and augmented reality

Figure 2.1   5G use cases (cont.)

quickly adapt to dynamic changes in the surrounding environments. Besides, high mobility is required for supporting possibly fast-moving vehicles, and full coverage is needed in case of completely unattended control [1]–[3][6].

2.1.1.2   Emergency communication

In an emergency situation it is crucial for a user to have a reliable network that can help the user to be rescued and thereby survive, as illustrated in Figure 2.1(b). Obviously, this need still holds even if parts of the network have been damaged in a disaster. In some cases, temporary rescue nodes may be brought in to assist in the damaged network. User devices may be given relaying functionality in order to support and/or assist other devices to reach the network nodes that are still operational. Here, it is typically of highest importance to locate the survivor positions in order to get them into safety [1][3].

A high availability and energy efficiency are the critical requirements in emergency communication. A high availability in the network enables a large discovery rate among the survivors. A reliable setup and call establishment ensures that one can maintain contact with the survivor after the discovery phase. In addition, throughout the search it is desirable to keep the energy consumption of the survivor's devices at a minimum to prolong the timespan in which it is still possible to detect survivors via the network [1][3].

2.1.1.3   Factory cell automation

In a factory, cell automation consists of devices in an assembly line communicating with control units with a sufficiently high reliability and sufficiently low latency to be able to support life-critical applications [7]; see Figure 2.1(c).

In fact, many industrial manufacturing applications require very low latency and high reliability [2][3]. Although only small payloads have to be sent and mobility is usually not the main issue, it is essential to satisfy such stringent latency and reliability requirements that are beyond the current wireless network capabilities. That is why the communication system for industry production is usually realized by a fixed line network today. However, this can be prohibitively expensive in many cases (e.g. when dealing with remote locations), and thus there is a clear need for 5G wireless technologies with very low latency and high reliability.

2.1.1.4   High-speed train

When traveling in a high-speed train (see Figure 2.1(d)), passengers would want to utilize the on-board time for their usual activities in a similar way as when they are at home. Examples are watching high-quality video, gaming or working via remote access to office clouds and virtual reality meetings. With trains traveling at a high speed, it may be challenging to satisfy requirements for these services without significant degradation of user experience [5][8].

The most relevant requirements for high-speed trains are experienced user throughput and end-to-end latency that are satisfactory for passengers to enjoy various services [5][8].

2.1.1.5   Large outdoor event

Some large-scale outdoor events that are held temporarily in a certain area can be visited by a significant amount of people in a limited time period; see Figure 2.1(e). Such events include, for example, sports, exhibitions, concerts, festivals, fireworks and so on. Visitors typically want to take high-resolution photos and videos and share them with their family and friends in real time. Since so many people are concentrated in a specific area of the event, the aggregated traffic volume can be enormously large. The network is highly under-dimensioned since the density of users in such an area is usually much lower unless there is such an event [1][8].

Thus, the critical requirements for crowded outdoor events are to provide average experienced user throughput that is sufficient for video data and to accommodate the large traffic volume density for the high connection density in this crowded setting that may correspond to multiple users per square meter. Besides, the outage probability should be as low as possible for realizing a highly satisfactory user experience [1][8].

2.1.1.6   Massive amount of geographically spread devices

A system that is able to collect relevant information to and from a massive number of geographically spread devices can make use of that information in various ways to improve the end-user experience; see Figure 2.1(f). Such a system can keep track of relevant data and perform tasks and make decisions based on the received and collected input, e.g. providing surveillance, monitoring critical components and assisting in information sharing [1][3].

One possible way to collect the information to and from the massive number of locations is by the use of sensors and actuators. However, it is challenging to make such solution feasible. As the number of devices will be very high, each device needs to be of very low cost and have a long battery life. Further, the generated traffic from all the massive number of locations with small amounts of data communicated at various occasions needs to be handled efficiently in order not to smother the system with interference [1].

2.1.1.7   Media on demand

Media on demand is simply about an individual user's desire to be able to enjoy media content (such as audio and video) at any preferred time and location; see Figure 2.1(g). The user location may range from various places in the city or at home, where one might want to select and see one of the latest movies online. At home, the movies may be viewed on a large TV screen where the wireless device is either a smartphone or a wireless router that forwards the video to the TV screen. A challenging situation appears when a large number of users located in a certain area want to watch their own unique media contents at the same time. For example, residential users in close proximity may want to watch individually selected movies during the evening hours at home [2].

Significantly high data rates are required in order to provide media contents with great user experience. This type of on demand media traffic is typically downlink-dominated, while the uplink is used mainly for application signaling. The absolute media starting delay, i.e. the delay from when the media content is requested to the point when the user can start consuming the media, may not be the most crucial requirement. Here, up to a few seconds may in fact be acceptable, though any reduction in delay is of course desirable. However, once the media application is up and running, the user is much more easily annoyed with interruptions. Therefore, a low latency is still required to be able to quickly get up to the speed after possible link interruptions. The high availability is required to widely provide services to as many users as possible, regardless of their location [2].

2.1.1.8   Remote surgery and examination

The examination and surgery of a patient can potentially be performed remotely; see Figure 2.1(h). In some crucial moments in life, the fraction of a second might make the difference between life and death. To be able to trust wireless technology in such moments, it is of outmost importance to have reliable connection. If, for example, a doctor is performing remote surgery on a patient, the system needs to be able to react almost instantly in order to save the patient's life. Furthermore, remote surgery also offers opportunities for patients in isolated areas to receive healthcare services in a timely and cost-efficient manner [2][3].

A very low end-to-end latency and ultra-reliable communications are required for enabling such critical healthcare services, since it is essential to instantly provide the condition of patients (e.g. through high-resolution images, accessing medical records), and to provide accurate feeling and tactile interaction (i.e. haptic feedback), in the case of remote surgery [2][3].

Although patients are in many cases stationary, such telemedicine services should be also provided, at least partly, in ambulances (eAmbulance) where the stringent requirements may be relaxed and traded off with the high mobility of the vehicle [2].

2.1.1.9   Shopping mall

In a large shopping mall (see Figure 2.1(i)), there are many customers looking for various kinds of personalized services. Access to mobile broadband enables traditional communication as well as other applications such as indoor guiding and product information. Surveillance and other security systems that e.g. enable fire and safety protection can be coordinated via the infrastructure. These services will involve both the traditional radio network and coordinated wireless sensors [1].

The main challenges in a shopping mall are to ensure available connection (upon request for all the users) and to provide secure communications for sensitive services e.g. related to financial aspects. Such a secure link typically does not have very challenging data rates or latency requirements for its protected messages, but may benefit from high availability, and might need reliability to not confuse the secure link with a possible intruder.

To enable these applications, the network needs to have high availability and reliability, especially for the safety-related applications. In addition, the experienced user throughput is of high importance in order to provide good end-user experiences to the customers [1].

2.1.1.10   Smart city

Many aspects from an urban inhabitant perspective will become smarter, e.g. the 'smart home', 'smart office', 'smart building', 'smart traffic control'. All of those together bring 'smart cities' into reality [1][4], see Figure 2.1(j).

Today, connectivity is mainly provided among people, but it will be significantly extended in the future to connect people also with their surrounding environments that can dynamically change as they move from one place to another; such as home, office building, shopping mall, train station, bus station and many others. The connectivity will enable the 'smartness' to life, in order to provide services that are personalized, context- and location-aware. Furthermore, the connectivity among 'objects' is expected to play an increasingly important role for enabling 'smart' services.

To accommodate an unprecedentedly wide range of services, the requirements for mobile wireless technology will be more diverse. For example, cloud services in a 'smart office' will require high data rates at low latency, whereas small devices, wearables, sensors and actuators usually need small payloads with moderate latency requirements, such as product information, electric payment in a shopping mall and temperature/ lighting control in a 'smart home' and 'smart building' context. Besides the diverse requirements, it is also challenging to support a large number of concurrently active connections and an overall high traffic volume in densely populated urban areas. Moreover, the requirements are dynamically changing due to the spontaneous crowd concentration both outdoors and indoors: for example, at train or bus stations when trains or busses arrive or leave, at road crossings when the traffic light changes and in certain rooms when meetings or conferences are held [1].

2.1.1.11   Stadium

A stadium gathers many people interested in the various events, such as sports and concerts, which take place there; see Figure 2.1(k). These spectators want to be able to communicate and exchange media content during the event in the densely crowded arena. This communication generates large amounts of traffic during the events, with highly correlated traffic peaks for instance during breaks or at the end of events, while the traffic is very low at other times [1][8].

The experienced user throughput is of high relevance for the spectators. On a network level, the traffic volume density is a major challenge due to the crowd of users wanting access at the same time [1][8].

2.1.1.12   Teleprotection in smart grid network

Smart grid networks (e.g. related to electricity, water and gas production, distribution and usage) need to be able to react fast to changes in the supply or usage of resources to avoid massive system failures with a potentially critical impact on society. For example, blackout could be a consequence in an energy distribution network when damage is caused by an unforeseen event such as a tree falling in a thunderstorm unless necessary reaction and countermeasures are taken promptly. Here, monitoring and controlling systems in conjunction with wireless communications solutions can play a vital role in providing teleprotection; see Figure 2.1(l). The timely exchange of critical information in a highly reliable manner plays a vital role for the system to be able to react immediately [1][4].

Thus, teleprotection applications require very low latency and high reliability. In the case of teleprotection in a smart grid network distributing electricity, when detecting a fault, alerting messages must be sent and relayed in the network with a very low latency and high reliability in order to take corrective actions for preventing the power system from cascading failures and a critical damage. Although only small payloads have to be sent and mobility is usually not a main issue in many cases, it is essential to satisfy such stringent latency and reliability requirements. The future wireless system satisfying the stringent requirements enables to provide such services in a wide area (nationwide, including rural areas) at a reasonable cost. Due to the critical nature of such infrastructure related applications, high security and integrity standards are commonly required [1][4].

2.1.1.13   Traffic jam

If caught in a traffic jam (see Figure 2.1(m)), many of the passengers would want to enjoy mobile services such as streamed media content. The sudden increase in data traffic demand poses a challenge on the network, especially if the location of the traffic jam is not well covered by the infrastructure, which has typically not been optimized for this case. From an end-user perspective, high experienced user throughput and high availability are important [1].

2.1.1.14   Virtual and augmented reality

Virtual reality is about users being able to interact with one another as if they were physically at the same location; see Figure 2.1(n). In a virtual reality scene, people from various places could meet and interact for a wide range of applications and activities that conventionally need physical presence and interactions, such as conferences, meetings, gaming and playing music. It enables people with specific skills located remotely to jointly perform complicated tasks [1][2].

While virtual reality resembles the reality, augmented reality enriches the reality by providing additional information that is relevant to the surrounding environment of the users. With augmented reality, the users are able to benefit from the additional contextual information that may be also personalized according to their interests [1][4][5][8].

A very high data rate and tight latency are required for enabling the virtual and augmented reality. In order to create the immersive feeling for virtual reality, all users must continuously be updated by streaming data to the others, since each member affects the virtual reality scene. Moreover, in order to enable high user experience of augmented reality, a significant amount of information should be exchanged between sensors/devices of the users and the cloud in both directions. The rich information of the surrounding environment is needed for the cloud to select the appropriate context information, which in turn has to be provided to the users in a timely manner. Further, it is known that if there is a delay between the 'real' reality and the augmented reality of more than a few ms, humans may experience so-called 'cyber sickness'. Multi-directional streams with very high data rates and low latencies are needed to maintain the high-resolution quality.

2.1.1.15   Other use cases: two examples

The fourteen presented 5G use cases capture the main 5G anticipated services, but the list is far from being exhaustive. In the following, two additional use cases are mentioned for completeness.

By providing connectivity to vehicles, **smart logistics** may enable cars and trucks to lower their fuel consumption and to reduce traffic congestion. Such potential benefits may be further enhanced when smart logistics are combined with smart traffic control in smart cities. Another emerging trend is Unmanned Aerial Vehicles (UAVs), which may be autonomously controlled to deliver packets to remote areas [1][2].

With **remote control**, industry applications and machines could be run and managed from other geographical locations than where they are physically located. This would enable tasks to be performed at one (or several) locations that are distant from the location where the task is being geographically executed, increasing productivity and reducing costs. Stringent security and privacy are also commonly required for industry applications [2].

2.1.2   Requirements and key performance indicators

This section provides a summary of the requirements for the use cases that were discussed in Section 2.1.1. A brief description of the Key Performance Indicators (KPIs) in 5G is given herein:

- **Availability**: Availability is defined as the percentage of users or communication links for which the Quality of Experience (QoE) requirements are fulfilled within a certain geographical area.
- **Connection density**: Connection density is defined as the number of simultaneous active devices or users in the considered area during a predefined time span divided by the area size.
- **Cost**: Cost typically arises from infrastructure, end-user equipment and spectrum licenses. A simple model could be based on the assumption that the total cost of ownership for an operator is proportional to the number of infrastructure nodes, the number of end-user devices and the spectrum.
- **Energy consumption**: Energy consumption is typically defined as energy per information bit (typically relevant in urban environments) and as power per area unit (often relevant in suburban/rural environments).
- **Experienced user throughput**: Experienced user throughput is defined as the total amount of data traffic (excluding control signaling) an end-user device achieves on the MAC layer during a predefined time span divided by that time span.
- **Latency**: It is the latency of the data traffic on the MAC layer of the radio interface. Two definitions are relevant: One-Trip Time (OTT) latency and Round-Trip Time (RTT) latency. The OTT latency is defined as the time it takes from when a data packet is sent by the transmitting end to when it is received by the receiving end. The RTT latency is defined as the time it takes from when a data packet is sent by the transmitting end to when an acknowledgement sent by the receiving end is received.
- **Reliability**: Reliability is generally defined as the probability that a certain amount of data has been successfully transmitted from a transmitting end to a receiving end before a certain deadline expires.
- **Security**: The security of a certain communication taking place is very difficult to measure. One possible way to quantify it would be to measure the time it would take for a skilled hacker to access the information.
- **Traffic volume density**: The traffic volume density is defined as the total amount of traffic exchanged by all devices in the considered area during a predefined time span divided by the area size.

Table 2.2 summarizes the main challenging requirements that characterize each use case [1]–[8].

Table 2.2 Summary of main challenging requirements for use cases.

Use cases	Requirements	Desired value
Autonomous vehicle control	Latency	5 ms
	Availability	99.999%
	Reliability	99.999%
Emergency communication	Availability	99.9% victim discovery rate
	Energy efficiency	1 week battery life
Factory cell automation	Latency	Down to below 1 ms
	Reliability	Down to packet loss of less than 10^{-9}
High-speed train	Traffic volume density	100 Gbps/km^2 in DL, and 50 Gbps/km^2 in UL
	Experienced user throughput	50 Mbps in DL, and 25 Mbps in UL
	Mobility	500 km/h
	Latency	10 ms
Large outdoor event	Experienced user throughput	30 Mbps
	Traffic volume density	900 Gbps/km^2
	Connection density	4 subscribers per m^2
	Reliability	Outage probability < 1%
Massive amount of geographically spread devices	Connection density	1,000,000 devices per km^2
	Availability	99.9% coverage
	Energy efficiency	10 years battery life
Media on demand	Experienced user throughput	15 Mbps
	Latency	5 s (start application) 200 ms (after possible link interruptions)
	Connection density	4000 devices per km^2
	Traffic volume density	60 Gbps/km^2
	Availability	95% coverage
Remote surgery and examination	Latency	Down to below 1 ms
	Reliability	99.999%
Shopping mall	Experienced user throughput	300 Mbps in DL, and 60 Mbps in UL
	Availability	At least 95% for all applications, and 99% for safety-related applications
	Reliability	At least 95% for all applications, and 99% for safety-related applications
Smart city	Experienced user throughput	300 Mbps in DL, and 60 Mbps in UL
	Traffic volume density	700 Gbps/km^2
	Connection density	200 000 users per km^2
Stadium	Experienced user throughput	0.3–20 Mbps
	Traffic volume density	0.1–10 Mbps/m^2
Teleprotection in smart grid network	Latency	8 ms
	Reliability	99.999%
Traffic jam	Traffic volume density	480 Gbps/km^2
	Experienced user throughput	100 Mbps in DL, and 20 Mbps in UL
	Availability	95%
Virtual and augmented reality	Experienced user throughput	4–28 Gbps
	Latency	10 ms RTT

2.2   5G system concept

This section describes a 5G system concept that meets the requirements described in the previous sections. To do so, it must provide a flexible platform. It should not be designed toward one single '5G killer application', but toward a multitude of use cases of which many cannot be foreseen today. Vertical industries (e.g. automotive, energy, manufacturing) in particular will require flexibility to obtain tailored solutions using a common network. Hence, the use cases have been utilized as guidance in the development of the 5G system concept, but the system concept is not limited to meet only the identified use cases.

2.2.1   Concept overview

Because of the wide range of requirements, the earlier generations' one-size-fits-all approach will not work for 5G. Therefore, the proposed 5G system concept generalizes key characteristics of the use cases and aligns the requirements, and combines technology components into three generic 5G communication services, supported by four main enablers, as shown in Figure 2.2. Individual use cases can be considered as a 'linear combination' of the 'basis functions'. Each generic 5G service emphasizes a different subset of requirements, but all are relevant to some degree. The generic 5G communication services include functions that are service-specific, and the main enablers include functions that are common to more than one generic 5G service. Further details are found in [9] and in the subsequent chapters.

The three generic 5G services are:

- **Extreme Mobile BroadBand** (xMBB) provides both extreme high data-rate and low-latency communications, and extreme coverage. xMBB provides a more uniform experience over the coverage area, and graceful performance degradation as the number of users increases. xMBB will also support reliable communication for e.g. National Security and Public Safety (NSPS).
- **Massive Machine-Type Communication** (mMTC) provides wireless connectivity for tens of billions of network-enabled devices, scalable connectivity for increasing number of devices, efficient transmission of small payloads, wide area coverage and deep penetration are prioritized over data rates.
- **Ultra-reliable Machine-Type Communication** (uMTC) provides ultra-reliable low-latency communication links for network services with extreme requirements on availability, latency and reliability, e.g. V2X communication and industrial manufacturing applications. Reliability and low latency are prioritized over data rates.

The generic 5G services will not necessarily use the same air interface. The preferred waveform depends on design decisions, and how the generic 5G services are mixed. A flexible OFDM-based air interface is the most suitable for xMBB, whereas new air interfaces as FBMC and UF-OFDM may be promising for uMTC where fast synchronization is necessary. Air interface candidates include e.g. OFDM, UF-OFDM and FBMC; cf. Chapter 7.

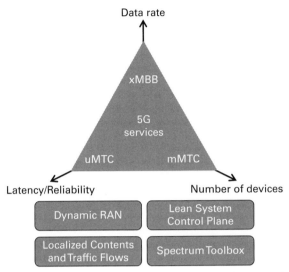

Figure 2.2　The described 5G system concept, showing the three generic 5G services emphasizing different 5G requirements, and the four main enablers [9].

The four main enablers are:

- The **Dynamic Radio Access Network** (DyRAN) provides a RAN that adapts to rapid spatio-temporal changes in user needs and the mix of the generic 5G services. The DyRAN incorporates elements such as

  - Ultra-Dense Networks,
  - Moving Networks (i.e. nomadic nodes and moving relay nodes),
  - Antenna beams,
  - Devices acting as temporary access nodes and
  - D2D communication for both access and backhaul.

- The **Lean System Control Plane** (LSCP) provides new, lean control signaling necessary to guarantee latency and reliability, supports spectrum flexibility, allows separation of data and control planes, supports a large number and variety of devices with very different capabilities, and ensures energy performance.
- **Localized Contents and Traffic Flows** allow offloading, aggregation and distribution of real-time and cached content. Localization reduces the latency and the load on the backhaul and provides aggregation of e.g. sensor information.
- The **Spectrum Toolbox** provides a set of enablers to allow the generic 5G services to operate under different regulatory frameworks, spectrum usage/sharing scenarios and frequency bands.

Overlaps between the services and enablers exist, and certain functions may end up in either category depending on final design decisions. However, it is desirable to make as many functions as possible common, without unacceptable performance degradation, to minimize complexity. Evolved LTE will play an important role in 5G to provide

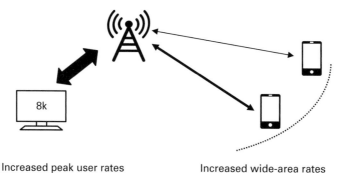

Increased peak user rates   Increased wide-area rates

Figure 2.3   Extreme Mobile Broadband (xMBB) addresses increased peak data rates and increased wide-area rates.

wide-area coverage, and can be considered as one additional generic 5G communication service.

A working definition of a 5G system is one common network that can provide all generic 5G services and is flexible enough to change the service mix dynamically. An operator should be able to change the offered service mix as customer needs change. The spectrum usage should not have to be planned for a certain type of service and should be re-farmed when not needed.

Supporting this 5G system concept requires an architecture that is flexible enough to emphasize different characteristics of the system, e.g. coverage, capacity and latency. The system architecture is described in Chapter 3.

2.2.2   Extreme mobile broadband

The Extreme Mobile Broadband (xMBB) generic 5G service extends today's MBB service and provides versatile communication that supports new applications requiring higher data rates, lower latency and a more uniform user experience over the coverage area; see Figure 2.3. xMBB will meet the expected increase of data volumes and data rates far beyond year 2020 use cases.

xMBB will provide extreme data rates, on the order of Gbps per user, to meet the requirements of high-demand applications such as augmented or virtual reality, or ultra-high-definition video streaming. In addition to high user data-rates, lower latency is also required, e.g. for Tactile Internet [10] combined with cloud computing. Because of the higher user data rates, the system peak rate must increase, often combined with a network densification.

Equally important for the end user is the reliable provisioning of moderate rates everywhere. This extreme coverage is manifested through very reliable support of 50 Mbps–100 Mbps everywhere in the intended coverage area. In crowded areas, xMBB provides a graceful decrease of rate and increase of latency as the number of users increases, instead of refusing service to some users.

The extreme coverage of xMBB and the DyRAN make it possible to establish reliable communication for NSPS as a mode of xMBB, and provide connectivity when the infrastructure is damaged, e.g. after a natural disaster.

The xMBB service will also exhibit robustness with respect to mobility and ensure seamless provision of high-demand applications with a QoE comparable to that of stationary users even while traveling at high speeds in e.g. cars or high-speed trains.

Some of the key solutions to realize the xMBB are access to new spectrum and new types of spectrum access, increased density of the network, improved spectral efficiency including localized traffic, and higher robustness for mobile users. This will require a new air interface suited for dense deployments and new spectrum bands. This should be the same air interface for access, D2D, and wireless self-backhaul in xMBB.

2.2.2.1 Access to new spectrum and new types of spectrum access

To meet the traffic requirements, access to more spectrum, and techniques for more flexible and efficient spectrum utilization are needed; cf. Chapter 12. Contiguous bandwidth is preferable since it allows for a simpler implementation and avoids carrier aggregation[1].

Both centimeter Waves (cmW) and millimeter Waves (mmW) are important to xMBB and 5G, but the solutions must be adapted to the frequency range and actual deployment. For example, as the frequency increases, beamforming becomes both necessary to counteract the decrease in received power due to the decrease in antenna aperture, and also practically feasible due to smaller antenna size. Here cmWs are good candidates for an attractive trade-off between coverage and ability to use multi-antenna systems.

xMBB must support spectrum-flexible operations in traditional spectrum, cmW and mmW bands using licensed access, Licensed Shared Access (LSA) and Licence-Assisted Access (LAA). Furthermore, multi-connectivity built on a close integration between the new air interface operating above 6 GHz and different existing systems, such as evolved versions of LTE, is necessary to provide a consistent user experience.

2.2.2.2 New radio interface for dense deployments

xMBB must consider the continued densification of the network into Ultra-Dense Networks (UDNs). As a consequence of the densification, the number of active devices per access node will decrease, and UDNs will less often be working under high-load conditions.

An air interface optimized for spectrum-flexible, short-distance communications for xMBB based on a harmonized OFDM is described in Chapter 7. The air interface is not only optimized for access in traditional cellular use, but also for D2D, and wireless self-backhaul. The air interface is harmonized for operations from 3 GHz to 100 GHz, and the cmW and mmW bands (cf. Chapter 6) are optimized for spectral efficiency in the UDN context.

---

[1] Carrier aggregation can lead to a huge number of band combinations globally.

2.2.2.3   Spectral efficiency and advanced antenna systems

The most promising technology to improve spectral efficiency is the use of advanced multi-antenna systems, e.g. massive Multiple Input Multiple Output (MIMO) and Coordinated Multi-Point[2] (CoMP); cf. Chapters 8 and 9, respectively.

Multi-antenna systems in xMBB support both extreme data-rates in a given area by improving the spectral efficiency, and extreme coverage and reliable moderate rates for users in crowded situations. For xMBB, OFDM-based solutions are preferred since they have a proven track-record with MIMO and they simplify the backwards interoperability.

Minor spectral efficiency gains are achieved by additional filtering, e.g. using UF-OFDM or FBMC, in xMBB applications. The main benefit of additional filtering comes when mixing services.

2.2.2.4   Number of users

To support a high number of users during initial access, xMBB can benefit from the methods to overload the physical resources proposed for mMTC; cf. Chapter 4. After initial system access, it is up to the scheduler to provide fairness. The use of DyRAN, D2D communication, and localized traffic can also improve the QoE for large numbers of users.

2.2.2.5   User mobility

Mechanisms for interference identification and mitigation, mobility management and prediction, handover optimization, and context awareness all provide benefits to xMBB; cf. Chapter 11.

2.2.2.6   Links to the main enablers

DyRAN provides short-distance communication in UDN which improves the Signal to Interference and Noise Ratio (SINR) and hence increases the data-rates and capacity. The network densification results in new three-dimensional and multi-layer interference environments, which must be addressed; cf. Chapter 11.

Localized Contents and Traffic Flows improve the system performance and reduce latency in xMBB using D2D communication to exchange information locally between devices in proximity; cf. Chapter 5.

The Spectrum Toolbox allows xMBB operations using licensed access, LSA, and LAA in traditional spectrum, cmW and mmW bands; cf. Chapter 12.

The Lean System Control Plane supports spectrum-flexible and energy-efficient operations; cf. Section 2.2.6.

2.2.3   Massive machine-type communication

Massive Machine-Type Communication (mMTC) provides efficient connectivity for a large number of cost- and energy-constrained devices. mMTC includes a very wide

---

[2] The performance of CoMP suffers from backhaul limitations in already deployed systems.

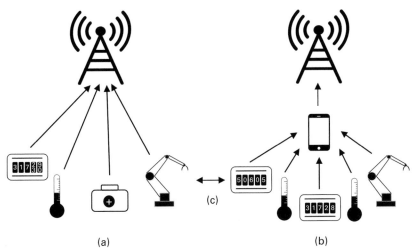

Figure 2.4   Massive Machine-Type Communication (mMTC) and its three access types, (a) direct network access (MTC-D), (b) access via an aggregation node (MTC-A) and (c) short-range D2D access for mMTC devices (MTC-M).

range of use cases, ranging from the wide-area use case with deployments of massive numbers of geographically spread devices (e.g. sensors and actuators) for surveillance and area-covering measurements, to more local cases connecting electronic devices in the smart home restricted to indoor environments in populated areas, or in close proximity of human users as in the case of body-area networks. Common for all these cases is that data payload is of small size and traffic is typically sporadic in comparison to xMBB. Since frequent battery charging and replacement is not feasible due to the large number of devices and the fact that devices may be deployed once and for all, particularly energy-consuming operations should be located on the infrastructure side, and operations on the device side should be kept as brief as possible to minimize the device on-time. This leads to an increase in asymmetry compared to today's networks, a trend which is exactly opposite compared to xMBB.

mMTC must be generic enough to support new, yet unforeseen use cases and not be restricted to what can be imagined today. To manage the highly heterogeneous mMTC devices, three mMTC access types are envisioned for mMTC: direct network access (MTC-D), access via an aggregation node (MTC-A), and short-range D2D access for mMTC devices (MTC-M) in case the traffic is end-to-end between nearby devices; see Figure 2.4. Ideally, the same air interface solution should be used for all three access types to minimize device costs. For the majority of machine devices, the access method will be MTC-D.

The main challenges for mMTC are a large number of devices, extended coverage, protocol efficiency and other mechanisms to achieve long battery lifetime, and limited capabilities of inexpensive devices; cf. Chapter 4.

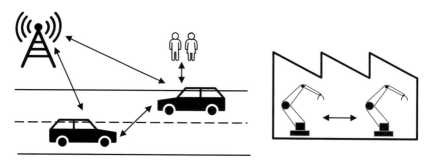

Road safety and traffic efficiency   Industrial manufacturing

Figure 2.5   Ultra-reliable Machine-Type Communication (uMTC) with applications to road safety and traffic efficiency and industrial manufacturing.

2.2.3.1   Links to the main enablers

For connection-oriented mMTC traffic, the DyRAN and Localized Contents and Traffic Flows can store context information in the network to reduce the amount of necessary transmissions. If context information is stored in the network, then transmissions can be reduced leading to longer battery lifetime. Relaying for coverage extension also affects the DyRAN.

mMTC can benefit from a closer integration of control- and user-planes, as opposed to xMBB, which affects the design of the LSCP.

2.2.4   Ultra-reliable machine-type communication

Ultra-reliable Machine-Type Communication (uMTC) provides ultra-reliable and low-latency communication for demanding applications. Two typical examples include road safety and traffic efficiency, and industrial manufacturing (see Figure 2.5), which both have stringent requirements on low latency and very high reliability.

In the road safety and traffic efficiency applications, information is exchanged between traffic participants using Vehicle-to-Vehicle (V2V), Vehicle-to-Pedestrian (V2P) or Vehicle-to-Infrastructure (V2I) communication. With a slight abuse of the terminology the term Vehicle-to-Anything (V2X) includes V2V, V2P and V2I, for traffic safety and efficiency applications.

V2X communication includes both periodic and event-driven messages. The periodic messages are transmitted to avoid the occurrence of dangerous situations. A traffic participant can broadcast its position, velocity and trajectory, etc. periodically (e.g. every 10 ms) to recipients within a certain range (e.g. 100 m). The event-driven messages are transmitted when an abnormal and/or dangerous situation is detected, e.g. oncoming vehicles or accidents. Though both kinds of messages should have high reliability, the event-driven messages are more critical and should be received in the proximity with very high reliability and almost no delay.

In the industrial manufacturing setting we consider three main categories;

- Stationary equipment, including rotating and moving parts, mostly indoors deployment. Sensors and actuators attached to the equipment are assumed to be part of the manufacturing process control loop.
- Autonomous transport robots, both indoors and outdoors. This category is similar to V2X applications but the expected speeds are lower, and the environment is not public.
- Sensors deployed on equipment and/or parts for monitoring purposes. The output of these sensors is not a part of the manufacturing process control loops.

In industrial manufacturing applications, the discovery and communication establishment requirements may be less stringent than for V2X but the reliability must still be high. Thus, many techniques applicable for V2X will also be useful for industrial manufacturing. Monitoring sensors can use solutions similar to mMTC but higher reliability at the cost of reduced battery lifetime can be desirable.

The main challenges for uMTC are fast communication setup, low latency, reliable communication, high availability of the system, and high mobility; cf. Chapter 4.

2.2.4.1　Links to the main enablers

uMTC will benefit from the interference identification and mitigation schemes in the DyRAN; cf. Chapter 11. For V2X applications, the interference environment will change rapidly and for industrial applications, the interference environment is typically not Gaussian [11]. Together with improved interference knowledge, context information and mobility prediction will play an important role for reliability in V2X communications.

Localized Contents and Traffic Flows are important to reduce latency and improve the reliability. The traffic status messages are local per se. For other applications (e.g. driver assistance and remote driving), it may be necessary to move application servers from central locations toward the radio edge to reduce latency, i.e. opposite to the cloud trend. This also affects the 5G architecture; cf. Chapter 3.

The fast connection set-up and low-latency communication affects the LSCP.

Multi-operator D2D operation including access to spectrum is covered by the Spectrum Toolbox; cf. Section 2.2.8 and Chapter 12.

2.2.5　Dynamic radio access network

To address the diverse requirements, the 5G Radio Access Network (RAN) will encompass different RAN enablers, or elements. Traditional macro-cellular networks provide wide-area coverage, Ultra-Dense Networks and Nomadic nodes provide local-area capacity increase, beamforming will be required for both wide-area coverage and SINR improvements at higher frequency, and D2D communication is applicable to both access and backhaul. However, each element by itself will not address the diverse needs and will not be able to adapt to the time- and location-varying requirements on capacity, coverage and latency. The Dynamic RAN (DyRAN) integrates all these elements in a dynamic manner for multi-RAT environments; see Figure 2.6. The DyRAN will adapt to rapid spatio-temporal changes in user needs and the mix of the generic 5G services.

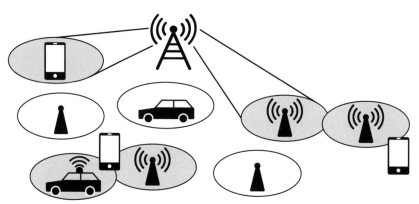

Figure 2.6   Illustration of a Dynamic RAN, including UDN nodes, nomadic nodes, antenna beams and backhaul. Shaded indicates activated, white indicates deactivated.

The different elements address the same fundamental technical requirement of increased SINR in the coverage area. E.g. massive MIMO beamforming and UDN can both be used to increase the average SINR in an area. The choice depends on technical and non-technical considerations. In dense urban environments, the UDN solution may be preferable whereas the massive MIMO solution may be preferable in suburban and rural environments.

Sections 2.2.5.1–2.2.5.4 describe the elements of the DyRAN, and Sections 2.2.5.5–2.2.5.8 describe some common functions in the DyRAN. Though the functions are common, the actual implementation may not necessarily be the same. The DyRAN is closely related to the system architecture, and supports different distributions of functions depending on the service and the computational capabilities of the network nodes; cf. Chapter 3.

2.2.5.1   Ultra-dense networks

Network densification is a straightforward way to increase the network capacity, and network densification will continue from macro-cellular networks through small cells to UDNs. UDNs will be deployed both outdoors and indoors, and can have inter-site distances down to a few meters.

UDNs target user data rates on the order of 10 Gbps, which translates to requirements on high (local) area capacity and high throughput. Providing this in an energy-efficient manner requires access to large, preferably contiguous, bandwidth, which is only realistic in the cmW and mmW bands. Communication in the cmW and mmW bands is treated in Chapter 6, and UDN air interface is treated in Chapter 7.

UDNs should be able to operate both stand-alone and capacity-boosting "islands" with tight interworking with an area-covering overlay network, e.g. evolved LTE. In stand-alone operation, the UDN must provide full functionality of a mobile communication network, including system access, mobility management, etc. In operation of UDN together with an area-covering overlay network, the UDN and the overlay network can share network functions. For example, the overlay network's C-plane can be common to the overlay network and the UDN, whereas the U-plane may be different for the overlay network and the UDN. A third-party-deployed UDN, e.g. in a building, can provide its capacity for indoors coverage and capacity to multiple operators, i.e. interworking with multiple overlay networks. Even user-deployed UDN access nodes may be supported.

The large number of UDN nodes prohibits traditional cell planning, and self-organization beyond today's self-organizing networks is instrumental. Novel methods for e.g. interference mitigation will be necessary; cf. Chapters 9 and 11.

The UDN network can be used for providing backhaul for a variety of different access technologies. Depending on the capabilities of the access nodes, the access link can be Wi-Fi, ZigBee, etc. This application is foreseeable in mMTC operations where devices access a UDN access node using an appropriate air interface.

2.2.5.2 Moving Networks

Consist of nomadic nodes and/or moving relays nodes.

- **Moving relay nodes** are wireless access nodes that provide communication capabilities to in-vehicle users, especially in high-mobility scenarios. Typical moving relay nodes would be trains, busses and trams, but possibly also cars. Moving relay nodes can overcome the outdoor to indoor penetration losses due to metalized windows.[3]
- **Nomadic nodes** are a new kind of network node, where the on-board communication capabilities of vehicles are utilized to make the vehicles serve as a temporary access node for both in-vehicle and outside users. Nomadic nodes enable network densification to meet traffic demands varying over time and space. Nomadic nodes resemble UDN nodes, but offer their services as temporary access nodes at non-predictable locations and at non-predictable times, and any solution must handle this dynamic behavior.

2.2.5.3 Antenna beams

Beamforming i.e. the forming of antenna beams, can be used for example to increase the SINR in a local area, in the context of massive MIMO or CoMP. Though the antenna site itself is fixed in location, the beam-direction is dynamic in space and time, and the illuminated area can be considered as a virtual cell. The virtual cell created by beamforming is more controllable than nomadic nodes.[4] Massive MIMO and CoMP are treated in detail in Chapters 8 and 9, respectively.

2.2.5.4 Wireless devices as temporary network nodes

High-end wireless devices, such as smartphones and tablets, have capabilities similar to inexpensive UDN nodes. A device equipped with D2D capability can act as a temporary infrastructure node for e.g. coverage extension. In this mode, a device may take certain network management roles, e.g. resource allocation between D2D pairs, or mMTC

---

[3] This outdoor-to-indoor penetration loss also occurs in energy-efficient buildings.
[4] The ability to fully control the antenna beams or select between a set of pre-defined beams depends on whether digital, analog or hybrid beamforming is used.

gateway functionality. However, admitting user devices into the RAN as temporary access nodes lead to trust issues that need to be resolved.

2.2.5.5 Device-to-device communication

Flexible D2D communication is a key element in the DyRAN where it can be used for access, offloading the U-plane to a D2D-link and backhaul. After device discovery, the most suitable communication mode will be selected based on various criteria, e.g. capacity needs and interference levels. D2D communication is also applicable in wireless self-backhauling. D2D communication is treated in detail in Chapter 5.

2.2.5.6 Activation and deactivation of nodes

As the number of candidate access nodes increases, so does the probability that an access node is idle. To minimize the network energy consumption and interference, the DyRAN employs activation/deactivation mechanisms to select which elements (nodes, antenna beams, D2D links and/or devices) should be activated at which times and locations to meet coverage and capacity demands. Nodes or beams not serving any users should be deactivated. Activation and deactivation also affect where in the DyRAN network functions are executed. This may trigger dynamic reallocation of network functions; cf. Chapter 3.

2.2.5.7 Interference identification and mitigation

The interference environment will become more dynamic. The interference does not only arise from users, as the activation and deactivation of nodes and antenna beams will also affect the interference environment. Hence, dynamic interference and radio resource management algorithms are necessary in the DyRAN. Methods for interference identification and mitigation are treated in Chapter 11.

2.2.5.8 Mobility management

In the DyRAN, mobility management is applicable to both terminals and MTC devices, and to access nodes. For example, a nomadic node may become unavailable, and a user could encounter handover decision, even if the user itself is stationary. Similarly, the wireless backhaul to moving nodes must be protected against sudden disruptions. Smart mobility management techniques are required that ensure seamless connectivity in the DyRAN; cf. Chapter 11.

2.2.5.9 Wireless backhaul

The nodes constituting the DyRAN are not always connected to a wired backhaul; moving nodes will never be connected to a wired backhaul, nomadic nodes rarely and UDN nodes likely. Hence, wireless backhaul is essential to leverage the gains of DyRAN. Wireless backhaul links can be arranged in a mesh topology utilizing D2D communication between nodes, and significantly improve the capacity and reliability of the whole network.

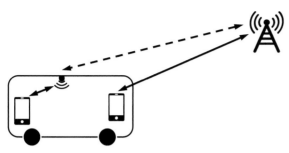

Figure 2.7   Similarity between access and backhaul links shown for a moving node. The left terminal is connected to the on-board access node which is using wireless backhaul, and the right terminal is connected to the macro base station.

For moving and nomadic nodes, predictor antennas, massive MIMO and CoMP techniques can be used to increase the robustness and throughput of wireless backhaul; cf. Chapters 8 and 9. Relaying and network coding (cf. Chapter 10) and interference-aware routing can increase the throughput.

The backhaul nodes are often assumed to be stationary. However, the backhaul to a moving node (e.g. bus or train) has very similar characteristics to the access link; see Figure 2.7. Hence, it is desirable to have a common air interface for access, backhaul and D2D links.

2.2.6   Lean system control plane

The control signaling must be fundamentally readdressed in 5G systems to accommodate the different needs of the three generic 5G services, spectrum flexibility and energy performance. The purpose of the Lean System Control Plane (LSCP) is to:

- provide a common system access,
- provide service-specific signaling,
- support C- and U-plane separation,
- integrate different spectrum and inter-site distance ranges (in particular for xMBB) and
- ensure energy performance.

Finally, the LSCP must provide enough flexibility to accommodate not yet foreseen services.

2.2.6.1   Common system access

The initial access to the 5G system is through a broadcast where the first signaling is common to all services; see Figure 2.8. The broadcast signal should be the minimally needed system detection signal that is always[5] transmitted. This common system access transmission integrates the generic 5G services and should also allow selected legacy technologies to be accessed through the LSCP.

---

[5] Here 'always' is not necessarily 'continuously', but with silent periods that are short enough so that the detection delay is not unacceptably large.

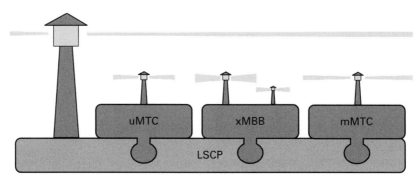

Figure 2.8   Illustration of the access broadcast signaling and the service-specific signaling of the Lean System Control Plane (LSCP).

2.2.6.2   Service-specific signaling

Additional service-specific signaling, see Figure 2.8, should be transmitted only when a user/device desires to transmit data using that service in order to avoid that service-specific reference signals are transmitted in empty areas.

To support extremely high data rates, xMBB requires service-specific signals to obtain precise channel state information and enable spectrally efficient transmissions. The realization of the service-specific signals depends on which frequency bands xMBB operates in.

mMTC requires optimized sleep mode solutions for battery operated devices, and mobility procedures with a minimum of signaling and measurements.

uMTC requires guaranteed latency and reliability, and here 'lean signaling' should also consider effects on the total latency budget for a given packet transmission. For critical uMTC applications, the signaling should be designed in a way that ensures connection resilience under practically all conditions.

2.2.6.3   Control and user plane separation

Different separation of the C- and U-planes can be considered for the generic 5G services. For xMBB, it is beneficial to separate the C- and U-planes to allow transmission at different frequencies, e.g. C-plane at lower frequencies for good coverage and U-plane at higher frequencies for higher data rates. In the case of network-controlled D2D offloading, the U-plane is transmitted over D2D links.

For mMTC, it may be advantageous to integrate the C- and U-Planes [12]; cf. Chapter 4. The current solution in LTE is not sufficiently good for mMTC concerning signaling overhead, energy performance and coverage.

The potential dual connectivity of uMTC implies even more C- and U-plane combinations.

2.2.6.4   Support of different frequency ranges

To realize xMBB, 5G systems will integrate nodes with large and small coverage areas operating in different frequencies, e.g. macro cells below 6 GHz and fixed and/or nomadic nodes in cmWs and/or mmWs. The LSCP provides a seamless mechanism for operation in different frequency ranges.

2.2.6.5 Energy performance

Energy performance is achieved by having separate signaling solutions to provide coverage and capacity. The coverage signaling is achieved through the common system access described above. The capacity signaling must be more adaptive than today's solutions, since different services will be used at different times and locations. This is achieved by the service-specific signaling.

Separation of the C- and U-planes minimizes the "always on" signaling, and supports discontinuous transmission and reception in the data plane, which improves the system energy performance. Activation and deactivation of network nodes (cf. Section 2.2.5.5) also improve energy performance.

2.2.7 Localized contents and traffic flows

One of the key challenges of 5G is to reduce latency. The largest delay contributions occur in the core network and Internet parts of the end-to-end communication. Data traffic offloading, aggregation, caching and local routing can be employed to meet the latency target [13]. Latency is also reduced, and reliability improved, by moving application servers toward the radio edge.

The increased amount of traffic will pose a challenge not only for the wireless access links but also for the backhaul and transport network. Some information is only of local interest, e.g. traffic safety information and proximity-based marketing. By identifying this kind of contents and keeping it close to the radio edge, the load on the transport network will be minimized.

Localized contents and traffic flows include functions to reduce latency and offload the transport network.

2.2.7.1 Anti-tromboning

Tromboning occurs when traffic between two nodes in close proximity is routed to a central location and then back toward the edge again [14]. Anti-tromboning techniques enable the traffic to be 'turned around' as early as possible in the network to minimize latency and transport load; see Figure 2.9. In addition to the technical challenge of identifying traffic intended for nodes in close proximity, this has regulatory and legal implications since inspection and analysis of the traffic is needed.

2.2.7.2 Device-to-device offloading

One anti-tromboning technique is to offload the traffic to a D2D communication link; see Figure 2.9. The U-plane is transmitted over the direct D2D link, whereas the C-plane remains under network control to e.g. ensure interference coordination, and provide authentication and security features. In a sense, D2D communication provides the maximum localization of traffic since the U-plane never enters the network.

Device discovery mechanisms can be utilized to identify suitable pairs for D2D offloading, both with and without network coverage; cf. Chapter 5. In some V2X

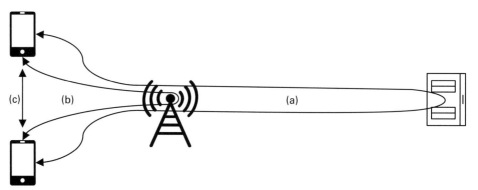

Figure 2.9   Illustration of how Localized Contents and Traffic Flows avoid tromboning. In a) the traffic is routed to the central office and back, in b) the traffic is turned in the base station and in c) the U-plane is offloaded to D2D communication.

applications, the message is broadcasted without the discovery phase in order to further reduce the latency.

In mMTC, the use of concentrators acting as local gateways could allow direct communication among sensors located in a local area without the need to reach the core network gateway cf. Figure 2.4. For mMTC, the localized traffic flows allow low-power access to the network. Further, the necessary context information for mMTC operations can be stored locally.

2.2.7.3   Servers and contents close to the radio edge

To meet the delay constraints of certain delay-sensitive services, e.g. autonomous vehicle control, it is necessary to move the application server close to the radio edge and perform critical computations close to the user. This is the opposite of the centralization often assumed in C-RAN, and has implications on system architecture; cf. Chapter 3. Moving application servers to the radio edge requires mobility management not only for the terminals but also for applications running in servers close to the edge.

Contents can also be distributed, and caching can be shifted toward the radio edge, including the access nodes. Devices can act as proxies in case they have the requested content in the memory. Storing contents on devices allows shifting the communication in time (preloading the expected contents), but digital rights management issues must be addressed.

2.2.8   Spectrum toolbox

The generic 5G services will support a wide range of use cases with different requirements on the employed spectrum, e.g. spectrum band, signal bandwidth and authorization schemes. Additionally, the mix of different services in a 5G system may be changed, and it is therefore necessary to reassign spectrum on a timescale of hours.

Hence, in addition to access to more spectrum, the spectrum usage in 5G systems must be highly flexible, and have the capability to operate under different authorization modes in various frequency bands. The Spectrum Toolbox provides the tools to meet these requirements.

This section describes the spectrum requirements of the three generic 5G services xMBB, mMTC and uMTC, and gives a brief overview of the Spectrum Toolbox. For further detail; cf. Chapter 12.

2.2.8.1   Spectrum needs for xMBB

xMBB addresses the increase in traffic volume and data rates, and reliable moderate rates. To meet the requirements on data rates, additional spectrum is necessary, most likely above 6 GHz to find wider contiguous bandwidths. In the cmW band, contiguous bandwidths of 100s of MHz are desirable, and in the mmW band bandwidths exceeding 1 GHz are desirable.

To meet the requirements on coverage with moderate rates, spectrum in lower frequency bands is essential. Hence, a mix of spectrum in lower bands for coverage purposes and higher bands for capacity, including wireless backhaul solutions, is required for xMBB.

Exclusive spectrum access is preferred to guarantee coverage and QoS, complemented by other licensing regimes, e.g. LAA, LSA or unlicensed access (e.g. Wi-Fi offload) to increase the overall spectrum availability and hence capacity.

In case of wireless self-backhaul, where the same resources are shared for both access and backhaul, a sufficient amount of spectrum must be available to allow for high-data rate in both access and backhaul links.

2.2.8.2   Spectrum needs for mMTC

mMTC requires good coverage and penetration conditions, while the bandwidth requirements are comparably small. For coverage and propagation purposes, frequency spectrum below 6 GHz is most suitable and spectrum below 1 GHz is needed. The available bandwidths in these frequency ranges are considered sufficient since the spectrum requirement for mMTC is relatively small; 1–2 MHz is currently expected to be enough [1]. However, it is important to be able to increase the system bandwidth for mMTC if needed in the future. Therefore, fixed band allocations should be avoided.

Sensors will be simple devices with no or very limited possibility for upgrades after deployment, and with a long expected lifetime. Therefore, a stable regulatory framework is needed. Exclusive licensed spectrum is the preferred option. Other licensing regimes can be considered depending on application-specific requirements and desire for global harmonization.

2.2.8.3   Spectrum needs for uMTC

uMTC requires high reliability and low latency. To realize low latency, the signal bandwidth can be increased to reduce the transmission time. Frequency diversity also increases the reliability.

Exclusive spectrum access or very high priority in spectrum access is essential to provide the reliability. For V2X communication, a harmonized band for Intelligent Transport Systems (ITS) exists [15]; cf. Chapter 12.

2.2.8.4   Properties of the spectrum toolbox

The Spectrum Toolbox enables flexible use of available spectrum resources aiming at increasing the efficiency in the use of spectrum. Thus, it is a fundamental enabler for multi-service operations and spectrum-flexible air interfaces. The toolbox provides tools to:

- Enable operation in widely distributed spectrum bands, both at high and low frequencies, by considering the suitability of different spectrum bands dependent on applications.
- Facilitate different sharing scenarios by applying respective mechanisms either solely or in combination.
- Facilitate operation by using small as well as large bandwidths, which enables spectrum-flexible air interfaces supporting higher data rates.
- Adopt different rules for different services, e.g. certain spectrum may only be used for specific services.

The functionality of the Spectrum Toolbox is divided into three domains: the regulatory framework domain, the spectrum usage scenario domain, and the enabler domain; cf. Chapter 12.

2.3   Conclusions

This chapter has summarized the key 5G use cases and their requirements, and the overall 5G system concept.

It was shown that the identified 5G use cases can be classified into three main categories representing the requirement extremes in 5G: extreme mobile broadband (xMBB), where ubiquitous high throughput is key, massive machine-type communications (mMTC), where coverage and device-side cost and power constraints are the key challenges, and ultra-reliable MTC (uMTC), which is characterized by stringent latency and reliability requirements.

The chapter has further described the four key system concepts required to provide the efficiency, scalability, and versatility to address the wide range of requirements associated with the aforementioned use cases: a Dynamic Radio Access Network, Lean System Control Plane, Localized Contents and Traffic Flows, and a Spectrum Toolbox.

As the ITU-R 5G requirements are yet to be finalized, the identified 5G use cases and system concept may have to be updated, but it is expected that the key considerations with respect to the classification of use categories and the identified key system concepts should still be valid.

References

[1] ICT-317669 METIS project, "Future radio access scenarios, requirements and KPIs," Deliverable D1.1, April 2013, www.metis2020.com/documents/ deliverables/

[2] ICT-317669 METIS project, "Updated scenarios, requirements and KPIs for 5G mobile and wireless system with recommendations for future investigations," Deliverable D1.5, April 2015, www.metis2020.com/documents/deliverables/

[3] NGMN Alliance, "NGMN 5G White paper," February 2015, www.ngmn.org/upl oads/media/NGMN_5G_White_Paper_V1_0.pdf

[4] 4G Americas, "4G Americas' recommendations on 5G requirements and solutions," October 2014.

[5] ARIB 2020 and beyond ad hoc group, "Mobile communications systems for 2020 and beyond," October 2014.

[6] GSMA, "Understanding 5G: Perspectives on future technological advancements in mobile," December 2014.

[7] Industrie 4.0 working group, "Recommendations for implementing the strategic initiative INDUSTRIE 4.0," April 2013.

[8] IMT-2020 (5G) promotion group, "5G vision and requirements," May 2014.

[9] ICT-317669 METIS project, "Final report on the METIS 5G system concept and technology roadmap," Deliverable D6.6, April 2015, www.metis2020.com/documents/deliverables/

[10] G. P. Fettweis, "The Tactile Internet: Applications and challenges," *IEEE Vehicular Technology Magazine*, vol. 9, no. 1, pp. 64–70, March 2014.

[11] P. Stenumgaard, J. Chilo, P. Ferrer-Coll, and P. Angskog, "Challenges and conditions for wireless machine-to-machine communications in industrial environments," *IEEE Communications Magazine*, vol. 51, no. 6, pp. 187–192, June 2013.

[12] G. Wunder, P. Jung, M. Kasparick, T. Wild, F. Schaich, Y. Chen, S. ten Brink, I. Gaspar, N. Michailow, A. Festag, L. Mendes, N. Cassiau, D. Ktenas, M. Dryjanski, S. Pietrzyk, B. Eged, P. Vago, and F. Wiedmann, "5GNOW: Non-orthogonal, asynchronous waveforms for future mobile applications," *IEEE Communications Magazine*, vol. 52, no. 2, pp. 97–105, February 2014.

[13] C. B. Sankaran, "Data offloading techniques in 3GPP Rel-10 networks: A tutorial," *IEEE Communications Magazine*, vol. 50, no. 6, pp. 46–53, June 2012.

[14] Y.-B. Lin, "Eliminating tromboning mobile call setup for international roaming users," *IEEE Transactions on Wireless Communications*, vol. 8, no. 1, pp. 320–325, January 2009.

[15] CEPT ECC, "ECC Decision (08)01: The harmonised use of the 5875–5925 MHz frequency band for Intelligent Transport Systems (ITS)," ECC/DEC/(08)01, March 2008, www.erodocdb.dk/docs/doc98/official/pdf/ECCDec0801.pdf

3 The 5G architecture

Heinz Droste, Icaro Leonardo Da Silva, Peter Rost, and Mauro Boldi

3.1 Introduction

The design of a mobile network architecture aims at defining network elements (e.g. Base Stations [BSs], switches, routers, user devices) and their interaction in order to ensure a consistent system operation. This chapter discusses basic considerations and provides an overview of current research activities. Network architecture can be considered from different angles that are needed in order to fulfill objectives like integration of technical components into an overall system, proper interworking of multi-vendor equipment and efficient design of physical networks from cost and performance point of view.

As 5G systems have to integrate a plethora of partly contradicting requirements, enablers such as Network Function Virtualization (NFV) and Software Defined Networking (SDN) are to be applied in order to provide the needed flexibility of future networks, especially for the core network. Applying these tools may require a rethinking of some traditional aspects of network architecture design. This chapter will give the reader an impression of the most important topics influencing architecture design of future networks.

3.1.1 NFV and SDN

Today's operator networks include a large and increasing variety of hardware appliances. Launching new services often requires integration of complex hardware dedicated to the service including costly procedure design and is associated with lengthy time to market. On the other hand, hardware life cycles become shorter as technology and service innovation accelerates.

At the end of 2012, network operators have started an initiative on NFV [1]. NFV aims at consolidating the variety of network equipment onto industry-standard high-volume servers. These servers can be located at the different network nodes as well as end-user premises. In this context, NFV relies upon but differs from traditional server virtualization. Unlike server virtualization, Virtualized Network Functions (VNF) may consist of one or more virtual machines running different software and processes in order to replace custom hardware appliances (Figure 3.1). As a rule, multiple VNFs are to be used in sequence in order to provide meaningful services to the customer.

5G Mobile and Wireless Communications Technology, ed. A. Osseiran, J. F. Monserrat, and P. Marsch. Published by Cambridge University Press. © Cambridge University Press 2016.

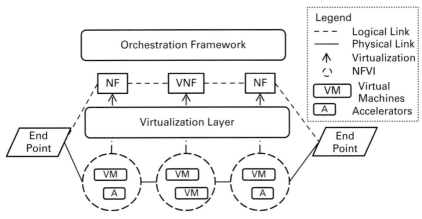

Figure 3.1   NFV framework.

Example: The air interface is arranged in different layers that build upon each other (cf. Figure 3.5). In order to provide connectivity, RF processing, physical layer, medium access control, radio link control and packet data convergence protocol layer are arranged in sequence.

NFV requires an orchestration framework that enables proper instantiation, monitoring and operation of VNFs and Network Functions (NFs) (e.g. modulation, coding, multiple access, ciphering, etc.). In fact, the NFV framework consists of software implementations of network functions (VNF), hardware (industry standard high volume servers) that is denoted as NFV Infrastructure (NFVI) and a virtualization management and orchestration architectural framework. In order to realize real time requirements some NF may need inclusion of hardware accelerators. The accelerators take over computation intensive and time critical tasks that still cannot be realized by NFVI. Hence, not only can traffic be off-loaded from NFVI but also adherence of latency requirements can be ensured. Opportunities and limitations for NFV are described in more detail in Section 3.4.1. As depicted in Figure 3.1 in context with virtualization physical and logical paths between the end points (e.g. devices) in the network are to be distinguished.

The most significant benefits of NFV [1] are reduction of capital and operational expenditures as well as increased speed of time to market. However, an important prerequisite for leveraging these benefits is that VNFs are portable between different vendors and that they can coexist with hardware-based network platforms.

As mentioned previously, in addition to NFV, SDN is another important enabler for 5G future networks. The basic principles of SDN are separation of control and data planes (also called infrastructure layer or user planes), logical centralization of network intelligence and abstraction of physical networks from the applications and services by standardized interfaces [2]. Further, the control of the network is concentrated in a control layer (control plane), whereas network devices like switches and routers that

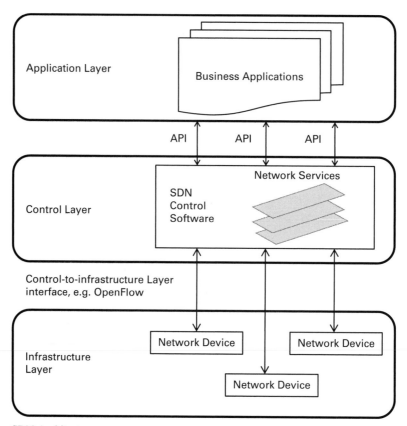

Figure 3.2   SDN Architecture.

handle data plane functionalities are distributed within the network topology of the infrastructure layer (Figure 3.2).

The control layer interacts at one side with the applications and services via standardized Application Programing Interfaces (API) and at the other side with the physical network via a standardized instruction set called OpenFlow. APIs make it possible to implement network services like routing, security and bandwidth management. OpenFlow allows direct access to the plane of network devices like multi-vendor switches and routers. Since it allows the network to be programmed on a per flow basis, it provides extremely granular control enabling the network to respond in real time to demand changes at the application layer and avoid cumbersome manual configuration of network devices. From a topology point of view, NFs belonging to the control and infrastructure layers may be arranged centrally, as well as in a distributed manner, based on the requirements explained in more detail in Section 3.3.

NFV and SDN do not rely on each other. However, as NFV is providing a flexible infrastructure on which the SDN software can run and vice versa, that is, SDN concept enables flow based configuration of network functions, both concepts are to be seen as highly complementary.

In 5G networks, both concepts will serve as key enablers to provide the needed flexibility, scalability and service-oriented management. As for economic reasons networks cannot be dimensioned for peak requirements, flexibility means that tailored functionalities will have to be made available on-demand. Scalability is to be supported in order to fulfill requirements of contradicting services like massive Machine-Type Communication (mMTC), ultra MTC (uMTC) and extreme Mobile BroadBand (xMBB), e.g. by inclusion of adequate access procedures and transmission schemes (see Chapter 4 for more information on MTC). Service-oriented management will be realized by flow-oriented control and user planes enabled by joint NFV and SDN frameworks.

3.1.2   Basics about RAN architecture

The design of network architectures aims firstly at integrating technical components into an overall system and making them properly interoperable. In this context, it is very important to define a common understanding on how components designed by different manufacturers are capable of communicating so that they can execute the needed functionalities. So far in standardization, this common understanding is achieved by the specification of a logical architecture consisting of logical Network Elements (NEs), interfaces and related protocols. Standardized interfaces allow for communication between NEs with aid of protocols including procedures, message formats, triggers and behavior of the logical network elements.

*Example: The E-UTRAN architecture defined in 3GPP for the 4th Generation radio access (*Figure 3.3*) consists of the NEs radio node (eNodeB (eNB)) and devices (User Equipment (UE)) [3]. eNBs are linked with themselves via an inter-node interface X2 and UEs are linked to eNBs via the air interface Uu. 4G systems consist of a flat architecture hence eNBs are linked directly to the core network (evolved packet core, EPC) via an S1 interface.*

Each NE accommodates a set of NFs that execute operations based on a set of input data. NFs generate a set of output data that is to be communicated to other NEs. Each of these NFs must then be mapped to NEs. The functional decomposition of technical components and the assignment of NFs to NEs are described by a functional architecture (Figure 3.4). An implementation of technical components may require the placement of their NFs at different places within the logical architecture.

Example: Channel measurements can only be done directly at the air interface of devices or radio nodes whereas resource assignment based on these measurements might be done in the radio nodes.

NFs impose different interface requirements with respect to latency and bandwidth. This implies the need for a view on how the boxes are arranged in a concrete

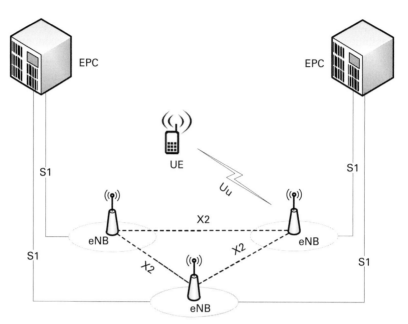

Figure 3.3   E-UTRAN architecture.

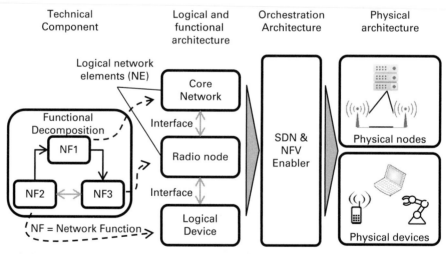

Figure 3.4   Relation between functional, logical, orchestration and physical architectures.

deployment. The physical architecture describes the assignment of NEs or NFs to physical nodes that are located at certain locations within the network topology. Physical architecture design has a big impact on network performance and network economics. Some NF may preferably be placed at central sites for, e.g. economic reasons in the case that statistical multiplexing of computing resources can be exploited. However, due to functional or interface requirements in terms of latency or bandwidth demand some functions may need to be operated close to the air interface or close together, which rather requires a distributed placement. In this case, both performance and cost may be affected. Technical and operational aspects in the context of function placement are elaborated in more detail in Section 3.3.

Traditionally the assignment of NFs to NEs as well as of NEs to physical nodes has been customized for each specific deployment. As it was seen from Chapter 2, diverse end-user requirements, services and use cases suggest the need for much more flexibility in 5G networks. Novel architectural enablers such as NFV and SDN aim to facilitate this increased flexibility [1][2]. An orchestration and control architecture as described in more detail in Section 3.4.1 will allow for significantly more flexible placement of NFs in future physical networks. To be more precise, the usage of SDN/NFV is already happening in 4G networks, mainly for core network functions. The 5G architecture will explore these technologies from the beginning. At this point, it is to be highlighted that in future networks the focus will be much more on NFs rather than on NEs.

Specification by standardization groups plays an important role to guarantee worldwide interoperability of equipment originating from different manufacturers. Even though traditional NEs, protocols and interfaces are specified, both network and device manufacturers still have considerable degrees of freedom. The first degree of freedom consists of how NEs are mapped to the physical network.

Example: Even though E-UTRAN is essentially a distributed logical architecture, a network manufacturer can still design a centralized solution where physical controller equipment, e.g. placed at a central access site, executes some eNB functions while other functions are executed at the distributed physical nodes closer to the radio heads. In that sense, a network manufacturer would be splitting a standardized NE into multiple physical nodes in order to enable a centralized deployment architecture. In another direction, the same vendor has the freedom to merge NEs in the same physical node, as it is the case in some core network nodes that are used in the market where manufacturers offer solutions of integrated Packet Data Network Gateway (P-GW) and Serving Gateway (S-GW) [3].

The second degree of freedom regards the hardware and software platforms architectures that are applied by the different manufacturers. So far, this has not been in the scope of 3GPP, which does not define any specific software or hardware architectures or platforms for the NEs.

The third degree of freedom relates to how manufacturers can implement the decision logic behind the different NFs.

Example: 3GPP has specified protocols for information exchange at the air interface. This defines the way how radio nodes (eNB) communicate, among others, scheduling information and the way devices (UE) interpret this information and how the UE should react. However, there exists some degree of freedom on how an eNB uses information in order to assign resources.

3.2   High-level requirements for the 5G architecture

Before the RAN logical architecture is specified, high-level principles should be defined. These principles take into account the 5G end-user requirements and envisaged services. In the following, the most important high-level design principles for the 5G architecture are listed.

PRINCIPLE I: *The 5G architecture should benefit from co-deployments with the evolution of LTE, but inter-system dependencies should be avoided. At the same time, all fundamental RAN functionalities should be designed having in mind the frequencies that the new air interface should operate at, e.g. system access, mobility, QoS handling and coverage.*

This principle has been derived from i) the acknowledgement of the proven success of LTE when it comes to Mobile Broadband (MBB) services and possibly other services such as mMTC [4] and ii) the fact that it is likely that at the time of the initial 5G deployments LTE will likely have wide coverage [5]. This principle is endorsed in [5], where enhanced multi-Radio Access Technology (RAT) coordination is stated as a design principle to be followed for the 5G architecture [4].

Inter-RAT coordination should also include non-3GPP technologies, e.g. IEEE 802.11 family, but the level of coordination may differ.

There may be no need to support handover or service continuity between 5G and 3G or 2G networks [5].

PRINCIPLE II: *The 5G architecture should enable multi-connectivity, including multi-layer and multi-RAT.*

It is expected that a device may be connected to several links of the same RAT (e.g. to macro and small cells), as well as to different RATs, including new RATs and LTE. This may leverage or extend existing technologies such as carrier aggregation and dual connectivity. This combination of RATs may involve also non-3GPP RATs, e.g. IEEE 802.11ax (High Efficiency Wi-Fi).

PRINCIPLE III: *The 5G architecture should support coordination features usable for nodes connected via different categories of backhaul.*

This means that the new air interface should be designed in a way that avoids unnecessary constraints such that deployments with different functional splits are possible. This is a very important principle since coordination for interference cancellation, for example, is part of the "design principles for 5G" [5], where massive MIMO and Coordinated Multipoint (CoMP) transmission and reception are given as examples of expected technologies [4]. This principle is also valid for non-collocated deployments of

LTE evolution and the new air interfaces. It would guarantee that operators with their existing backhaul should be able to deploy the 5G technology.

PRINCIPLE IV: *The 5G architecture should have embedded flexibility to optimize network usage, while accommodating a wide range of use cases, and business models.*

This principle implies that the same RAN logical architecture, specified by 3GPP, should be sufficiently flexible to address MBB and non-MBB use cases, e.g. uMTC, and a diversity of business models, e.g. network sharing. When it comes to the RAN and CN architecture, it implies that the protocol design is flexible enough to support the different requirements.

PRINCIPLE V: *The 5G architecture should have a programmability framework to enable innovation.*

In order to support the envisaged wide range of requirements, address many use cases (not clear in the time frame 5G is implemented) and allow for fast business innovation, 5G devices should have a high degree of programmability and configurability, multi-band multi-mode support and aggregation of flows from different technologies, device power efficiency and service aware signaling efficiency.

3.3   Functional architecture and 5G flexibility

In traditional networks, the assignment of NFs and NEs to physical nodes is designed for a specific deployment. SDN and NFV are novel architectural enablers that allow for a new way of deploying a mobile network. Hence, recent 5G research projects have addressed the logical architecture design by defining NFs and inter-function interfaces, instead of NEs and inter-node interfaces [3][6], except for the air interface, for obvious reasons. This implies a number of potential benefits such as

- NFs can be placed at optimal locations in a flexible way considering opportunities and limitations of the transport network.
- Only necessary NFs are applied to avoid overhead.
- NFs can be optimized through dedicated implementations.

However, this approach would require a plethora of interface definitions to enable multi-vendor interoperability. Hence, operators must be enabled to define and configure flexibly their own interfaces based on the functions that are used. A potential challenge that will concern mobile network operators is the increased complexity of such a system where many interfaces would need to be managed. As it is elaborated further in Section 3.4.1, software interfaces instead of inter-node protocols may be a solution but the 5G architecture design must carefully take into account the trade-off between complexity and flexibility.

This section provides an introduction on criteria for splitting functionality between NEs, an overview of exemplary functional splits, and examples for optimizing the operation of a mobile network. It is worth mentioning that the analysis not only supports

Table 3.1 Assessment of centralized versus distributed architectural approaches. Capacity is compared to required rates in the case of fully distributed operation.

Approach	Expected benefit	Requirements	Physical Constraints
Full centralization	Cloud and virtualization enabler Coordination schemes simplified Routing optimization	Latency[1]: <100 μs, Capacity: ≈20x [7]	Limited set of suitable BH technologies
Partial centralization	Both centralized and distributed approaches available	Latency: <1 ms (HARQ constrained); <10 ms (frame constrained) Capacity: ≈1-5x	Many interfaces need to be standardized
Fully distributed operation	Distributed processing simplified	Synchronization among nodes	Inter node connections necessary

the shift from inter-node to inter-function interfaces but also might be used to understand potential RAN functional splits with inter-node interfaces.

3.3.1   Functional split criteria

During the logical architecture design, the so-called "functional split" allows mapping of NFs to protocol layers and defining the placement of these layers within different NEs. There are different possibilities for implementing the functional split in 5G and they will mainly be driven by the following two factors:

- a distinction between NFs that operate synchronous or asynchronous with respect to the radio frames. Depending on this distinction there exist stronger or looser timing constrains on the interfaces,
- backhaul (BH) and fronthaul technologies which may be used to operate the 5G system. Depending on the technology, there might be latency or bandwidths limitations on the interfaces.

In particular, when it comes to functional split, the following aspects should be carefully taken into account [8]:

- **Centralization benefits:** Defining whether the architectural approach would imply benefits in case it is centralized with respect to the case of distributed implementation (see Table 3.1).
- **Computational needs and diversity:** Some functions may require high computation capabilities that should be provided centrally, at the same time at these locations applications with very different types of traffic demands may be implemented.

---

[1] Latency refers to the RTT latency between radio access point and central processor.

- **Physical constraints on the link:** With particular reference to the latency and bandwidth requirements on the connections between central unit pool and remote units.
- **Dependencies between different NFs in terms of synchronicity and latency toward the air interface:** NFs running at higher network layers in the OSI model are considered to be asynchronous. Two NFs should not be split if one of them depends on time-critical information of the other.

Table 3.1 summarizes benefits, requirements and constraints related to the functional decomposition from a fully centralized approach to a completely distributed positioning of NFs.

3.3.2   Functional split alternatives

As previously mentioned, 5G is characterized by the flexibility of placing NFs at any location within the network topology. This flexibility potentially introduces the two options of a Centralized RAN (C-RAN) and a Distributed RAN (D-RAN). Traditionally, C-RAN primarily aims at centralizing (pooling) base band processing resources. With the aid of NFV and using industry standard high volume server hardware for the baseband signal processing, the C-RAN approach can be extended to the so-called Cloud-RAN where NFs are deployed in a virtualized manner. For legacy physical architectures where mainly D-RAN is operated, C-RAN as well as Cloud-RAN architectures represent a kind of paradigm change.

So far, only fully centralized RAN architectures have been implemented, which require that the digitized receive signal (I/Q samples, one stream per antenna) is communicated via a fronthaul link between radio access point and central baseband pool, e.g. using an interface such as CPRI [9] or ORI [10]. The flexibility in 5G networks refers to the extension of this concept to a generic split of the NFs. A classical representation of this functional split is reported in [8]. Figure 3.5 shows four different options to split the functionality between the local radio access point and the central processor, the split line identifying what is in the central location (above the line) and what is locally placed (below):

- **Split A:** Lower Physical layer split. Similar to the currently deployed CPRI/ORI based functional split, where highest centralization gains are achieved at the expense of strong fronthaul requirements.
- **Split B:** Upper Physical layer split. Similar to the previous option, but only user-based NFs are centralized while cell-specific functions are remotely managed. For instance, Forward Error Correction (FEC) coding/decoding may be centralized. Its processing and fronthaul requirements scale with the number of users, their occupied resources and data rates. Hence, multiplexing (MUX) gains are possible on the fronthaul link and centralization gains are slightly reduced.
- **Split C:** MAC centralization. Time-critical centralized processing is not needed but also less centralization gains are exploitable. This implies that scheduling and Link-Adaptation (LA) must be divided into a time-critical (locally performed) and less time-critical part (centrally performed).

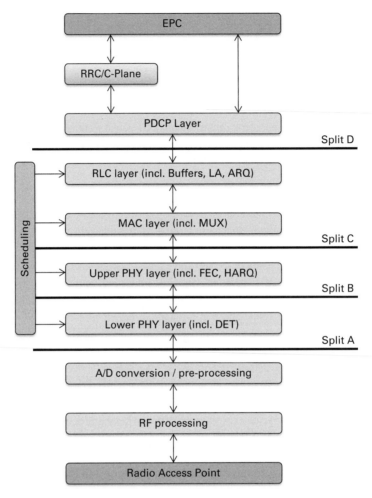

Figure 3.5   Overview of four exemplary functional splits [7].

- **Split D:** Packet Data Convergence Protocol (PDCP) centralization. Similar to existing dual connectivity mechanisms in 3GPP LTE.

Functions that operate asynchronously to the air interface frames are the ones with the least restrictive requirements on centralization and virtualization. These are the ones typically assigned to PDCP and RRC protocols. It has also been pointed out that the functions running in the lower layer must be performed synchronously to the air interface frames, i.e. part of the functionality which is centralized in splits A and B. This imposes strong requirements on their interfaces, which makes centralization and virtualization very challenging. On the other hand, core network functions, not explicitly addressed here are the ones benefiting the most from both centralization and virtualization. As it is discussed in detail in Section 3.4, the actual choice of functional split highly depends on the physical deployment and specific applications.

In addition, the functional split could be arranged in different options regarding control and user plane. Three models are envisioned [8]:

- **Straight flow:** Packets from the core go to the central entity that afterwards sends them to the remote units. This option is viable with centralized higher layers and distributed lower layers.
- **Forward-backward flow:** Packets from the core are sent directly to the remote units that decide what must be processed by the central unit. Afterward, the central unit NFs perform required processing and send the packets back one more time to the remote units. This option is viable when some higher-layer NFs are managed in a distributed way.
- **Control/user plane separation:** The previous two models can be further split in the case that central units perform only control plane processing and remote units only user plane processing.

3.3.3 Functional optimization for specific applications

5G networks will provide more degrees of freedom to optimize the mobile network operation, e.g. based on a specific purpose, dedicated software may be deployed and only a subset of the whole RAN protocol stack is implemented. Some factors that should be considered for optimizing mobile network functionality are listed in Table 3.2.

Functionality that may be optimized based on the scenario can be identified on all RAN protocol layers. On physical layer, coding plays an important role, e.g. block codes for mMTC and turbo-codes for xMBB, hard-decision decoding for resource limited nodes, carrier modulation, e.g. single-carrier for latency-critical applications and multi-carrier for high-throughput services, or channel estimation, which may be performed differently depending on the scenario.

Table 3.2 Influence factors on functional composition.

Factor	Impact	Example
Structural properties	Interference pattern, shadowing, deployment limitations	High buildings, streets or pedestrian area
User characteristics	Multi-connectivity need, D2D availability, handover probability	Mobility, user density
Deployment type	Local breakout, cooperative gains, dynamic RAN	Stadium, hot spot, airport, mall, moving/nomadic nodes
Service pattern	Local breakout, latency and reliability requirements, carrier modulation	mMTC, MBB
RAN technology	Backhaul connectivity, coordination requirements	Massive MIMO, CoMP, Inter-Cell Interference Coordination (ICIC)
Backhaul network	Centralization options, coordination opportunities	Optical fiber, mmWave, In-band

On MAC layer, among others Hybrid ARQ may be differently optimized depending on latency requirements, mobility functions highly depend on the actual user mobility, scheduling implementations must take into account user density, mobility, and QoS requirements and random access coordination may be optimized for MTC if necessary.

Furthermore, also functionality on network level can be optimized based on the actual **deployment type** and **service pattern**. Local break-out functionality depends on whether local services are to be provided, i.e. in the case that localized services are offered, internet traffic may be handled locally at the radio access point. Multi-cell cooperation and coordination depend on network density, **structural properties** and **user characteristics** like interference pattern and user density, respectively. Dual connectivity features depend on which multi-RAT coordination feature is applied (see Section 3.3.5).

Example: Consider a wide-area deployment where massive MIMO and ultra-dense networks (UDNs) (of small cells) are deployed. As UDN (see Chapter 11) and massive MIMO could operate at higher frequencies due to small cell areas and narrow beams robust mobility may not be guaranteed. Hence, multi-RAT connectivity for C-plane diversity is needed.

The possible degree of centralization will depend heavily on the envisioned **backhaul network**.

Example: Macro-cells with optical fiber connectivity can be deployed more centrally while, for economic reasons, UDN nodes are equipped with wireless backhaul and due to bandwidth limitations less NFs can be centralized.

Finally, the applicability of NFs depends on scenario and the deployed **RAN technology**.

Example: For densely UDNs, inter-cell interference coordination or multi-cell processing algorithms are essential, while massive MIMO will require pilot coordination algorithms. Furthermore, UDNs deployed in a pedestrian area with low mobility requirements allow for application of different interference mitigation schemes than wide area nodes at railway lines. Finally, the use of massive MIMO for backhauling will not require mobility management. In the case of a stadium, content will be provided locally and therefore core-network functionality as well as information and telecommunication services should be provided locally. Similarly, at hot spots local services may be offered, which requires again local core-network functionality.

For each of the above examples, dedicated software may be deployed which is optimized for the particular use case.

3.3.4   Integration of LTE and new air interface to fulfill 5G requirements

The integration of new air interfaces with legacy one(s) has always been an important task during the introduction of a new generation of mobile systems. Up to the introduction of 4G, the main goal of such integration was the provision of seamless mobility over the whole network, the smother introduction of new services in particular areas provided by the new generation and the maintenance of the services supported by the previous generation such as voice, supported by UTRAN during initial LTE deployments via circuit-switched fall-back. Among the different 3GPP systems this integration has been typically achieved via inter-node interfaces between the different core network nodes such as S11 (between MME and Serving Gateway) and S4 (between Serving Gateway and SGSN) [11].

For the transition to 5G a tight integration of the new air interface with LTE (compared to the integration between current legacy systems) will be from day one an essential part of the 5G RAN architecture. A tight integration in this context basically means a common protocol layer for multiple accesses residing above access-specific protocol layer(s). The demand for this tight integration comes from data rate requirement in 5G (up to 10 Gbps), which together with lower latency, drives the design of a new air interface(s) to be optimized to operate in higher frequencies above 6 GHz. Under these frequency bands, propagation is more challenging and coverage can be spotty [12].

In parallel with the 5G research activities, 3GPP is continuously adding new features to LTE and it is likely that at the time 5G reaches market LTE should be capable of addressing many of the 5G requirements, such as the ones related to MTC and MBB. At that time, LTE is also expected to be heavily deployed and, the fact that it operates in frequency bands with better propagation properties, makes the integration of LTE and the new interface operating in higher frequency bands very appealing [4][5][6][12].

This kind of tight integration of multiple accesses has been previously investigated [13], where a common RRM-based architecture for GSM, UTRAN and WLAN has been introduced for service-based access selection. In the Ambient Networks project [14], different tight integration architectures have been discussed and an architecture, relying on a multi-radio resource manager and a generic link layer, has been proposed. More recently alternatives for a tightly integrated architecture have been evaluated taking into account the LTE protocol architecture and aspects that are an important part of the new air interface [12]. Further, according to [12], at least PDCP and RRC layers should be common for LTE and the new air interface supporting the 5G requirements. This preferred option leads to a protocol architecture somehow similar to the one standardized in LTE Release 12 to support dual connectivity. The various options are the following (and shown in Figure 3.6):

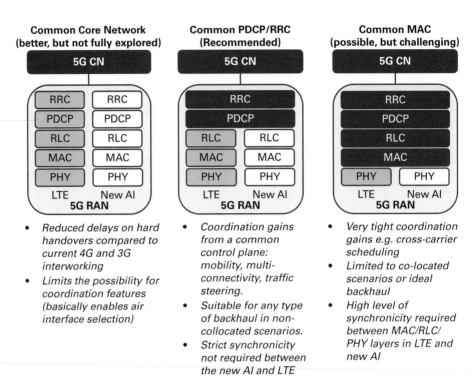

Figure 3.6   Different protocol architectures for the tight integration of LTE and new air interface.

Inter-connected core networks or a common core network

In this case, each RAT has its own RAN protocol stack and its own core networks where both core networks are linked via inter-node interfaces. The current solution integrates UTRAN (3G) and E-UTRAN (4G), where an inter-node interface exists between Mobility Management Entity (MME) and S-GW for the control plane. When it comes to the integration between 5G and LTE, this is unlikely to be the way forward since it would be challenging to fulfill the requirements of seamless mobility and transparent connectivity.

In the case that each RAT has its own RAN protocol stack but the core network is common, new 5G core NFs can be used by both LTE and the new air interface. This has the potential to reduce hard handover delays and enable more seamless mobility. On the other hand, potential multi-RAT coordination features might not be possible.

Common physical layer (PHY)

The LTE PHY layer is based on OFDM. It provides services to the MAC layer in the form of transport channels and handles the mapping of transport channels to physical channels. OFDM-based transmission will most likely remain as a good baseline also for the new air interface, that will likely have quite different characteristics compared to LTE, e.g. in terms of OFDM numerology, which means numbers for carrier spacing, symbol length, guard intervals and cyclic prefix length (cf. Chapter 7). Hence, the introduction of a common PHY may be very challenging. In addition, this architecture would impose limitations in terms of deployments since non-collocated operation of multi-RAT radios would likely not be possible due to the high level of synchronicity needed between the physical layers of LTE and the new air interface.

Common medium access control (MAC)

The LTE MAC layer provides services to the RLC layer in the form of logical channels, and it performs mapping between these logical channels and transport channels. The main functions are: uplink and downlink scheduling, scheduling information reporting, Hybrid-ARQ feedback and retransmissions, (de)multiplexing data across multiple component carriers for carrier aggregation. In principle, the integration of the new air interface and LTE on the MAC level can lead to coordination gains, enabling features such as cross-carrier scheduling across multiple air interfaces. The challenge to realize a common MAC comes from the assumed differences in the time- and frequency-domain structures for LTE and the new air interface. A high level of synchronicity would be needed between the common MAC layer and underlying PHY layers, including LTE and novel air interfaces. In addition, harmonized numerology for the different OFDM-based transmission schemes is needed. This challenge would likely limit this level of integration of the MAC layer level of integration to co-located deployments in which this high level of synchronicity could be achieved.

Common RLC

In LTE, the RLC layer provides services for the PDCP layer. The main functions for both user and control plane are segmentation and concatenation, retransmission handling, duplicate detection and in-sequence delivery to higher layers. RLC integration is likely to be challenging due to the required level of synchronicity between PHY, MAC and RLC. For example, in order to perform fragmentation/reassembly, the RLC needs to know the scheduling decisions in terms of resource blocks for the next TTI, information that has to be provided in time by the PHY layer. A joint fragmentation and reassembly for multiple air interfaces would likely not work unless a common scheduler is deployed. Similarly to the previous alternative (common MAC), a common RLC would only properly operate in co-located deployments of LTE and the new air interface.

Common PDCP/radio resource control (RRC)

In LTE, PDCP is used for both control and user planes. The main control plane functions are (de)ciphering and integrity protection. For the user plane, the main functions are (de)ciphering, header (de)compression, in-sequence delivery, duplicate detection and retransmission. In contrast to PHY, MAC and RLC functions, the PDCP functions do not have strict constraints in terms of synchronicity with the lower layers. Hence, a specific design for PHY, RLC and MAC functionalities for both air interfaces would likely not impose any problems for a common PDCP layer. In addition to this, such integration would work in both co-located and non-collocated network deployment scenarios, making it more general and future proof.

The RRC layer is responsible for the control plane functions in LTE. Among these, the broadcast of system information for non-access stratum and access stratum, paging, connection handling, allocation of temporary identifiers, configuration of lower layer protocols, quality of service management functions, security handling at the access network, mobility management, and measurement reporting and configuration.

The RRC functions do not require synchronization with functions in lower layer protocols, which makes it quite likely that they can be common to multiple air interfaces in order to exploit potential coordination gains from a common control plane. As in the case of common PDCP layer, both co-located and non-collocated network deployment scenarios would be allowed.

3.3.5   Enhanced Multi-RAT coordination features

Different multi-RAT coordination features can be envisioned thanks to the recommended protocol architecture alternative relying on a tight integration with common PDPC/RRC, as shown in the previous section. Some of these options are shown in Figure 3.7.

Control plane diversity

A common control plane for LTE and the new air interface would allow a dual-radio device to have a single control point for dedicated signaling connected via the two air interfaces. An equivalent concept has been developed as part of the dual connectivity concept for LTE Release 12 in order to improve mobility robustness [15].

With such a feature, no explicit signaling would be needed to switch the link and the receiver should be capable of receiving any message on any link including the same message simultaneously on both air interfaces. This might be the main benefit of the feature, which might be important to fulfill the ultra-reliability requirements for certain applications in challenging propagation conditions. In addition, a common control plane is also an enabler for user-plane integration features, as discussed in the following.

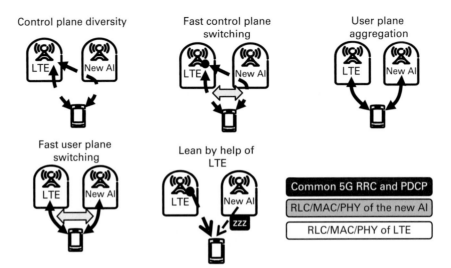

Figure 3.7   Different multi-RAT coordination features.

Fast control plane switching

With such a feature relying on a common control plane, the device would be capable of connecting to a single control point via any of the air interfaces and switch very fast (without the need of core network signaling, context transfers, etc.) from one link to another without requiring extensive connection setup signaling. The reliability might not be as high as with applying control plane diversity and additional signaling would be needed.

User plane aggregation

One variant of the user plane aggregation is called flow aggregation, which allows a single flow to be aggregated over multiple air interfaces. In another variant, defined as flow routing, a given user data flow is mapped on a single air interface, such that each flow of the same UE may be mapped on different air interfaces. The benefits of this feature is increased throughput, pooling of resources and support for seamless mobility. The flow aggregation variant may have limited benefits when the air interfaces provide different latency and throughput.

Fast user plane switching

Here, instead of aggregating the user plane, the user plane of devices uses only a single air interface at a time, but a fast switching mechanism for multiple air interfaces is provided. In this case, a robust control plane is required. Fast user plane switching provides resource pooling, seamless mobility and improved reliability.

Lean by help of LTE

This feature relies on a common control plane. The basic idea is to make 5G "lean" by transmitting all control information over LTE that will anyway be transmitted for backwards compatibility purpose (cf. Chapter 2). Information to idle mode devices, e.g. system information, is transmitted over LTE. The main benefit is that it likely reduces overall network energy consumption and "idle" interference in 5G. Even though the transmitted energy is just moved from one transmitter to another, substantial energy can be saved when the electronic circuitry associated with a transmitter can be turned off.

3.4   Physical architecture and 5G deployment

3.4.1   Deployment enablers

The logical architecture enables specification of interfaces and protocols whereas a functional architecture describes the integration of NFs into an overall system. The arrangement of functions in a physical architecture is important for practical deployment. NFs are mapped to physical nodes trying to optimize cost and performance of the whole network. In that sense, 5G will follow the same design principles as previous generations. However, in 5G networks the introduction of NFV and SDN

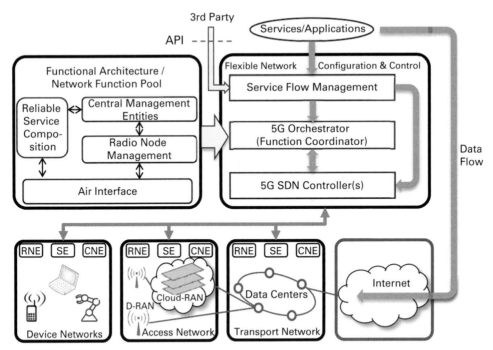

Figure 3.8   Relation of logical, functional, physical and orchestration architecture.

concepts will cause also a rethinking of imaginations in the context of traditional protocol stack methodologies. There could be interfaces directly between NFs rather than between NEs. Interfaces between functions not necessarily have to be protocols but may be software interfaces.

The idea around SDN and NFV are mainly driven by flexibility requirements at the core network. However, an extension of both enablers to RAN architectures has been developed [6]. A relation of the logical, functional, physical and orchestration architecture is shown in Figure 3.8.

NFs are compiled in a Network Function Pool. The function pool collects data processing and control functions, and allows them to be available centrally. It includes information on the interfaces, function classification (synchronous vs. asynchronous) and placement options as well as input and output relations. At high level, RAN related functions can be assigned to the following building blocks:

- Central management entities include overarching network functions that mainly are to be deployed at some central physical nodes (data centers). Typical examples are context and spectrum management.
- Radio Node Management provides functions that usually affect more than one radio node to be operated at selected physical radio node sites (D-RAN or Cloud-RAN).
- Air Interface functions provide functionalities directly related to the air interface in radio nodes and devices.

- Reliable service composition[2] represents a central C-plane integrated into service flow management that interfaces to the other building blocks. This function evaluates the availability or enable provisioning of ultra-reliable links applied for novel services requiring extremely high reliability or extremely low latency.

The task of the flexible network configuration and control block is to realize an efficient integration of functions according to service and operator requirements by mapping elements of the logical topologies of data and control plane to physical elements and nodes as well as configuration of the NFs and data flows as shown in Figure 3.8. Thereby, in a first step, the Service Flow Management is analyzing customer-demanded services and outlines requirements for data flows through the network. Requirements from 3rd party service providers, e.g. minimum delay and bandwidth, can be included through a dedicated API. These requirements are communicated to the 5G orchestrator and 5G SDN controller. The 5G orchestrator is responsible for setting up or instantiating VNFs, NFs or logical elements within the physical network. Radio Network Elements (RNEs) and Core Network Elements (CNEs) are logical nodes that can host virtualized functions (VNF) or hardware (non-virtualized) platforms (NF). Logical Switching Elements (SEs) are assigned to hardware switches. In order to guaranty sufficient performance required by some synchronous NFs, the RNEs will include a mixture of software and hardware platforms in the physical network – especially at small cells and devices. Hence, the flexibility with respect to deployment of VNF in radio access is limited. As most of the respective NFs act asynchronously to the radio frames and hence are less time critical to the air interface, CNEs allow more degrees of freedom to apply function virtualization. The 5G SDN Controller flexibly configures the elements set up by the 5G Orchestrator according to service and operator needs. Thereby, it sets up the data flow through the physical nodes (U-plane) and executes the C-plane functionalities including scheduling and handover functions.

At high level, the physical network consists of transport networks, access networks and device networks. The transport network realizes interconnection between data centers by high-performance link technology. Transport network sites (data centers) host physical elements dealing with big data streams including the fixed network traffic and core network functionalities. RNEs may be collocated realizing centralized base band processing (Cloud-RAN). In radio access, 4G base station sites (sometimes referred as D-RAN) as well as sites hosting Cloud-RAN connected via fronthaul to pure antenna sites will coexist. In other words, the flexible functional placement will lead to deployments where traditional core network functions could be instantiated closer to the air interface. The need for local break out, for instance, will cause a coexistence of RNE, SE and CNE even at radio access sites. SDN concepts will allow for creation of customized virtual networks using shared resource pools (network slices). Virtual networks may be used to realize optimized resource assignment to diverse services such as mMTC and MBB. It also allows for resource sharing between operators.

---

[2] Reliable service composition has been highlighted since it is expected to be one the new 5G services. In fact, service composition can be as well about any new expected service.

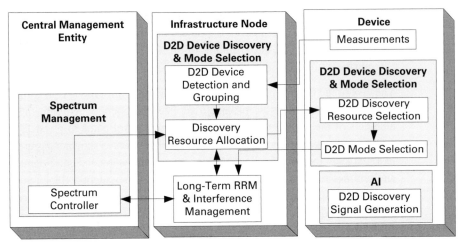

Figure 3.9   Assignment of network functions to logical nodes for device-to-device communication (D2D).

With some limitations, 5G architectures will permit device networks where devices act as part of the network infrastructure enabling other devices to get access to the network e.g. by D2D communication. Even at this device networks, RNEs will coexist with SEs and CNEs.

An example for assignment of network functions to logical nodes is given in Figure 3.9. Enabling D2D Type 2B (see Chapter 5) network functions are interworking at three different logical nodes denoted as device, infrastructure node and central management entity. Functions enabling device discovery are located at devices and infrastructure nodes. Device discovery is based on measurements that are executed by devices at certain radio resources where D2D discovery signals are transmitted over the Air Interface (AI). Responsible infrastructure nodes execute device grouping and resource allocation based on information about network capabilities, service requirements and measurement reports of the devices. Network capabilities include options for sharing D2D frequency resources with the cellular infrastructure (underlay D2D) or partitioning spectrum dedicated to cellular and D2D (overlay D2D). Discovery resource allocation is prepared by the infrastructure based on load situation and density of devices. Devices have to initiate selection between infrastructure or D2D mode (mode selection). The long-term radio resource and interference management considers allocation of D2D resources during resource assignment. Multi-operator D2D can be enabled by out-band D2D at dedicated spectrum resources. In this case, functionalities of a centrally operated spectrum controller are needed. In a physical network the central management entity will be located at central data centers in the transport network whereas logical infrastructure nodes are to be located in the access network e.g. at Cloud-RAN or D-RAN locations. As all described network functions operate asynchronous to the radio frames the infrastructure node functions provide potential for centralization meaning that not all RNEs located at BS sites need to host D2D detection and mode selection functionalities.

3.4.2   Flexible function placement in 5G deployments

A physical architecture determines a set of characteristics of the radio access such as network density, radio access point properties (size, number of antennas, transmit power), propagation characteristics, expected number of user terminals, their mobility profile and traffic profile. It also determines the backhaul technology between radio access nodes and transport network, which may be a heterogeneous technology mix composed of fixed line and wireless technologies. Furthermore, the physical deployment defines the technology toward the core network and its logical elements. All these characteristics imply the physical properties and limitations, which apply to the interaction of functional and logical mobile network components.

The impact of these limitations and the way how to cope with them may differ significantly depending on the data rate requirements, network status and service portfolio.

The choice of functional split options and the physical deployment conditions are tightly coupled, i.e. a decision of a certain functional split option determines the logical interfaces that must be carried over the physical infrastructure, which imposes constraints on this interface. Consider first the network density. The more radio access points are deployed per unit area, the more backhaul (BH) traffic must be supported. Figure 3.10 shows the number of supported base stations depending on the BH data rates and functional split [7]. The higher the functional split is located within the RAN protocol layers, the more access points can be supported. Split A (cf. Figure 3.5) implies

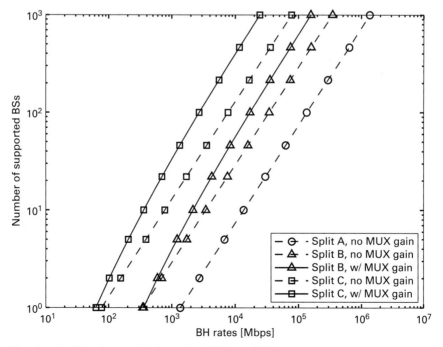

Figure 3.10   Functional split options and their required BH rates [16].

a static data rate per radio access point while in the case of split B and C, data rates vary with the actual data rates toward users. Hence, these two splits are able to exploit a statistical multiplexing gain[3] in the transport network, which may be up to a factor of 3. By contrast, split A will always induce the same data rate per access point independent of the actual load and therefore no multiplexing gain can be exploited.

Backhaul technologies not only determine possible data rates but also influence end-to-end latencies that can be realized. Split A requires optical fiber or mmWave backhauling technologies with either wavelength-switching or daisy-chaining of mmWave links. Low latency is very critical for split A as the physical transmission is implemented by CPRI, which obtains its time and frequency synchronization from the CPRI data stream.

Split B and C could also tolerate higher latencies in the order of a few milliseconds, which allows for using higher layer switching technologies such as MPLS or Ethernet. This increases the degrees of freedom to design the backhaul network significantly. The main difference between splits B and C is that split B performs central encoding and decoding. In current 3GPP LTE, this may imply stringent timing requirements because the Hybrid ARQ process requires that code words are processed within 3 ms after reception. If the backhaul latency is in the order of a few milliseconds, this constraint would not be met. Hence, either workarounds are required which relax this requirement [17] or 5G mobile network must be sufficiently flexible to scale its latency requirements. However, also split C (and inherently split B) has to cope with latency requirements for instance for scheduling and link-adaptation. The latter is very critical as a sub-optimal link-adaptation due to imperfect channel knowledge may severely deteriorate the performance [18]. The impact of this latency is mainly determined by the user mobility and changing interference patterns.

Both network density and user density have an inherent impact on the choice of the functional split as well as its gains. In the case that each cell has to serve a large number of users, one can expect scenarios where almost all resources are occupied. Hence, significant inter-cell interference is caused, which must be mitigated through cooperative algorithms. In this case, functional splits at lower RAN protocol layers are preferable. Such a scenario would occur, for instance, in hot spots, stadium or indoor deployments such as malls and airports. By contrast, if the number of users per cell is rather low or significantly varying due to the user's traffic profile, the number of occupied resources per cell may be lower. This increases the degrees of freedom for inter-cell coordination, which could be performed with higher layer functional splits and may be similarly efficient as cooperative algorithms.

Finally, the service profile has a strong impact on the choice of functional split as well as the deployment. Splits A and B offer more optimization opportunities compared to split C because more functionality can be implemented in software and optimized for the actual purpose as discussed in Section 3.3.3. For instance, split B allows for using coding technologies specific for the actual service, e.g. block-codes for MTC and LDPC codes for MBB services. Furthermore, split B allows for joint decoding algorithms in order to

---

[3] Statistical multiplexing gain in this context refers to the gain achieved by mixing independent, statistical bandwidth demand sources. Due to their random nature, the sum-rate of the multiplexed demand streams (effective bandwidth) is lower than the sum of the individual rates of each stream.

mitigate interference efficiently. Hence, if high service diversity can be foreseen, it may be worth to increase the degree of centralization. However, there also may be services that must be processed locally, e.g. vehicular traffic steering. Hence, the network may need to selectively apply the degree of centralization.

The next three examples describe how the placement of functionality may be determined by the type of deployment.

3.4.2.1   Wide-area coverage with optical fiber deployment

In this case, all radio access NFs are centralized. This imposes the strongest requirements on the network in terms of transport network capacity and latency. However, since all radio NFs are executed in a data center, preferably co-located with core NFs, also the maximum cooperative diversity gains as well as software virtualization gains can be exploited. Furthermore, other RAT standards could easily be integrated by providing for each a specific implementation running at the data center. However, relying on optical fiber backhaul also limits the flexibility and increase deployment costs, e.g. for small-cell networks where all nodes need to be connected either via optical fiber or line-of-sight (LOS) mmWave backhaul technologies.

3.4.2.2   Wide-area coverage with heterogeneous backhaul

This scenario is illustrated in Figure 3.11 [19] and would allow for different backhaul technologies, which are chosen based on available backhaul links as well as structural limitations, e.g. the usage of multi-hop mmWave technologies as well as Non-Line-of-Sight (NLoS) backhauling. This mix of backhaul technologies enables different degrees of centralization. Hence, the ability to cooperate among radio access points and the flexibility to adapt to changing network parameters may vary. If, for instance, two radio access points would apply functional split B and C, respectively, then both could coordinate their resources through ICIC and split B could apply advanced and tailored coding algorithms. This deployment scenario is optimal in terms of capital expenditures [20] as it exploits a large portion of cooperative gains and reduces deployment costs

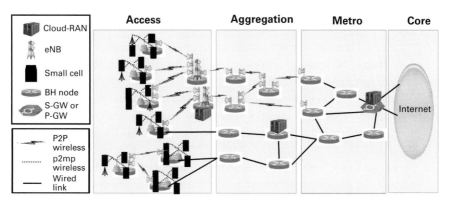

Figure 3.11   Illustration of a heterogeneous wide-area deployment [19] including point-to-point (p2p) and point-to-multi-point (p2mp) transmissions.

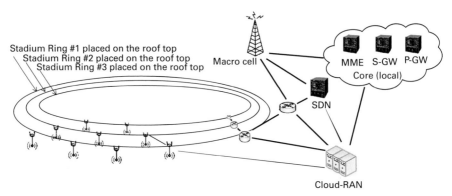

Figure 3.12   Deployment within a stadium [7].

compared to conventional deployments. However, it is also very challenging from many perspectives such as cooperation among radio access points, placement and dimensioning of data processing elements, deployment of software, and management of network elements, e.g. by means of SDN.

3.4.2.3   Local-area stadium

A stadium deployment is illustrated in Figure 3.12 and a very good example for deployments where the infrastructure is owned by the operator of the venue. Similar deployments are airports or malls. In this case, the operator of the venue provides the connectivity while mobile network operators must share the facilities. Furthermore, those deployments are very well planned and dimensioned in order to fit with the expected traffic demands. Finally, the deployed hardware will be very similar to wide-area or other hot-spot deployments but the applied software may vary significantly, not only in the radio access but also core-network functionality. For instance, core-network functionality may be placed right at the stadium in order to allow for local services such video streaming.

3.5   Conclusions

As next generation radio access has to fulfill a broad range of requirements, the design of future networks architectures will be driven by demand for flexibility, scalability and service-oriented management. Even though not directly associated with 5G, NFV and SDN will complement each other and enable the implementation of these basic requirements. 5G networks respond to changing market conditions will be much faster compared to legacy networks e.g. 3G or 4G. By fulfilling high-level requirements like co-deployments of 5G with LTE evolution and provisioning of multi-RAT connectivity, high capacity islands as well as ultra-reliable radio links can be enabled without additional economic effort. Flexible placement of network functions paves the way for better matching of functional split to service requirements, user density, propagation conditions as well as mobility and traffic profiles. To enable all these benefits, it will be fundamental to provide a compromise between the needed flexibility of communication among the arranged network functions and the number of standardized interfaces that allow for interworking of multi-vendor equipment.

References

[1] AT&T et al., "Network Function Virtualization: An Introduction, Benefits, Enablers, Challenges & Call for Action," White Paper, October 2012, http://portal.etsi.org/NFV/NFV_White_Paper.pdf

[2] Open Networking Foundation, "Software-Defined Networking: The New Form for Networks," ONF White Paper, April 13, 2012, www.opennetworking.org/images/stories/downloads/sdn-resources/white-papers/wp-sdn-newnorm.pdf

[3] 3GPP TS 36.300, "Overall description; Stage 2 (Release 12)," Technical Specification TS 36.300 V11.7.0, Technical Specification Group Radio Access Network, September 2013.

[4] ICT-317669 METIS project, "Final report on the METIS 5G system concept and technology roadmap," Deliverable 6.6, Version 1, May 2015.

[5] NGMN, "5G Whitepaper," February 2015, www.ngmn.org/uploads/media/NGMN_5G_White_Paper_V1_0.pdf

[6] ICT-317669 METIS project, "Final report on architecture," Deliverable 6.4, Version 1, January 2015.

[7] ICT-317941 iJOIN project, "Final definition of iJOIN architecture," Deliverable 5.3, Version 1, April 2015.

[8] ICT-317941 iJOIN project, "Revised definition of requirements and preliminary definition of the iJOIN architecture," Deliverable 5.1, Version 1, October 2013.

[9] Common Public Radio Interface (CPRI), "Interface Specification," CPRI Specification V6.0, August 2013.

[10] Open Radio equipment Interface (ORI), "ORI Interface Specification," ETSI GS ORI V4.1.1, June 2014.

[11] M. Olsson, S. Sultana, S. Rommer, L. Frid, and C. Mulligan, *SAE and the Evolved Packet Core, Driving the Mobile Broadband Revolution*, 1st ed. Academic Press, 2009.

[12] I. Da Silva et al., "Tight integration of new 5G air interface and LTE to fulfill 5G requirements," in 1st 5G architecture Workshop, IEEE Vehicular Technology Conference, Glasgow, May 2015.

[13] IST-2002–001858 Everest Project. [Online] www.everest-ist.upc.es

[14] M. Johnsson, J. Sachs, T. Rinta-aho, and T. Jokikyyny, "Ambient networks: A framework for multi-access control in heterogeneous networks," in IEEE Vehicular Technology Conference, Montreal, September 25–28, 2006.

[15] 3GPP TR 36.842, "Study on Small Cell Enhancements for E-UTRA and E-UTRAN – Higher layer aspects," Technical Report TR 36.842 V12.0.0, Technical Specification Group Radio Access Network, January 2014.

[16] ICT-317941 iJOIN project, "Final definition and evaluation of PHY layer approaches for RANaaS and joint backhaul-access layer," Deliverable 2.3, Version 1, April 2015.

[17] P. Rost and A. Prasad, "Opportunistic hybrid ARQ: Enabler of centralized-RAN over non-ideal backhaul," *IEEE Wireless Communications Letters*, vol. 3, no. 5, pp. 481–484, October 2014.

[18] R. Fritzsche, P. Rost, and G. Fettweis, "Robust rate adaptation and proportional fair scheduling with imperfect CSI," *IEEE Transactions on Wireless Communications*, vol. 14, no. 8, pp. 4417–4427, August 2015.

[19] ICT-317941 iJOIN project, "Final definition and evaluation of network-layer algorithms and network operation and management," Deliverable 4.3, Version 1, April 2015.

[20] V. Suryaprakash, P. Rost, and G. Fettweis, "Are heterogeneous cloud-based radio access networks cost effective?," *IEEE Journal on Selected Areas in Communications*, vol. 33, no. 10, pp. 2239–2251, October 2015.

4 Machine-type communications

Joachim Sachs, Petar Popovski, Andreas Höglund, David Gozalvez-Serrano, and Peter Fertl

4.1 Introduction

Machine-Type Communication (MTC) denotes the broad area of wireless communication with sensors, actuators, physical objects and other devices not directly operated by humans. Different types of radio access technologies are targeting MTC (see [1]). For Long Term Evolution (LTE), it has emerged as an important communication mode during the recent standard evolution. The research and development efforts made to enhance LTE in a way to support MTC clearly indicate the need for the wireless system architecture to address MTC. As the role of MTC is expected to grow in the future, there is a good opportunity in the development of a 5G wireless system to address MTC from the very beginning in the system design.

This chapter is organized in the following way. Section 4.1 outlines some of the most important use cases for MTC and categorizes MTC into the groups of massive MTC (mMTC) and ultra-reliable and low-latency MTC (uMTC). The requirements for these two MTC categories are defined. Section 4.2 describes some fundamental techniques for MTC. Sections 4.3 and 4.4 address mMTC and uMTC respectively and explain the corresponding design principles and technology components. Section 4.5 summarizes the chapter.

4.1.1 Use cases and categorization of MTC

4.1.1.1 The general use case of low-rate MTC

MTC use cases exist in a wide range of areas. They are mainly related to large numbers of sensors monitoring some system state or events, potentially with some form of actuation to control an environment. One example is automation of buildings and homes, where the state e.g. of the lighting, heating, ventilation and air condition, energy consumption, are observed and/or controlled. There are also wide area use cases, such as environmental monitoring over larger areas, monitoring of some infrastructure (e.g. roads, industrial environments, ports), available parking spaces in cities, management of object fleets (e.g. rental vehicles/bicycles), asset tracking in logistics,

5G Mobile and Wireless Communications Technology, ed. A. Osseiran, J. F. Monserrat, and P. Marsch. Published by Cambridge University Press. © Cambridge University Press 2016.

monitoring and assistance of patients. There are use cases that comprise remote areas, such as in smart agriculture. In the context of the use cases described in Chapter 2, MTC appears as an important, if not the crucial, element in (1) autonomous vehicle control, (3) factory cell automation, (6) massive amount of geographically spread devices, (10) smart city, (12) teleprotection in smart grid network and (15) smart logistics/remote control of industry applications.

A commonality in these use cases is that the reporting of sensory information is typically delay-tolerant and sent from a sensor to some (cloud) service. Furthermore, the information could be correlated across the sensors, so that it is not crucial that each individual sensor sends the data. For example, it may not matter if a single temperature sensor fails if others that measure a correlated temperature are successfully transmitting. An exception is the reporting of alarms, for which some delay constraint exists [2]. Most data transmissions take place from the devices toward some centralized service function. The communication in the reverse direction is often more rare, e.g. some simple form of actuation, but mostly configuration of the devices and services or confirmations about transmitted messages; communication toward the devices is delay-tolerant. In many use cases, the number and density of devices can be very large. An extremely high level of reliability for individual sensor reports is only sometimes needed and simplicity is an important target.

4.1.1.2   Use case: the connected car

The connected car has gained a lot of attention during the recent years, as it enables new services and functionalities for the automotive industry based on the use of wireless communications, and, most particularly, cellular systems. Only these systems are capable of providing the wide area coverage and performance demanded by automotive applications, including both human and machine type of communication. For Human-Type Communication (HTC), the challenge is to provide to passengers in the vehicle with comparable mobile broadband connectivity performance as can be found in stationary environments. In the automotive context, MTC refers to the exchange of information between machines that can be located in vehicles, user devices or servers, with little or no human interaction. The scope of MTC in the automotive domain encompasses a wide range of applications including road safety and traffic efficiency (e.g. highly autonomous driving), remote processing or remote diagnostics and control, among others. Some automotive applications in the area of MTC, such as road safety and traffic efficiency, require ultra-reliable connections with stringent requirements for latency and reliability, as the timely arrival of information can be critical for the safety of passengers and vulnerable road users. Furthermore, a highly reliable and widely available connectivity to the cloud can allow some functions, such as video processing, audio recognition or navigation systems to be carried out remotely by cloud servers instead of the central processing unit in the vehicle. Remote processing has not only the potential to increase the processing power beyond the vehicle capabilities but also to enable a continuous service improvement during the vehicle´s lifetime. Other applications such as remote diagnostics and control are based on the transmission of small telemetry and command messages, and therefore, do not possess stringent requirements in terms of latency or data rate. Nevertheless, they must operate when the vehicle is turned off and even in reception scenarios with very high attenuation values, like in the case of underground parking places. These characteristics demand low power consumption and a significant coverage extension.

4.1.1.3   Use case: the smart grid

The smart grid represents an evolution of the electric power grid into an immensely complex cyber-physical system that will rely on decentralized energy production, as well as near-real-time control and coordination between the energy production and consumption. A fundamental enabler of the smart grid is the reliable, two-way wireless MTC. In the downlink, the smart grid should be able to send commands and polls. The communication design is more challenging in the uplink, as it needs to coordinate a large set of partially or fully uncoordinated transmissions and therefore research attention is more focused on the uplink. An exemplary MTC device in the smart grid is the smart electricity meter. At present, smart electricity meters are primarily used by electricity providers only for availability monitoring and billing. However, as the Distributed Energy Resources (DERs), such as wind turbines and solar panels, increase their share of energy generation, the role of the smart meter is expected to become more complex and communication-intensive [3]. Specifically, there can be increased need to grid state estimation, where the meter should frequently monitor and report the power quality parameters, such as e.g. power phasors, which enables real-time estimation and control of the grid state. Due to their sheer number within a given region, the smart meters represent a showcase of massive MTC. The Smart Grid also features instances of ultra-reliable MTC, as many of the devices should reliably and very timely report critical events in the grid, such as outage or islanding of a micro-grid.

4.1.1.4   Use case: factory cell automation

Wireless communication in factory cell automation systems provides connectivity of movable machine parts or mobile machines integrated in distributed control systems. The advantage over wired connection is the low installation cost, as well as avoidance of the mechanical/weight problems that the cables may introduce. The typical application is closed-loop, real time control of interconnected sensors and actuators. The performance requirements fall often in the area of ultra-reliable MTC: low latency, strong determinism (low jitter), and high reliability. For example, the latency requirements can go down to 1 ms, while the packet loss probability requirement may reach down to very low values, possibly even extremes such as 10^{-9} [4]. Since the interconnected devices are often constrained to a small geographical area, unlicensed wireless systems (Wireless HART, Wi-Fi, Bluetooth) and fixed communication systems have dominated industrial wireless systems during the past decade. On the other hand, cellular systems in the context of industrial automation have been used in remote service applications and alert systems. However, unlicensed spectrum is not suitable for very high reliability, which puts forward ultra-reliable cellular MTC in a licensed band as a candidate solution for future applications. Due to the stringent latency requirements, the use of small cells (see e.g. [5]) and network-controlled

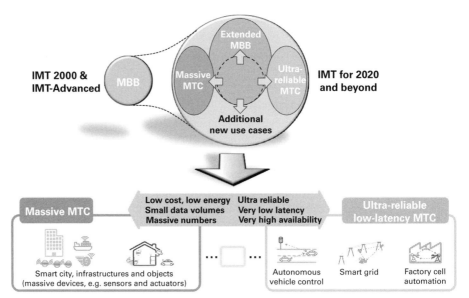

Figure 4.1   Categorization of MTC within the 5G wireless systems.

Device-to-Device (D2D) communication are interesting technologies for MTC in the area of industrial automation.

4.1.1.5   Categorization of MTC

The current view on the 5G wireless systems categorizes the broad area of MTC into two groups, as shown in Figure 4.1: massive MTC (mMTC) or ultra-reliable and low-latency MTC (uMTC). Sometimes uMTC is referred to as mission-critical MTC. mMTC assumes delay-tolerant data services for typically infrequent data transmissions, with massive amounts of devices that are sometimes even battery-operated. In contrast, uMTC tends to be of very high reliability, very low latency and real-time control of objects and processes. However, this categorization is not very strict and there are also use cases that do not fit perfectly into this categorization, e.g. there can be massive sensing use cases that do actually require very high reliability and devices maybe do not need to operate on a battery.

Another important observation is that the content of the chapter is more extensive on mMTC technology assessment than for uMTC. The reason is that research on mMTC has been going on for several years, while uMTC has emerged more recently.

4.1.2   MTC requirements

4.1.2.1   Massive MTC

The nature of MTC is quite different from that of human-oriented traffic from e.g. a smart phone. For example, many MTC devices are expected to be less mobile, implying that there will be a limited need for handover of ongoing transmissions with short delay-tolerant transmissions. Further, to make a massive number of devices feasible the device cost must be low and the need to frequently re-charge devices must be eliminated. The system must also scale with the number of devices, so that large numbers do not become limiting. There is a need for ubiquitous coverage, as devices should also be reachable in isolated locations, e.g. basements or outskirts.

The mMTC requirements listed below are considered important for 5G (see [6]–[7] and Chapter 2 for more information):

- **10 years device battery life**: A general guideline that omits any charging during the life span of the device.
- **Coverage enhancements of 20 dB**: 3GPP LTE Release 13 is targeting 15 dB coverage enhancement for MTC and the 5G requirement is a notch higher.
- **300,000 devices per cell**: The 3GPP requirement states the capability to support 30,000 devices per radio cell and for a 5G system a 10 times higher capacity should be envisioned. The requirement is stated per cell, as it is not necessary to reach the capacity only by densification. Note that 300,000 is a rather extreme upper bound; most of the cells will have a number of devices that is lower by orders of magnitude.
- **Low device complexity**: This requirement addresses the need to enable simple smart devices to become connected. Low device complexity enables low device cost, which is a prerequisite for connectivity in many mMTC use cases. The objective is that low-complexity communication modes for mMTC can be used by devices of low complexity, even if the communication system supports also high-performance communication modes for other devices, which have significantly higher complexity.

Adding to the above MTC-specific requirements is the overall requirement that a 5G system should be able to flexibly support a large variety of fundamentally different services. That is, an operator should not have to license a spectrum band and deploy a dedicated communication system solely for the use of massive-MTC, but should be able to devote the band to all types of 5G services and allocate the resources to massive-MTC according to the amount of MTC traffic, which may vary over location and time.

In the following discussions, 4G LTE Release 10 is used as a reference or baseline when needed (e.g. for the 20 dB improved coverage requirement). Note that the evolution of 4G, that is LTE Releases 11–13, addressed some of these requirements, and that work will continue in future releases. It can be noted that also 2G is addressing mMTC, under the Extended Coverage GSM for IoT (EC-GSM-IoT), in its evolution (see [8][9] [10] and Chapter 1). In addition, a new narrowband radio interface for mMTC, called NarrowBand-IoT (NB-IoT) was adopted in 3GPP. It assumes 180 kHz radio frequency bandwidth. Further, it follows similar design principles as presented in this chapter and enables e.g. deployments in narrow spectrum allocations like a single 2G carrier [11]. Some proprietary radio technologies for long-range transmission at very low data rates, such as LoRa or Sigfox, have also recently been developed, but are not further considered in this chapter. Those operate in unlicensed spectrum, where they need to co-exist and share their spectrum with many other radio systems [1][12].

4.1.2.2   Ultra-reliable MTC

Ultra-reliable communication is seen as one of the new features of 5G wireless systems [13][14], offering stable wireless connections, consistent experience for the users and always-on connectivity for IoT devices. Specifically, ultra-reliable MTC refers to wireless communication links that require unprecedented levels of reliability, often supplemented with a strict latency requirement. For example, in some industrial applications the latency requirement can be such that the packet should be successfully received with a probability higher than 99.9999% within a time of 1 ms–2 ms. The successful delivery of the packet also implies that the control information of that packet has been received correctly. Hence, uMTC poses new challenges on design and transmission of control information.

A definition of link reliability that is commonly adopted is that reliability is the probability that a certain amount of data (e.g. number of bytes) can be successfully decoded by the intended receivers within a certain deadline (e.g. number of seconds or ms). Availability of a given connection can be defined as the capability to support the minimal required level of reliability at a minimal required data rate. For example, when the link is not available, it cannot be guaranteed with sufficiently high likelihood that data messages and associated control messages are successfully exchanged between the two communicating parties within the given delay bound.

The provided link reliability definition is rather rigid, as it implies that if the specified amount of data does not arrive at the destination with the required reliability, the service using this data transmission should completely fail. Therefore, an important concept related to uMTC is Reliable Service Composition (RSC), which is a way to specify different versions of a service, such that when the communication conditions are worsened, the Quality of Service (QoS) gracefully degrades to the service version that can be reliably supported, instead of having a binary decision "service available/not available". The concept of graceful degradation of a service is not new, it has been used in e.g. scalable video coding. However, video and its perception naturally allows for graceful degradation. In RSC, the objective is to design services that offer certain level of functionality when it is not possible to get the full one. For example, the uMTC for reliable Vehicle-to-Vehicle (V2V) communication can be designed in a way that if only a part of the data can be reliably decoded, then the service runs at its basic version, and hence, only the most critical information for the safety of road users is transmitted.

Achieving ultra-high reliability in a given scenario requires a careful analysis of the risk factors or reliability impairments that are dominant in that scenario. There are at least five different reliability impairments [13]: (1) *Decreased power of the useful signal* – for example, coverage extension is directly combating this impairment; (2) *Uncontrollable interference*, which occurs in unlicensed bands, but also among uncoordinated small cells; (3) *Resource depletion due to competition* – e.g. when multiple devices send concurrently to the same receiver; (4) *Protocol reliability mismatch* – e.g. when the control information in the protocol is not designed to attain high reliability levels and (5) *Equipment failure*. The design of protocols and transmission schemes for uMTC in a given scenario should carefully assess the impact of each of the reliability impairments in that scenario.

4.2 Fundamental techniques for MTC

Traditionally, the evolution of cellular wireless communication systems has been centered around broadband communication and provision of increasingly high data rates. The emergence of mMTC and uMTC changes that focus, as the target scenarios do not require excessively high data rates, but rather new modes of connectivity to a massive number of simple devices and/or support extremely reliable connections. Although the performance requirements for mMTC and uMTC are vastly different, they point toward the revision of the same set of communication-theoretic mechanisms. In this section, two of those mechanisms that are promising for the radio access part are discussed: (1) creation of short packets, where the data and the associated control information are of comparable size, and (2) non-orthogonal protocols for distributed access.

4.2.1 Data and control for short packets

The success of broadband wireless communication systems is largely based on the methods for reliable transmission that follow the principles of information theory. Those principles are applicable when each transmitted packet contains a large amount of data, due to the following two features: (1) large data means that one can use methods (codes, modulation) that are applicable in an asymptotic case to guarantee reliable transmission under a constraint of the total energy used for transmission; (2) the size of control information is small compared to the size of data, as shown in Figure 4.2(a) such that, even if the control information is sent suboptimally (e.g. repetition coding), its overall effect on the system performance is negligible. These features have led to a common approach in designing broadband communication, in which data is transmitted using optimized and sophisticated methods, while the transmission of control information has been largely left to a heuristic design. This approach for creating packets for broadband communication needs to be revised when the amount of transmitted data

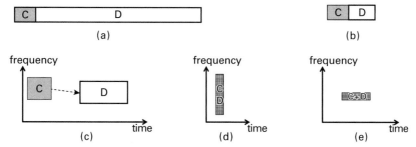

Figure 4.2   Structure of control information (C) and data (D) in a packet. (a) Packet for broadband communication D>>C. (b) Short packet for machine-type communication D≈C. (c) Common causal relationship in receiving the control information and data. (d) Low-delay transmission spread in frequency. (e) Short packet with blended data and control.

is small and comparable in size to the associated control information, see Figure 4.2(b). In order to do that, it is needed to look in the anatomy of a data packet.

The packet structure that is commonly used for broadband transmission of large data portions is based on separation of the control information and the data, as shown in Figure 4.2(c). Each packet is sent by being coded and modulated into N transmitted symbols. Each possibility to send one of those symbols can be seen as a Degree of Freedom (DoF), which is the total number of independent communication resources available to send the packet. A communication resource can be e.g. a specific part of the time-frequency grid or a spreading code in CDMA systems used at a given time. The N symbols could be sent equally spaced in time, using a single-carrier frequency, but they can also be sent through a combination of different time and frequency resources, as actually depicted on Figure 4.2(c). A common way to send the packet is to use N_C DoFs to send the control information that are separated/orthogonal from the N_D DoFs that are used to send the data. Figure 4.2(c) illustrates the well-established causal relationship between the reception of the control information and the data: the successful reception of control information is a condition to receive the data. Usually, the control information is a small sub-packet that should be received correctly with very high probability to avoid a false positive, i.e. if the packet is intended to a person called Alice, but another person called Bob erroneously decodes the header and thinks that the packet is intended for him. Furthermore, decoding the control information is a hint for Alice whether the data that follows is intended for her; if not, Alice can turn off the receiver, which is an essential principle for designing energy-efficient wireless networks.

Let us now look at the low-latency communication where the relationship between the size of the data D and the size of the control information C is arbitrary, i.e. it can be as in Figure 4.2(a) or Figure 4.2(b). If the packet should be sent with a low latency, then the transmission is confined in time, while the required DoFs are gathered in frequency. For example, the packet can use a single OFDM symbol that consists of many subcarriers; see Figure 4.2(d). In such a setting, it is not possible to condition the reception of the data on the correct reception of the control information, as they need to be decoded simultaneously. In other words, control information cannot be used to make a decision whether to invest energy to decode the data. This simple example illustrates that the communication protocols feature a trade-off between latency and energy efficiency.

Finally, Figure 4.2(e) illustrates a scenario in which the short packet from Figure 4.2(b) should be sent with a limited number of DoF. This situation is typical for mMTC, where a massive number of devices share the communication resources. Nevertheless, this case will also occur in uMTC, where low-latency transmissions of short packets coexist with other broadband traffic, such that there is no opportunity to use frequency resources in abundance, as in Figure 4.2(d). The transmission on Figure 4.2(e) is done by assuming that data and control are combined and sent by using the same set of DoF. This is following the recommendations from the recent fundamental result in information theory [15], which states that in the region of short packet lengths going to several hundreds of bits, the reliability of the coding is very sensitive to the packet length, which is not the case for the very long packets. This implies that it may be beneficial, instead of complete separation of data and control, to have at least part of the control information jointly encoded with the data. Similar to the case on Figure 4.2(d), the causality between the control information and data is lost, and thereby the possibility to use the corresponding energy-efficient mechanism. However, one should not haste to the conclusion that the joint encoding of data and control is not energy efficient: due to the increased reliability, if an ARQ protocol is applied, the number of required retransmissions decreases, which may eventually result in better energy efficiency.

4.2.2   Non-orthogonal access protocols

The field of wireless access protocols has been, for a long time, dominated by protocols that require the transmissions from different nodes to be received in orthogonal frequency resources. The non-orthogonal use of the same resource by two or more terminals is treated as a collision, in which all the packets of the involved terminals are lost. However, both the use cases for mMTC/uMTC as well as the recent trends in access protocols point toward methods in which the receiver utilizes the collisions through advanced processing and Successive Interference Cancellation (SIC). Specifically, non-orthogonal access and SIC become significant for uMTC in scenarios where devices use random access to send data within a given deadline, such that the controlled utilization of collisions can improve the overall reliability.

Let us consider the case in which a massive amount of sensors attempts to report event-driven and correlated information, e.g. occurrence of an alarm. In this case, orthogonal transmission would assume that the reading of a particular sensor is acquired only if the packet of that sensor is received by the Base Station (BS) without experiencing collision with another packet. However, the BS may use the fact that the information carried in the packets is correlated, apply advanced processing on the received signals, which features collisions from many packets, and send feedback to the sensors to stop the transmission when the BS has extracted sufficient information about the event. This is an example of joint source coding, channel coding and protocol design.

The use of SIC at the BS leads to a new class of coded random access protocols [16], which are suitable for massive coordinated access. Differently from the classical ALOHA approach, in coded random access each device repeats its packet multiple times, which sets the stage for the use of SIC. Figure 4.3 illustrates a simple example of coded random access, in which three devices send their packet in four slots. In classical

Figure 4.3   Illustration of coded random access.

ALOHA, only device 2 would successfully send its packet within the four slots, as in slot 1 and slot 3 there are collisions. If the receiver applies SIC, then it *buffers* the collided packets from slot 1 and 3. The decoded packet of device 2 from slot 4 contains pointer to where else device 2 has transmitted. The receiver then cancels the interference of device 2 from the buffered reception from slot 1 and thus recovers the packet of device 3. Finally, it cancels the packet from device 3 from the buffered reception in slot 3, thereby recovering the packet of device 1. In this specific example, the throughput increases three times compared to classical ALOHA.

The use of non-orthogonal transmission for uMTC is motivated by spectrum efficiency. Namely, a straightforward approach to support ultra-reliable communication is to allocate a dedicated spectrum for it. If the low-latency packet is spread in frequency, as on Figure 4.2(d), then there is a risk that a large frequency band is not used most of the time, being reserved for uMTC. One therefore has to look for spectrum access methods in which uMTC and non-uMTC traffic use the spectrum in a non-orthogonal, concurrent way, while the receiver applies some form of SIC in order to extract the interfering packets. A discussion on non-orthogonal access can also be found in Chapter 7. Random access design for MTC is discussed more in depth in [17][18].

4.3   Massive MTC

4.3.1   Design principles

The basic design principle of massive MTC is to exploit that mMTC services are delay-tolerant and consist of transactions with small amounts of data. These relaxed requirements can enable extensive sleep cycles for devices (to enable long battery lifetimes), define low-complexity transmission modes (to enable low device costs) and define extra-robust low-rate transmission (to enable extended transmission range). Since the total data volume of massive MTC is rather small (compared to e.g. multimedia services like video), even a very large number of devices is expected to generate (on average) manageable traffic volumes for a mobile network that is also dimensioned for mobile broadband services. However, still considerations have to be given for mMTC with high density of devices when it comes to handling the control signaling, context handling in the network, as well as overload of system resources in access peaks when large device populations try to access the network simultaneously.

4.3.2   Technology components

As mentioned previously, the desired features for a massive MTC system are low device complexity, long battery lifetime, and scalability and capacity. In the following the technology components that address these features are presented.

4.3.2.1 Features for low device complexity

The complexity of a device is related to the performance that is expected for the communication. Massive MTC services transmit typically infrequently small amounts of data and have relaxed requirements in terms of required data rate and transmission reliability. This provides opportunities to exploit the relaxed performance requirements to simplify the transmission mode and reduce device complexity. A significant evaluation on how device complexity can be simplified has been provided in [7] for LTE, and the features listed below are already addressed for LTE evolution in Releases 12–13. However, the general findings are independent from the specific radio access technology and are explained in the following.

Transmitting at wide bandwidth can provide high peak data rate at the costs of device complexity. When the transmission and reception bandwidth used by the device is bounded, the costs can be reduced compared to wide-bandwidth devices. Therefore, for mMTC devices it is desired to have a transmission mode with a limited device bandwidth. Already at bandwidths in the order of 1 MHz, as it is used in e.g. Bluetooth design, very low device complexity is achievable. It shall be noted that the total system bandwidth provided by the 5G system can be much wider, and may be used by other devices targeting e.g. high peak rates. A further cost reduction can be achieved by limiting the peak data rate in order to limit the amount of allocated buffer. The number of antennas that a device includes directly affects the device complexity, so that a low-complexity transmission mode should not depend on multiple device antennas being present. Further, a device that needs to transmit and receive simultaneously requires a duplex filter to separate the transmit signal from the receiver. If a device is alternately transmitting or receiving, like in time-division duplex or half-duplex frequency division duplex, the costs of a duplex filter can be avoided. Finally, it is desirable to limit the transmit power of a device, so that the power amplifier can be embedded onto the integrated circuit, thereby avoiding the need for a separate external power amplifier. For this purpose, for LTE Release 13 a new device power class is defined, where the device output is limited to around 20 dBm.

4.3.2.2 Features for service flexibility

MTC services comprise typically only small amounts of data that are transmitted per transaction. However, it is not easy to define the maximum transaction size for MTC services. Even more, services can be easily updated during the lifetime of a device, e.g. the monitoring of some process may be based on infrequent status reports. However, after a few years in operations, the service may be updated via over-the-air configuration or software update. As a result, the amount of data transmitted per device, the frequency of transmissions, and the priority of message may change over time. Therefore, a flexible access design is needed in order to enable flexible service provisioning. Even if some upper bound in capabilities may be set by the category of the device, some flexibility in service provisioning shall be catered for.

4.3.2.3   Features for coverage extension

Coverage is normally defined as the maximum range or path loss at which a certain throughput limit can still be upheld. With the latency tolerance of mMTC, this must not any longer be the case but a degradation of the throughput is acceptable. In fact, LTE Release 13 is specifying 15 dB coverage gain by means of time repetition. The resulting low bit rate is not a problem in itself, however, since the device energy consumption is very dependent on the time during which the device cannot stay in sleep mode, the longer transmission times are affecting the battery life. Therefore, the coverage extension and battery life requirements are somewhat contradictory and it is difficult to meet them at the same time. One way of doing this is to make use of the massive number of devices and allow for some of them to function as simple relays and greatly improve the link budget for devices in challenging coverage. Clearly, the relays dissipate more power, but improve the overall connectivity. This has been evaluated in the Madrid propagation map developed in METIS [19] at a 2 GHz carrier frequency and a bandwidth of 1 MHz, where the devices acting as relays are operating in the same frequency band and with the same device output power of 23 dBm, and still send their own traffic (for more details see [20]). Further, the simulations are static, limiting to two-hop relaying and the uplink and downlink considered separately. In Figure 4.4 it is seen that the drop rate is greatly reduced when MTC devices are enabled to act as relays to other MTC devices. The drop rate refers to those devices that are dropped by the network if a certain throughput cannot be upheld. Therefore, this relates to the coverage since devices are dropped if located outside a certain cell radius. Although this evaluation does not give the coverage enhancement in number of dB, MTC device relaying is obviously very promising for improving

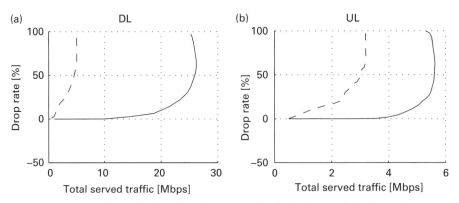

Figure 4.4   Drop rate versus total served traffic (number of MTC devices) with MTC devices acting as relays (continuous line) compared to LTE Release 10 case without relaying (dashed line): (a) for the downlink and (b) for the uplink.

the coverage. More importantly, it does so without extending the transmission times, which is beneficial for the device battery lifetime, as discussed below.

4.3.2.4 Features for long battery lifetime

In broad terms, the device energy consumption is proportional to the transmission and reception time of the device during which it cannot power down to a conservative sleep state [21]. For very long battery lifetimes, also the self-discharge rate of the battery plays a role. The discontinuous reception (DRX) technique was developed to reduce the reception time of the device. In LTE Release 10 DRX cycle lengths are configurable up to a maximum value of 2.56 s, which means that the device is only required to listen for paging e.g. once per such interval and not continuously. To reduce the transmission time, many proposals for 5G rely on the working assumption that contention-based transmission of data is beneficial [22]. That is, omitting the RRC Connection Setup procedure and transmitting the payload (and associated control overhead) at once. In the most favorable case, the device would not even have to obtain uplink synchronization before transmission. This is applicable to waveforms that require no or relaxed synchronization such as FBMC or UFMC (see Chapter 7), as an unsynchronized uplink will cause no or little inter-subcarrier interference, or to cases where the timing advance could be estimated, e.g. reusing a previous value for stationary devices. Figure 4.5 depicts in different curves different reporting periodicities for the uplink payload of 125 Bytes. From Figure 4.5 it can be seen that the largest gain is achieved by extending DRX cycles beyond 2.56 s, while contention-

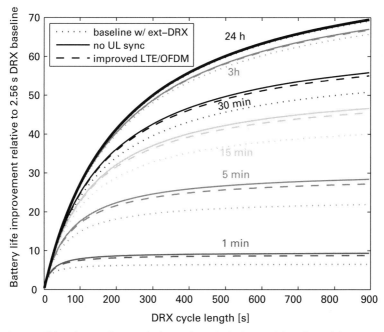

Figure 4.5 Battery life gain as a factor relative to the LTE Release 10 baseline with DRX cycle of 2.56 s.

based transmission (labeled "no UL sync") provides an additional, but significantly smaller, extra gain. The gains of extended DRX cycles are especially for longer uplink reporting periodicities (for more details see [22]). The longer the reporting periodicity, the smaller the gain from contention-based transmission and the larger the gain from longer DRX cycles since the paging monitoring is the dominant part of the device energy consumption. For too frequent reporting, every minute in this case, the 10 years battery lifetime requirement is simply not fundamentally possible. The largest gains of contention-based transmissions are therefore found for the case of 5-minute reporting. In this case, the extension of the DRX cycle from 2.56 s to 300 s gives an improvement of 20 times longer battery life (see [21] for a calculation of the battery lifetime), and the contention-based transmission can further increase this up to a factor of 25 times. Note, however, that in practice these results represent an upper limit, as one also needs to account for the overhead in the RRC Connection Setup (addresses, security, etc.), which also consumes transmission resources [23]. To further separate out the gains of omitting the RRC Connection Setup signaling from those of not having to obtain uplink synchronization, a third alternative is included in which Random Access is used only to obtaining the Timing Advance in Random Access Response, which is sent with a fixed timing (this alternative is labeled "improved LTE/OFDM"). It is seen that the majority of the additional gain (i.e. increase from 20 to 25 times) comes from omitting the initial RRC signaling, since in this case the gain is 24 times of that of the reference case. Still, the largest gain (20 times) overall comes from monitoring paging that is performed less often.

The above results are for a typical cell-edge throughput of 23 kbps. The cell edge users have the lowest battery lifetime and one should seek to fulfill the 10-year battery lifetime for those devices. Devices in the cell center or other favorable locations benefit from short transmission times and can obtain significantly longer battery lifetimes then devices at the cell edge. If extended coverage modes (e.g. by means of time repetition) is considered, with a data rate of around 1 kbps, the gain of contention-based transmission is always below 1% and therefore insignificant.

Note that LTE Release 12 introduces a so-called Power Saving Mode (PSM). In this mode, the device resides in a sub-state of the idle state where it is not reachable by paging. Periodically, the device performs Tracking Area Update (TAU) through exchanging signaling information with the network and is afterwards reachable for paging in a pre-defined time period. PSM enables improvements that are close to those of the above DRX cycle extension, but also has several shortcomings over extending DRX cycles: (1) it has rather extensive signaling which makes it less beneficial for shorter sleeping cycles, (2) it is most suitable for periodical and predictable traffic since TAU keep-alive will be transmitted even when there is no data payload to transmit, and (3) it is mostly suitable for mobile originated traffic since downlink reachability comes at the price of uplink transmissions (TAU). In general, the extended DRX cycle solution is therefore a better and more general solution; DRX extensions are addressed in LTE Release 13.

Coverage enhancements are based on lower data rates and lead to longer transmission times and considerably higher device energy consumption. This makes it difficult

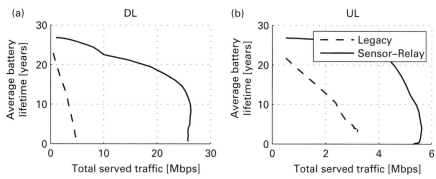

Figure 4.6 The average battery lifetime, for the set of users that benefit from the sensor-relays, presented as a function of the total served traffic/number of MTC devices: (a) for the downlink and (b) for the uplink.

to meet both the coverage enhancement and battery life requirements at the same time. One approach to address very long battery lifetimes in extended coverage is to use MTC device relaying, where devices with decent coverage forward messages to and from devices in extended coverage. In Figure 4.6, it is seen that the battery life is considerably better upheld with an increasing amount of devices for the case where MTC devices can act as relays to others. For the MTC use case 6 (called "Massive amount of geographically spread devices" in Chapter 2) with a data rate of 2 Mbps (including 50% control overhead), the uplink battery life can be improved from 12 to 26 years for the devices served by relays. Note, however, that this has an equalizing effect of the battery life for other devices in the cell. Devices in bad coverage will get much better energy consumption, whereas devices acting as relays will naturally get worse.

4.3.2.5 Features for scalability and capacity

Besides the problem of improvement of coverage and battery life, another problem that needs to be addressed is the scaling of the protocols with the predicted massive number of MTC devices. Specifically, there is a need for protocols to tackle massive contention and distributed sharing of the wireless medium. The most common are the quasi-orthogonal techniques that are used to overload the physical resources and allow for a higher number of simultaneously transmitting devices. This is typically applicable to uplink transmissions since it requires a more advanced receiver, which is only feasible to implement in the infrastructure, but not the low cost devices. One example is coded random access with SIC over time slots, already discussed in Section 4.2.2. Another example is multi-user detection based on compressed sensing, where the sparsity is given by the sporadic access of MTC devices [24]. Coded random access was successfully combined with compressed sensing to obtain a potential gain of factors of 3–10 in the number of servable devices as compared to idealized LTE. Another technique is Sparse Coded Multiple Access (SCMA), where several OFDM time-frequency resources are used to define the contention space and different devices have different sparsity patterns over these resources on which transmissions are directly coded (see [25] and Chapter 7). SCMA with a contention-based data transmission can double the uplink capacity of device transmissions [6] compared to LTE Release 11. Somewhat closer to legacy operation, [26] is expanding the Random Access (RA) by coding over multiple subframes in time and obtaining a larger contention space without increasing the number of physical resources. Compared to LTE Release 11, this would increase the RA capacity at the price of higher complexity and latency.

Regarding the uplink transmission bandwidth for MTC, the coverage is not improved by having narrowband transmission for a given device. However, when the path loss becomes sufficiently large, a larger transmission bandwidth only gives unnoticeable higher throughput as the maximum transmit power of the device becomes limiting. Therefore, a more narrow transmission bandwidth can improve the capacity by allowing coverage-limited devices not to use more bandwidth than required. In LTE, this could be done by scheduling transmissions over a subcarrier of 15 kHz instead of a physical resource block of 180 kHz. Note that this would not even require any physical layer changes. Alternatively, code-division multiplexing for multiple devices could be added for transmitting on the same physical resource block. Figure 4.7 shows the uplink capacity gains associated with this (for more details see [20]). Different coverage extension options are considered: LTE Release 10 cell edge coverage ("0 dB"), e.g. cell edge indoor ("20 dB") and an extreme coverage extension ("40 dB"). In the

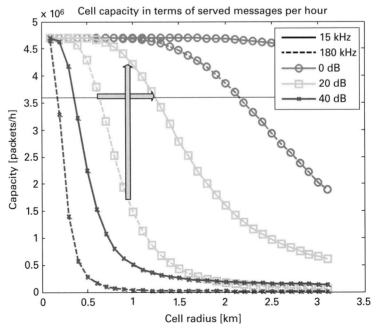

Figure 4.7   Uplink capacity with 1.4 MHz system bandwidth and QPSK rate limitation.

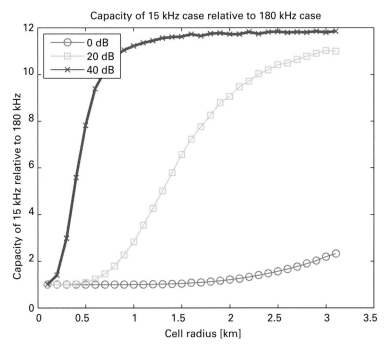

Figure 4.8   Relative uplink capacity increase of 15 kHz transmissions relative to 180 kHz.

figure, solid lines indicate 15 kHz transmission bandwidth, dashed lines indicate 180 kHz.

First of all, it is seen that if the device throughput is upwards limited by QPSK modulation, the required capacity of 3.6 million packets per hour (as defined in [6] and indicated by the horizontal dotted line) can be met by a system bandwidth of 1.4 MHz (0.5 MHz if limited by 64QAM).

For indoor devices with an assumed additional path loss of +20 dB, it is seen that at a cell radius of 1.1 km the capacity increases by a factor of three by going to the more narrowband 15 kHz transmissions (full curve) as compared to the 180 kHz LTE Release 10 reference case (dashed curve). This is the largest absolute capacity increase but the relative capacity increase can be up to almost twelvefold as seen in Figure 4.8. Note however that the same gains should be possible to achieve also by code multiplexing and that these results are for single cell.

As mentioned, the MTC device relaying will also give capacity gains, but as opposed to the solutions mentioned above there will also be gains for the downlink. As seen from Figure 4.9, when MTC devices are acting as relays five times more traffic can be served, which corresponds to a fivefold increase in number of devices (see [6]). For the uplink, the corresponding gain is only a factor of two, and the reason is that in the very dense Madrid grid fewer devices are being served by relays in the uplink case. However, the capacity limit can alternatively be given by a maximum acceptable drop rate. Assuming that this is e.g. 4%, it can be seen from Figure 4.4 that the capacity increases from 1 Mbps

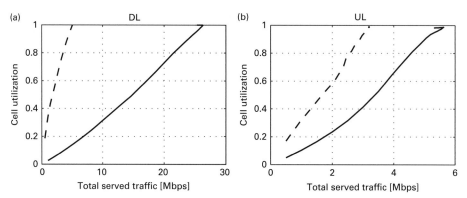

Figure 4.9   The utilization of the macro cell layer as a function of the total served traffic/number of devices for the LTE Release 10 baseline system (dashed line) and the MTC device-relay system (continuous line): (a) for the downlink and (b) for the uplink.

to 16 Mbps with the MTC device relaying. That is a capacity improvement by a factor of 16. For the uplink, the corresponding value is a factor of five.

Further capacity improvements could be obtained by taking advantage of the transmissions between MTC devices of the same type or in the same group. As suggested by [27] compression could be applied to similarities and redundant information can under some circumstances be removed.

4.3.3   Summary of mMTC features

The main mMTC features as well the techniques that allow enforcing those desired features are briefly summarized in Table 4.1.

4.4   Ultra-reliable low-latency MTC

4.4.1   Design principles

Reliable low-latency design is often needed in a control-related communication context. This can be remote control of machinery (e.g. tele-surgery or operations in hazardous environments) or factory cell automation. In these use cases, data messages are typically short control messages, e.g. 100–1000 bits that need to be transmitted within very strict delay bounds. For example, for industrial automation requirements can be as strict as requiring guaranteed end-to-end packet transmissions within 1 ms. Required reliability levels can be in the 99.999th percentile (i.e. $(1 - 10^{-5})$ but may need to even reach levels up to 99.9999999th percentile (i.e. $(1 - 10^{-9})$)) in extreme cases. The following Figure 4.10 shows the design objective of ultra-reliable low-latency communication in contrast to typical mobile broadband objectives. The mobile broadband systems commonly focus on metrics related to median and peak

Table 4.1 Techniques that allow enforcing main mMTC features. The gains are relative to LTE.

Feature	Techniques	Desired values/Gain
Low device complexity	Limited BW transmission	~1 MHz/n.a (not applicable).
	Limited peak rate	~2 Mbps/n.a.
	Limited output power	20 dBm/n.a.
Service flexibility	Flexible data access design	scalable up to a few Mbps/n.a.
Coverage extensions (by 15 dB)	Time repetition and relaying	Scenario dependent
Long battery lifetime (5–10 years)	Extended DRX and potentially relaying	DRX cycle of 300 s/scenario dependent
Scalability	Coded random access	Large number of devices/3–10 times gain
	Multi-user detection based on compressed sensing	
Capacity	SCMA	n.a./2 times UL capacity
	Narrow band transmission i.e. 15 kHz	n.a./3 to 12 times DL capacity gain
	Relaying	n.a./16 times gain for DL (4 for UL)

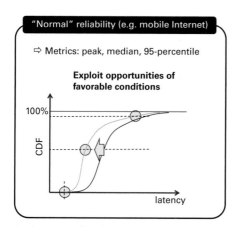
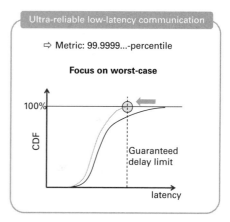

Figure 4.10   Design target for ultra-reliable low-latency communication.

performance, as well as certain modest percentile (e.g. the 95th or 99th percentile) in the performance distribution. For uMTC the focus is rather on a very high percentile, given by the reliability requirement, which ensures that at this level the required delay can be met. Improving the transmissions that are already within the delay bound, which is equivalent to a higher data rate, is not a goal per se.

When ultra-high reliability needs to be attained at a system level, then one way to proceed is to derive reliability requirements for each of the modules that constitute the system. For a data packet, such requirements can be sublimed as, e.g. "transfer of data packets that have at most B bytes with a delay D less than L seconds in 99.99% of the attempts". This creates a rather simple criterion to see whether the system meets the requirement or not. However, the problem with this criterion is that the data transmission model needs to report a failure whenever this simple and rigid criterion is not met. The data transmission module may still be able to send *something* and the overall service at the system level does not need to fail. In order to achieve this type of operation, one needs to reconsider the way in which a certain communication service is composed. The concept of *Reliable Service Composition* [13], discussed in Section 4.1.2.2, enables different versions of a service according to the reliability at which the connectivity can be provided.

In order to illustrate the idea, let us consider RSC in the case of V2V communication. It is noted that the percentages used in the example are provisional, only for illustration. The basic version of the service is available 99.999% of the time. In the V2V setting, the basic version could involve transmission of a small set of warning/safety messages without certification. The fact that the set of messages transferable in the basic mode is limited can be used to design efficient transmission mechanisms that use low rate. An enhanced version of the service is available 99.9% of the time, includes limited certification and guarantees the transfer of a payload of size D1 within time T1 with probability 99.9%. The full version is available 97% of the time, includes full certification and guarantees for transfer of payload of size D2 > D1 within time T2 < T1 with probability 97%. The key issue in making RSC operational is to have reliable criteria to detect which version the system should apply at a given time, i.e. have suitable indicators of availability and reliability.

4.4.2   Technology components

The desired features for an uMTC system are reliable low latency, and availability indication. In the following the technology components, among them D2D communications, that address these features are presented.

4.4.2.1   Features for reliable low latency

In wireless transmission, Rayleigh fading adds significant signal fluctuations, which increases the risk of temporary outage and packet losses. This can be compensated with a fading margin added to the average SNR. In order to achieve high levels of reliability, a significant margin needs to be added, e.g. 50 dB–90 dB for reliability levels of $1-10^{-5}$ to $1-10^{-9}$ (see Figure 4.11). Adding diversity to the transmission with independently faded signal components provides robustness against fading losses. With higher level of diversity, the fading margin can be significantly reduced. With a diversity order of 8 or 16, the fading margin for a reliability of $1-10^{-9}$ can be reduced from 90 dB to 18 dB or 9 dB respectively. Diversity can be achieved in dimensions space, frequency and time. Due to the low-latency requirement, the time dimension cannot be exploited. Overall, diversity is one of the key technology components to enable reliability on short time scales in wireless communication system with fluctuating channel properties [28][29]. Other sources of uncertainty that challenge reliability is interference caused by other transmissions. Orthogonal

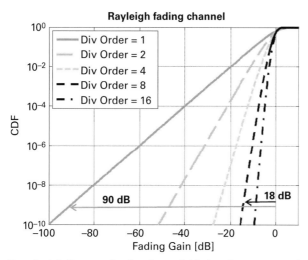

Figure 4.11 Required fading margins for ultra-reliable low-latency transmission.

Figure 4.12 Slot structure and access delay.

multiple access (like e.g. OFDM) with coordinated channel access enables robustness against interference.

In order to achieve low latencies, short transmission time intervals are desirable. A Transmission Time Interval (TTI) or access slot is the atomic unit in which access to the radio channel is enabled. An arriving packet has to wait until the next access opportunity before the data transmission can start (see Figure 4.12). Short transmission time intervals reduce this access delay. The transmission time interval also defines the resource blocks into which data is packaged during transmission; such that an entire transmission time interval needs to be received before data can be delivered at the receiver. Short transmission time intervals, e.g. in the order of 100 μs are needed if end-to-end latencies of around 1 ms shall be enabled.

Another important design choice for reliable low-latency communication is to enable fast receiver processing. Channel coding is playing a crucial role in providing reliability for data transmissions; at the same time channel code design has an impact on the receiver processing delay. For the transmission of short control messages at very high reliability, convolutional codes have benefits over iterative codes, like turbo codes or low-density parity check codes [29]. For short messages up to a few hundred bits the performance of convolutional codes is roughly on par with turbo and convolutional codes. At the same time, iterative codes may have an error floor, which is prohibitive to reaching very low packet error rates (e.g. 10^{-9}). Convolutional codes do not have an error floor and the receiver has a lower complexity. Besides that, a receiver can already start decoding data as the data is being received. In iterative codes, first an entire data block needs to be received before it is iteratively decoded. As a result, convolutional coding enables shorter receiver processing times. A prerequisite to enable on-the-fly channel decoding is that the channel can be early estimated. Therefore, reference symbols should be placed at the beginning of a transmission time interval and not be spread over the transmission time interval. This allows that a receiver can estimate the channel after receiving the first symbols in a TTI and then start decoding the code words. For very short TTIs, channel variations within a TTI should be very modest. While most wireless communication systems use hybrid ARQ with incremental redundancy in order to achieve high spectral efficiency, it is shown in [29] that potential benefit of HARQ for better resource utilization is very limited for very low transmission delays.

Reliable low-latency design has the fundamental property that the reliability cannot be improved by spreading in time domain in order to achieve diversity. Therefore, a general design choice is to make adaptive transmissions, such as link adaptation, very robust and apply conservative channel estimation.

4.4.2.2   Feature for reliability: availability indication

The application of wireless communication systems for ultra-reliable communication use cases depends on the capability to provide reliable connectivity. The failure to comply with the reliability requirements can render the service useless as it can lead to costly damage (e.g. as in the case of failure of a smart grid or an industrial process) or cause even harm (e.g. to vehicle passengers and other traffic participants in case of road safety). At the same time, wireless communication systems are inherently subject to uncertainties of the radio environment and are today typically designed to provide only modest reliability levels. A design to provide ultra-high reliability levels at all times and in every reception scenario may lead to an overdesign of a system with little commercial viability. In order to cope with this problem and enable the provision of uMTC applications with strict reliability requirements, future 5G systems should be able to warn the application about the presence or absence of reliability according to its requirements. In this manner, the applications would be capable of using the wireless communication system only in those instances in which the reliability can be guaranteed. For example, highly autonomous driving systems could reduce the velocity in case of insufficient reliability on the wireless interface, or even prompt the driver to take control of the vehicle.

In order to enable the use of wireless communications for the provision of critical applications, such as those in the area of road safety and traffic efficiency, it is of paramount importance to ensure high reliability. Within this context, two different aspects must be taken into account:

- The probability of false alarm, that is, the probability that the link is indicated as reliable when it is not, must be kept below a maximum as specified by the application. This is determined by the precision of the channel estimation and prediction method used for the computation of the availability. More accurate channel estimation and prediction methods result in a better knowledge of the propagation channel, and therefore, improve the accuracy of the availability estimation. The results in [30] illustrate how it is possible to ensure availability estimation at user velocities with Doppler shift up to 200 Hz as long as the propagation channel can be predicted perfectly.

- The availability of the wireless communication link, that is, the probability that the link is indicated as reliable, should be maximized under the false alarm. In general, the provision of link reliability depends on the robustness of the physical layer (i.e. channel coding, diversity schemes, etc.), the amount of available radio resources, etc. In this sense, the utilization of high orders of diversity is fundamental to improve the availability of the wireless communication link.

One possible implementation for the availability indication of reliability consists of two main components: a Reliable Transmission Link (RTL) that is optimized to transmit packets successfully and within a predefined deadline, and an Availability Estimation and Indication (AEI) function that is able to reliably predict the availability of the RTL under given conditions (see Figure 4.13). The AEI function receives Availability Requests (AR) from the application, which include relevant information about the service such as the packet size, the maximum latency, the required reliability, and other implementation specific information. Upon an AR, the AEI returns an Availability Indicator (AvI), which in its simplest form is a binary variable. In case the AvI = 0 (link is declared as not available), the application need to use some fall-back mechanism to gracefully degrade the application performance or even refrain from using the wireless communication link. In case AvI = 1, then it is possible for the application to use the link with the required reliability.

4.4.2.3 Features enabled by D2D communications
D2D communication is an important enabler for uMTC applications from the automotive and industrial domain that have very stringent latency requirements (in the order of

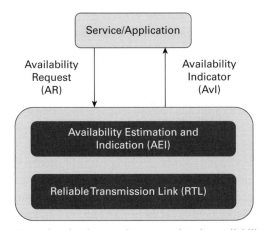

Figure 4.13 Exemplary implementation supporting the availability indication for reliability.

milliseconds). Exploiting the D2D capabilities, two physically close-located communicating peers can take advantage of their proximity to exchange information over a direct link so that messages do not need to be relayed by a central BS. Especially for short-range communication requirements, direct communication between peers by means of D2D reduces the end-to-end latency. It is spectrally more efficient than communication through the cellular infrastructure, as this avoids the redundancy in the use of resources due to the uplink and downlink and it enables a spatial reuse of resources by reducing the interference in the service area. Furthermore, the combination of D2D with cellular infrastructure-based communication can lead to increased reliability by means of multi-path diversity.

As an example, V2X communication – that is, the direct exchange of messages between vehicles (V2V), between a vehicle and a device carried by an individual (V2D) and between a vehicle and the infrastructure (V2I) – has the potential to significantly improve the provision of road safety and traffic efficiency services, including highly autonomous driving. While cloud connectivity is necessary in order to allow vehicles to download high definition digital maps as well as real-time traffic information, direct V2X communication based on D2D improves the awareness of the vehicle beyond the capabilities of sensor technology. Modern vehicles are equipped with a variety of sensors including cameras and radars that allow them to recognize objects in their environment. Nevertheless, the range of such sensors is quite limited and insufficient for the recognition of most hazards on the road. Combining the information gathered by multiple vehicles and fixed infrastructure (e.g. traffic surveillance cameras at intersections), leads to an extended perception horizon reaching far beyond the limited field of view of a single vehicle or its driver [31]. This would enable drivers and systems for autonomous driving to recognize hazards in advance and take preventive actions much earlier, but requires cooperation (i.e. information exchange) between the traffic participants. Note that the V2X information is only relevant for other traffic participants in the proximity, but not to those located beyond a certain distance. Therefore, D2D is an important key technology for automotive uMTC services.

Nevertheless, it is important to note that the provision of V2X communications presents some particularities that might condition the D2D operation:

- In the context of V2X communications, each traffic participant regularly sends updated information regarding its position, velocity, etc. to all the other traffic participants located within a certain range. The use of broadcasting transmissions (i.e. one-to-many) is therefore expected to be more efficient than unicast communication (i.e. one-to-one) from a resource utilization point of view and to scale better with an increasing number of users in the service area.
- V2X communications generally involve the exchange of information between fast moving terminals (e.g. vehicles) with relative velocities that can exceed 300 km/h in some situations (for vehicles moving in opposite directions). Compared to the case of static or slowly moving communication partners, the presence of highly mobile D2D users poses new challenges, especially in regards to Radio Resource

Management (RRM), as a result of the fast changing interference conditions. This calls for RRM solutions that are resistant to imperfect channel knowledge, as the collection of accurate channel information for all the relevant channels in the system could be extremely costly or even infeasible in a practical deployment scenario.

- Compared with some other uMTC applications such as the factory cell automation use case, the provision of V2X communications is not confined to a certain geographical area (e.g. an industrial complex) but extends along the entire road infrastructure. Since 100% network coverage is probably not achievable due to economic reasons, ad-hoc D2D might be used to enable V2X communications in areas with insufficient network coverage.

Based on the above particularities, it is easy to see that both network-assisted and ad-hoc D2D could play an important role for the provision of V2X communications in future 5G networks. While the coordination of transmissions performed by a central entity (i.e. a base station or a cluster head) in the case of network-assisted D2D allows for a better resource allocation and interference management (see Chapter 5 and [32]), ad-hoc D2D is fundamental in order to enable the exchange of data between traffic participants even in locations with insufficient network deployment (i.e. out of coverage). Future 5G systems could combine the superior performance of network-assisted D2D with the coverage extension of ad-hoc D2D based on the concept of reliable service composition, described in Section 4.4.1. As an example, the data packets that are exchanged between traffic participants could be divided in two different categories or classes. Data packets of class 1 might contain information that is critical for the safety of the traffic participants, such as their position, velocity or direction (e.g. 300 bytes per packet). Data packets of class 2 might contain non-critical information that nevertheless might contribute to the safety of traffic participants (e.g. 1300 bytes per packet). This solution is based on the idea that data packets of class 1 have to be transmitted at all times, even in situations with insufficient network deployment by means of ad-hoc D2D, whereas data packets of class 2 have to be transmitted only by means of network-controlled D2D and with a lower priority (e.g. they can be discarded in favor of data packets of class 1) in order to exploit the additional performance. In this manner, the system has to satisfy significantly lowered requirements in terms of user data rate and traffic volume density while operating in ad-hoc D2D mode as compared to network-controlled D2D mode.

For further details on the D2D technology and transmission schemes, we refer the reader to Chapter 5 of this book.

4.4.3 Summary of uMTC features

The techniques that will allow addressing the main uMTC features are briefly summarized in Table 4.2.

Table 4.2 Techniques addressing uMTC features.

Feature	Techniques	Desired values/Gain
Reliability (from $1-10^{-5}$ to $1-10^{-9}$)	Diversity (e.g. in frequency, space)	Order 4 to 16/35 dB to 72 dB
Delay (1 ms)	Short TTI	100 μs/n.a.
	No Retransmission	n.a.
	Convolutional Code	Scenario dependent
Delay or Reliability (generic)	D2D	Scenario dependent
Reliability (generic)	Availability indication	n.a.

4.5   Conclusions

Historically, the major improvement in each new mobile wireless generation, up to 4G, has been the increased data rate and spectral efficiency. While 5G is also poised to bring a significant improvement in the data rates over 4G, it will also address new use cases beyond personal mobile broadband services, including the field of MTC. Two distinct areas of MTC are of particular interest: massive Machine-Type Communication (mMTC) and ultra-reliable MTC (uMTC). This chapter has been dedicated to discuss uMTC and mMTC, describing use cases and requirements as well as 5G features to address those services. The requirements for a device in mMTC, such as low energy consumption and long battery lifetime, are vastly different from a device in uMTC, where the exemplary requirements include low latency and extremely high reliability of the packet delivery. Despite that, there are common fundamental communication theoretic principles that can be used in the design of both mMTC and uMTC, such as the transmission of the short packets. After describing those principles, the specific design principles for mMTC and uMTC are presented. The area of mMTC is more mature, as its development has already started within LTE. On the other hand, uMTC poses new research challenges in order to attain unprecedented levels of reliability that will enable new applications in 5G.

Concerning the specific conclusions of mMTC, mMTC devices can leverage on their very low requirements in terms of latency and rate to enable a device simplicity that is supported by the network with a MTC-specific transmission mode. Features of such a transmission mode are transmissions with limited peak-rate and limited bandwidth for the device. Furthermore, device design can be kept simpler with half-duplex transmission (avoiding duplex filters) and limited peak power (integration of power amplifier onto the integrated circuit).

As mMTC devices can be embedded with different types of physical objects and in different types of environments, they may be at locations where connectivity is hard to achieve from the mobile network infrastructure. Exploiting the delay-tolerant low-rate requirements, transmission modes with very low data rates and extra robust control channel design can be applied, thus providing significantly extended range of mobile networks. Another complementary approach toward extended coverage is to enable relaying of traffic via intermediate nodes between the mobile network infrastructure and a remote mMTC device.

In many cases it is desirable to deploy mMTC devices that are powered by batteries and still have them running for many years. The infrequent transmission of data per device and the delay tolerance enable long battery lifetime. In the downlink, the network has to buffer data while the device is in sleep mode. The battery-saving potential is limited by the frequency of data transmissions and also the reachability requirement of the device. Devices that need to be reachable within short times have to provide sufficient transmission opportunities, which limits the periods that the device remains in sleep mode. Devices that make use of extended range communication modes transmit at very low data rates, which reduces their battery lifetimes. Long battery lifetimes together with very low data rates can be provided by relaying, where the very large path loss is split into two parts of lower loss.

As the number of mMTC devices can be very large, with possibly several hundreds of thousands of devices per radio cell, a scalable transmission is required. In particular, for the random access, efficient multiple access schemes are needed for large device populations. This can be achieved by increasing the contention space e.g. by expanding contention signals in the time-frequency domain, which can be further enhanced with successive interference cancellation. For data transmissions, it is possible to improve the uplink capacity for devices that have very large path loss. For those devices, fine-granular resource allocations or overloading resource usage e.g. by code-division multiplexing is beneficial. Furthermore, the usage of relaying can provide a significant capacity increase for both uplink and downlink.

Concerning the specific conclusions of uMTC, the key requirement is to enable that all transmissions up to the required reliability meet the delay limit in order to make the transmission characteristics as deterministic as possible to the higher-layer application.

Design principles include high levels of diversity to combat uncertainties of the radio channel. Furthermore, short radio frames are needed. Frame structures should be so that receiver processing can be minimized, by enabling early channel estimation and on-the-fly decoding at the receiver rather than buffering larger amounts of data prior to receiver processing. For road safety and traffic efficiency applications, direct communication between traffic participants by means of network assisted as well as non-assisted D2D is an important component.

uMTC applications benefit from a reliable service composition framework, where service operations can be designed for different reliability levels. This should be combined with availability and reliability indications that expose the supported reliability to the service layer.

References

[1] S. Andreev, O. Galinina, A. Pyattaev, M. Gerasimenko, T. Tirronen, J. Torsner, J. Sachs, M. Dohler, and Y. Koucheryavy, "Understanding the IoT connectivity

landscape: A contemporary M2M radio technology roadmap," *IEEE Communications Magazine*, vol. 53, no. 9, pp. 32–40, 2015.

[2] M. Condoluci, M. Dohler, G. Araniti, A. Molinaro, and J. Sachs, "Enhanced radio access and data transmission procedures facilitating industry-compliant machine-type communications over LTE-based 5G networks," *IEEE Wireless Communications Magazine*, vol. 23, no. 1, pp. 56–63, 2016.

[3] J. J. Nielsen, G. C. Madueño, N. K. Pratas, R. B. Sørensen, Č. Stefanović, and P. Popovski, "What can wireless cellular technologies do about the upcoming smart metering traffic?," *IEEE Communications Magazine*, vol. 53, no. 9, pp. 41–47, 2015.

[4] A. Frotzscher, U. Wetzker, M. Bauer, M. Rentschler, M. Beyer, S. Elspass, and H. Klessig, "Requirements and current solutions of wireless communication in industrial automation," in IEEE International Conference on Communications Workshops, Ottawa, June 2014.

[5] M. Condoluci, M. Dohler, G. Araniti, A. Molinaro, and K. Zheng. "Toward 5G densenets: Architectural advances for effective machine-type communications over femtocells," *IEEE Communications Magazine*, vol. 53, no. 1, pp. 134–141, 2015.

[6] ICT-317669 METIS Project, "Scenarios, requirements and KPIs for 5G mobile and wireless system," Deliverable 1.1, April 2013.

[7] 3GPP TR 36.888, "Study on provision of low-cost Machine-Type Communications (MTC) User Equipments (UEs) based on LTE Equipments (UEs) based on LTE," Technical Report TR 36.888 V12.0.0, Technical Specification Group Radio Access Network, June 2013.

[8] VODAFONE Group Plc, "Update to study item on cellular system support for ultra low complexity and low throughput internet of things," Study item, GP-150354, 3GPP **TSG**-GERAN Meeting #66, May 2015.

[9] G. C. Madueño, C. Stefanovic, and P. Popovski, "How many smart meters can be deployed in a GSM cell?," in IEEE International Conference on Communications Workshops, Budapest, June 2013.

[10] 3GPP TR 45.820, "Cellular system support for ultra-low complexity and low throughput internet of things (CIoT) (Release 13)," Technical Report, TR 45.820 V13.1.0, Technical Specification Group GSM/EDGE Radio Access Network, December 2015.

[11] Qualcomm Incorporated, "New work item: NarrowBand IOT (NB-IOT)," Work Item RP-151621, 3GPP TSG RAN Meeting #69, September 2015.

[12] ETSI GS LTN 003, "Low throughput networks (LTN): Protocols and interfaces," September 2014.

[13] P. Popovski, "Ultra-reliable communication in 5G wireless systems," in International Conference on 5G for Ubiquitous Connectivity, Levi, November 2014.

[14] E. Dahlman, G. Mildh, S. Parkvall, J. Peisa, J. Sachs, Y. Selén, and J. Sköld, "5G wireless access: Requirements and realization," *IEEE Communications Magazine*, vol. 52, no. 12, December 2014.

[15] Y. Polyanskiy, H.V. Poor, and S. Verdu, "Channel coding rate in the finite block-length regime," *IEEE Transactions on Information Theory*, vol. 56, no. 5, pp. 2307–2359, May 2010.

[16] E. Paolini, C. Stefanovic, G. Liva, and P. Popovski, "Coded random access: Applying codes on graphs to design random access protocols," *IEEE Communications Magazine*, vol. 53, no. 6, pp. 144–150, 2015.

[17] A. Laya, L. Alonso, and J. Alonso-Zarate, "Is the random access channel of LTE and LTE-A suitable for M2M communications? A survey of alternatives," *IEEE Communications Surveys and Tutorials*, vol. 16, no. 1, pp. 4–16, 2014.

[18] K. Zheng, S. Ou, J. Alonso-Zarate, M. Dohler, F. Liu, and H. Zhu, "Challenges of massive access in highly dense LTE-advanced networks with machine-to-machine communications," *IEEE Wireless Communications*, vol. 21, no. 3, pp. 12–18, 2014.

[19] ICT-317669 METIS Project, "Intermediate system evaluation results," Deliverable 6.3, August 2014.

[20] ICT-317669 METIS Project, "Report on simulation results and evaluations," Deliverable 6.5, March 2015.

[21] T. Tirronen, A. Larmo, J. Sachs, B. Lindoff, and N. Wiberg. "Machine-to-machine communication with long-term evolution with reduced device energy consumption," *Transactions on Emerging Telecommunications Technologies*, vol. 24, no. 4, pp. 413–426, 2013.

[22] ICT-317669 METIS Project, "Components of a new air interface: Building blocks and performance," Deliverable 2.3, April 2014.

[23] 3GPP TR 37.869, "Study on enhancements to machine-type communications (MTC) and other mobile data applications," Technical Report TR 37.869 V12.0.0, September 2013.

[24] F. Monsees, C. Bockelmann, and A. Dekorsy, "Reliable activity detection for massive machine to machine communication via multiple measurement vector compressed sensing," in IEEE Global Communications Conference Workshops, Austin, December 2014.

[25] K. Au, L. Zhang, H. Nikopour, E. Yi, A. Bayesteh, U. Vilaipornsawai, J. Ma, and P. Zhu, "Uplink contention based SCMA for 5G radio access," in IEEE Global Communications Conference, San Diego, March 2015.

[26] N. K. Pratas, H. Thomsen, C. Stefanovic, and P. Popovski, "Code-expanded random access for machine-type communications," in IEEE Global Communications Conference Workshops, Anaheim, December 2012.

[27] Z. Chan and E. Schulz, "Cross-device signaling channel for cellular machine-type services," in IEEE Vehicular Technology Conference, Vancouver, September 2014.

[28] O. N. C. Yilmaz, Y.-P. E. Wang, N. A. Johansson, N. Brahmi, S. A. Ashraf, and J. Sachs, "Analysis of ultra-reliable and low-latency 5G communication for a factory automation use case," in IEEE International Conference on Communication, London, June 2015.

[29] N. A. Johansson, Y.-P. E. Wang, E. Eriksson, and M. Hessler, "Radio access for ultra-reliable and low-latency 5G communications," in IEEE International Conference on Communication, London, June 2015.

[30] H. D. Schotten, R. Sattiraju, D. Gozalvez-Serrano, R. Zhe, and P. Fertl, "Availability indication as key enabler for ultra-reliable communication in 5G," in European Conference on Networks and Communications, Bologna, June 2014.

[31] A. Rauch, F. Klanner, R. Rasshofer, and K. Dietmayer, "Car2X-based perception in a high-level fusion architecture for cooperative perception systems," in IEEE Intelligent Vehicles Symposium, June 2012.

[32] ICT-317669 METIS Project, "Final report on network-level solutions," Deliverable D4.3, February 2015.

5 Device-to-device (D2D) communications

Zexian Li, Fernando Sanchez Moya, Gabor Fodor, Jose Mairton B. Da Silva Jr., and Konstantinos Koufos

Direct Device-to-Device (D2D) communication, which refers to direct communication between devices (i.e. users) without data traffic going through any infrastructure node, has been widely foreseen to be an important cornerstone to improve system performance and support new services beyond 2020 in the future fifth generation (5G) system. In general, the benefits resulting from D2D operation include, among others, highly increased spectral efficiency, improved typical user data rate and capacity per area, extended coverage, reduced latency, and enhanced cost and power efficiency. These benefits are resulting from the proximity of the users employing D2D communication (proximity gain), an increased spatial reuse of time and frequency resources (reuse gain) and from using a single link in the D2D mode rather than using both an uplink and a downlink resource when communicating via the base station in the cellular mode (hop gain). The chapter starts with an overview of the fourth generation (4G) D2D development. Afterward, the challenges to be addressed in the context of 5G D2D and related key enablers are discussed. In particular, this chapter covers Radio Resource Management (RRM) for mobile broadband applications, multi-hop D2D communication, especially for public safety and emergency services, and multi-operator D2D communication.

5.1 D2D: from 4G to 5G

In the future 5G system, it is predicted that network-controlled direct D2D communication offers the opportunity for local management of short-distance communication links and allows separating local traffic from the global network (i.e. local traffic offloading). By doing this, it will not only remove the load burden on the backhaul and core network caused by data transfer and related signaling, but also reduce the necessary effort for managing traffic at central network nodes. Direct D2D communication therefore extends the idea of distributed network management by incorporating the end devices into the network management concept. In this way, the wireless user device with D2D capability can have a dual role: either acting as an infrastructure node and/or as an end-user device in a similar way as a traditional device.

5G Mobile and Wireless Communications Technology, ed. A. Osseiran, J. F. Monserrat, and P. Marsch. Published by Cambridge University Press. © Cambridge University Press 2016.

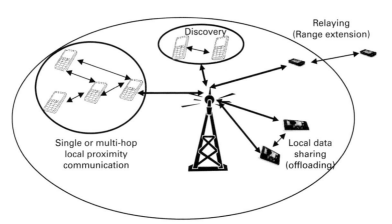

Figure 5.1   Typical use cases of D2D communication in cellular networks.

Further, direct D2D facilitates low-latency communication due to the local communication link between users in proximity. In fact, direct D2D has been seen as one of the necessary features to support real-time services in the future 5G system [1][2]. Another important aspect is reliability, where an additional D2D link can be employed to increase reliability through a larger extent of diversity. Moreover, due to the short-distance transmission, the device power consumption can be reduced significantly. Figure 5.1 illustrates typical use cases of D2D communication. A more detailed discussion on different 5G use cases can be found in Chapter 2. Four D2D scenarios are shown. The first one is about local data sharing where data caching in one device can be shared with other devices in proximity. In the second scenario, called relaying, D2D communication can play a key role to improve network availability (i.e. to extend the coverage area) via a D2D based relay. This is especially important for the use cases related to public safety and those including both indoor and outdoor users. The third scenario, called single or multi-hop local proximity communication, is the one considered in the 3rd Generation Partnership Project (3GPP) Release 12. In this scenario, the devices within proximity can set up a peer-to-peer link or multicast link that does not use the cellular network infrastructure. One of the particular applications is the public safety service. The last scenario is D2D discovery (considered in 3GPP Release 12 as well), which refers to a process that identifies whether a UE is in proximity of another UE.

Considering D2D air interface design, it is usually assumed that the air interface for D2D communication is derived from the cellular air interface in order to simplify the design and implementation. For example in 3GPP Release 12, Single-Carrier Frequency Division Multiple Access (SC-FDMA) based D2D signaling is employed for all data-carrying physical channels, and the structure of the Physical Uplink Shared CHannel (PUSCH), as defined in 3GPP, is re-used (with limited changes) for the D2D communication channel as well. Regarding spectrum usage, D2D can operate, depending on the scenario, in licensed spectrum and/or unlicensed spectrum.

Table 5.1 D2D scope in LTE Releases 12 and 13.

	Within LTE network coverage	Outside LTE network coverage
Discovery	Non-public safety & public safety requirements	Public safety
Direct Communication	At least public safety requirements	Public safety

When talking about cellular network-controlled D2D, it is necessary to mention the standardization progress especially in 3GPP on Long Term Evolution (LTE) D2D (also known as ProSe: Proximity Services). It is worthwhile to note that Wi-Fi Direct and Wi-Fi Aware are relevant as well, although they are not addressed here since the focus of this chapter is on cellular technology-based D2D.

In the following, the current D2D development in 4G LTE is examined. Thereafter, the 5G D2D concept is introduced in order to have a full picture on D2D concept development.

5.1.1 D2D standardization: 4G LTE D2D

Although, in principle, D2D can offer various promising benefits as discussed previously, in 3GPP LTE D2D work, the main driver is public safety in Releases 12 and 13 [3]. In addition, commercial discovery is supported as well, as can be seen from Table 5.1.

LTE D2D can be seen as an add-on feature in a 4G LTE system, hence allowing legacy cellular User Equipment (UE) to operate on the same carrier. In LTE, D2D is operated in a synchronous way, where the synchronization source can be an eNode-B[1] (in case of UEs being under network coverage) or a UE (in case at least one of the UEs is not under network coverage or in case of inter-cell operation). Either uplink (UL) spectrum (in case of Frequency Division Duplexing (FDD)) or UL subframes (in case of Time Division Duplexing (TDD)) can be used for D2D transmission. One interesting feature is the interference management among D2D links and cellular links. This feature has not been discussed in 3GPP, since in practice it is assumed that D2D is running within a dedicated resource pool (i.e. certain physical resource blocks in specific subframes), where the D2D-enabled UEs will get the resource pool configuration information from the eNode-B. In addition, the transmission signals are based on the UL signal design to avoid introducing a new transmitter at the UE side. Further compared to OFDM signaling, SC-FDMA can provide better coverage due to the lower Peak to Average Power Ratio (PAPR). The major features of the 4G LTE D2D concept are listed in the following, where it should be pointed out that the D2D link is referred to as sidelink in the 3GPP Radio Access Network (RAN) Working Groups (WGs).

5.1.1.1 D2D synchronization

The sidelink synchronization signal (i.e. D2D synchronization signal), which is transmitted by the D2D synchronization source (either eNode-B or UE), is used for time and frequency synchronization to facilitate synchronous D2D operation. In order to achieve

---

[1] The eNode-B term is used to refer to a base station when talking about LTE specific aspects.

synchronization, at least the following issues need to be solved: synchronization signal design, entities acting as synchronization source, and criteria to select/re-select the synchronization source.

The sidelink synchronization signal is composed of the primary sidelink synchronization signal and the secondary sidelink synchronization signal. Assuming the UEs have network coverage, then the eNode-B transmits primary and secondary synchronization signals (specified in LTE Release 8) that are reused for D2D synchronization. New sidelink synchronization sequences, which are transmitted by a UE acting as synchronization source (such UE can be in or out of network coverage), have been specified in 3GPP as well.

Both eNode-B and UEs can act as synchronization sources. It is easy to understand that the eNode-B can act as a synchronization source. However, in order to facilitate inter-cell D2D operation, under certain conditions, for example at the cell edge, UEs with network coverage can transmit synchronization signals as well. In case of partial coverage where some D2D UEs are with network coverage and the rest are without network coverage, synchronization signals transmitted by UEs within network coverage can also help the out-of-coverage synchronization by aligning the out-of-coverage transmission to cellular network timing. In this way, the possible interference from D2D transmission to cellular links can be reduced.

In order to solve the potential issue of synchronization source selection and re-selection, different types of synchronization sources are specified with different priority levels. The eNode-B has the highest priority order followed by in-coverage UEs, and then out-of-coverage UEs that are synchronized to in-coverage UEs. The out-of-coverage UEs not synchronized to any in-coverage UEs have the lowest priority.

5.1.1.2   D2D communication

In LTE Release 12, D2D communication is based on physical layer broadcast communication, i.e. a physical layer broadcast solution is used to support broadcast, multicast and unicast services at application layer. In order to support multicast or unicast, the targeted group ID (for multicast) or user ID (for unicast) is indicated in the higher layer message. Since by construction it is a broadcasted information, no physical layer closed control loop exists, i.e., no physical layer feedback, no link adaption, and no HARQ is supported for D2D links. The air interface is based on the Uu interface and the UL channel structure is extended to D2D communication. In particular, for D2D data communication related physical channels, the PUSCH structure (as defined in [4]) is reused whenever possible. Considering resource usage, D2D communication is based on a resource pool concept as illustrated in Figure 5.2, where certain time/frequency resources (called resource pool) are configured for D2D usage. The D2D resource pool is configurable within one cell and there are separate resources for D2D control information transmission and D2D data transmission. The resource pool information is carried over broadcast messages, i.e. *SystemInformationBlockType18*.

Before the D2D data transmission, every transmitter sends out a control signal with information on the data transmission format and the occupied resource. This applies to the scenario where the network is assigning resources to the D2D transmitter and the

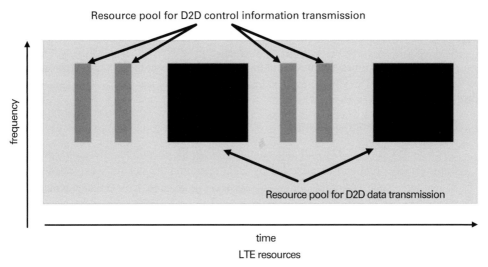

Resource pool for D2D control information transmission

Resource pool for D2D data transmission

frequency

time

LTE resources

Figure 5.2 D2D resource pool.

scenario where the transmitter selects the resource by itself. At the receiving side, it is not necessary to listen to the cellular control channels in order to find out where the D2D data is located. Just based on the content of the D2D control channel, the receiving devices can find out the right location of the relevant resources. As to the resource usage for D2D communication, two different modes were specified:

- **Mode 1:** An eNode-B or relay node schedules the exact resources used by a UE to transmit D2D data and D2D control information. Obviously, Mode 1 can be only applied to the scenarios where the transmitting UEs are within network coverage.
- **Mode 2:** A UE by itself selects resources from the configured resource pools to transmit D2D data and D2D control information. Mode 2 can be applied no matter whether the transmitting UE has network coverage or not.

5.1.1.3 D2D discovery

In LTE Release 12, discovery is applicable only to the UEs with network coverage. The concerned UEs can be in either RRC_IDLE state or RRC_CONNECTED state. Similar to the resources for D2D communication, the D2D discovery resources are arranged as resource pools as well, which are indicated by the eNode-B via *SystemInformationBlockType19*. The resource pools are defined with the parameters including *discoveryPeriod*, *discoveryOffsetIndicator* and *subframeBitmap*. The frequency resources within a D2D subframe are given by the parameters *startPRB, endPRB* and *numPRB*. There are two ways specified for a transmitting UE to get the resources for discovery message transmission:

- **Type 1:** The UE selects autonomously the resource for transmission from the discovery pools (independent of the UE RRC state).
- **Type 2B:** The UE transmits on resources allocated for it by the network (only applicable to RRC_CONNECTED UEs).

The 3GPP RAN WGs specified further D2D enhancements in Release 13 as described in [5]. These enhancements, for the public safety use cases, aimed to solve out-of-coverage discovery, layer 3 based UE-to-network relays, enhancement of D2D communication to support group priorities and group call functionality. However, these are different from the challenges of 5G D2D that aims to address a wider range of use cases.

5.1.2   D2D in 5G: research challenges

Since in 4G LTE D2D communication the focus is on public safety, the potential improvements that can be provided by D2D operation are not fully exploited. In the 5G system, such restriction does not exist anymore, and it is predicted that D2D operation will be natively integrated as part of the future 5G system. Main potential gains that can be achieved include:

- **Capacity/throughput gain:** Because the involved devices are in close proximity with potentially better propagation conditions comparing to the propagation conditions toward the Base Station (BS), link throughput can be improved due to e.g. better Modulation and Coding Scheme (MCS) level. In addition, there is the possibility of sharing the same radio resources among cellular users and D2D users, which can improve the overall spectrum usage. System capacity can be improved due to off-loading and local content sharing gain from D2D communication.
- **Latency gain:** The End-to-End (E2E) latency may be reduced due to a short distance with less propagation delay, and no involvement of infrastructure network entities resulting in reduced transport delay and processing delay.
- **Availability and reliability gain:** D2D can be used to extend network coverage with one hop or multi-hop. Network coding and cooperative diversity via D2D can be used to enhance link quality as well. Furthermore, a D2D ad-hoc network can provide a fall back solution in case of a failure of the infrastructure or in case the infrastructure cannot be easily established.
- **Enabling new services:** Full-blown D2D has great potential to enable new services and applications not only in the telecommunication area, but also in vertical industries, as for example Vehicle-to-X (V2X) communication as discussed in Chapters 2 and 4. The extension of D2D solutions for Vehicle-to-Vehicle (V2V) communication is part of LTE Release 14.

However, as discussed in [6][7], fully utilizing potential D2D gains poses new challenges in terms of device discovery, communication mode selection, co-existence, interference management, efficient multi-hop communication support and multi-operator support among others.

- **Device discovery**: Efficient network-assisted D2D discovery, which is used to determine the proximity between devices and the potential to establish a direct D2D link, is a key element in order to enable D2D communication and possible new applications.
- **Communication mode selection**: Mode selection is another core function that controls whether two devices will communicate to each other in direct D2D mode or in

regular cellular mode (i.e. via a BS). In direct D2D mode, the devices can take advantage of their proximity and may reuse cellular resources for the direct communication link. In cellular mode, the devices communicate through a common or separate serving BS by means of regular cellular links in orthogonal resources with cellular users. How to select the most appropriate communication mode in different scenarios is an important issue to be solved, as discussed in Sections 5.2.3, 5.3.4 and 5.4.2.

- **Co-existence and interference management**: Considering co-existence and related interference issues, at least two different aspects should be taken into account: (1) co-existence among a large number of D2D links, and (2) co-existence among D2D links and regular cellular links. Efficient schemes to handle the interference are of importance in order to achieve the potential D2D benefits.
- **Multi-operator or inter-operator D2D operation**: Inter-operator D2D is a clear requirement resulting from e.g. V2X communication, and supporting inter-operator D2D operation is essential for the 5G D2D concept. Without multi-operator D2D support, the applicability of the future D2D solution to e.g. Cooperative Intelligent Traffic Systems will be quite limited. Considering inter-operator D2D operation, issues to be solved include, for example, spectrum usage and how to control and coordinate UEs in D2D communication across multiple operators' networks.

Clearly, the above bullets are only a subset of the challenges related to D2D operation. In this chapter, the focus is on the challenges related to radio resource management with the proposal of one example of a 5G RRM concept in Section 5.2 followed by multi-hop D2D operation in Section 5.3. Finally, in Section 5.4, multi-operator D2D is addressed, including discovery support, distributed mode selection and spectrum for multi-operator D2D.

5.2   Radio resource management for mobile broadband D2D

In this section, the key aspects related to D2D RRM both from a state of the art and future research perspective are covered. The focus is on mobile broadband D2D scenarios, i.e. scenarios with typically low mobility where offloading of the cellular network, enhancement of system capacity and improvement of user experience in terms of reduced latency and increased data rates play a dominant role [8]. The focus will be on in-band underlay D2D, in which D2D communication uses the same spectrum and resources as cellular communication.

The section is structured as follows. Firstly, a brief overview of RRM techniques for mobile broadband D2D is presented. It is followed by some of the most significant RRM and system design challenges to be solved in order to make D2D a native and efficient technology in 5G systems. Finally, an example of a 5G RRM concept based on flexible TDD is described and performance numbers illustrating the user experience are provided.

5.2.1   RRM techniques for mobile broadband D2D

The addition of the D2D layer as an underlay to cellular networks poses new challenges in terms of interference management in comparison with traditional cellular communication. These challenges come from the reuse of resources between cellular and D2D users, which creates intra-cell interference [9][10]. Therefore, in order to exploit the benefits of D2D communication and achieve an improved system performance over baseline cellular-only systems, careful resource management that takes into account both cellular and D2D users is essential.

RRM algorithms and techniques for D2D underlay communications can be classified depending on the optimization metric and the tools used to achieve that optimized or improved performance. The most common objectives or optimization metrics of RRM algorithms and techniques are spectral efficiency, power minimization and performance with Quality of Service (QoS) constraints [11]. The basic toolbox of available RRM techniques commonly agreed in the literature, such as mode selection, resource allocation and power control [12][13], is described in the following.

- **Mode Selection (MoS):** Several factors influence the MoS decision such as distance between devices, path loss and shadowing, interference conditions, network load, etc. and the time scale on which MoS should be operated. A MoS decision can be made before or after D2D link establishment, while operating on a slow time scale, e.g. based on distance or large-scale channel parameters [14]. Further, a MoS can be done on a faster time scale [15][16], based on changing interference conditions coupled with the resource allocation phase.
- **Resource Allocation (ReA):** ReA determines which particular time and frequency resources should be assigned to each D2D pair and cellular link [9][17]. ReA algorithms can be broadly classified according to the degree of network control, e.g. centralized versus distributed, and the degree of coordination between cells, e.g. single-cell (uncoordinated) versus multi-cell (coordinated).
- **Power Control (PC):** In addition to MoS and ReA, PC is another key technique to deal with the interference, both intra- and inter-cell, that results from underlay D2D operation [18][19]. The focus is mostly on limiting the interference from D2D to cellular transmission, in order to improve the overall system performance while ensuring that the cellular user experience is not degraded. The applicability of LTE power control mechanisms to efficiently support D2D, and optimizations that rely on a practical distributed scheme, have been extensively studied in [20].

It is worth mentioning that the different algorithms do not rely on just one RRM component or isolated technique, but normally combine several of them to achieve better performance [19].

5.2.2   RRM and system design for D2D

Complementing evolved legacy standards with non-backward compatible radio interfaces in 5G will allow designing a radio technology that natively and efficiently supports

D2D from the onset. In Section 5.1.2, some of the general challenges to support D2D in 5G systems, with its broad scope of use cases and scenarios, were highlighted. The focus here is to specifically address some of the fundamental RRM and system design questions to be answered for an efficient support of mobile broadband D2D, for instance:

- How valuable is the usage of D2D across multiple cells, and does this justify the additional coordination and signaling burden introduced? Enabling inter-cell D2D requires some kind of basic conflict prevention of RRM decisions between the serving BSs of the devices involved in D2D communication, even if not targeting optimally coordinated resource allocation. It could be the case that, in a half-duplex system (e.g. a 5G system with flexible TDD optimized for dense scenarios), one BS schedules one of its assigned D2D users for UL transmission (cellular mode selection) while another BS schedules a direct D2D transmission toward the same user, violating the half-duplex constraint. Solutions to prevent this issue may include: exchange of scheduling information between BSs (or via a centralized coordination entity); protocol-level solutions that orchestrate the order of the transmissions; or simply disabling inter-cell D2D, i.e. only allowing intra-cell D2D and routing the inter-cell D2D traffic through the infrastructure to avoid the coordination burden.
- Does sophisticated D2D (e.g. fast joint MoS and ReA with flexible TDD) require centralized radio resource management, or can this be done in a decentralized or distributed manner? Apart from the multi-cell D2D aspect, it is possible to question whether centralized RRM can bring substantial benefits to the challenging interference conditions of D2D scenarios at a reasonable signaling and computational complexity cost.
- How should MoS between D2D communication and device-infrastructure-device (DID) be performed, and on which time scale should this be conducted? The possibility to make use of fast, instantaneous SINR-based MoS against a simpler path-loss based slow MoS will have a major impact on the protocol stack design. It is needed to carefully evaluate the trade-off between achievable gains, complexity and signaling overhead.
- Is instantaneous Channel State Information (CSI) of all potentially interfering cellular and D2D links needed for scheduling purposes, or is the statistical CSI knowledge enough? In general, D2D communication requires information on the channel gain of D2D pairs (i.e. the quality of the direct links), the channel gain among D2D pairs (i.e. generated/received interference to/from other D2D pairs), the channel gain between D2D transmitters and cellular UEs, and the channel gain between cellular transmitters and D2D receivers, in addition to the CSI information of cellular-only systems. The exchange of such extra channel information can become an intolerable overhead to the system if instantaneous CSI feedback is needed.

5.2.3   5G D2D RRM concept: an example

In this section, an example of a 5G D2D RRM concept in the context of a flexible TDD air interface is described. The seamless integration of D2D in the flexible UL/DL TDD

frame structure is presented and the joint multi-cell D2D and cellular resource allocation is explained for the case of centralized and decentralized schedulers. Afterward,
adequate mode selection schemes for D2D are analyzed. Finally, some performance
numbers showing the gains of D2D with flexible TDD over fixed TDD and centralized
over decentralized scheduling are provided. The performance and the implementation
implications of two MoS algorithms that operate on different time scales are also
compared.

5.2.3.1 Flexible uplink and downlink TDD concept for D2D

The UL and DL dynamic TDD concept for D2D is based on a MIMO-OFDMA air
interface, similar to the proposal in [21]. The TDD optimized radio has a flexible
frame structure that enables fast TDD access and fully flexible UL/DL switching, in
addition to support for non-conventional type of communications such as D2D and
self-backhauling (cf. Chapter 7 and [21][22]). Each cell can flexibly switch the data
frames to UL or DL within a scheduling slot based on short-term traffic requirements,
without requiring clustered TDD.

D2D communication is natively integrated into the flexible TDD frame by considering the D2D users in addition to the cellular users. The scheduler decides among UL, DL
and D2D (with simultaneous reuse of resources between cellular and D2D users
allowed) for that cell, taking into account both favorable transmission conditions and
user fairness [23].

Figure 5.3 illustrates the challenges and opportunities presented by multi-cell D2D
communication in scenarios with flexible TDD. The focus is on a specific scheduling
slot and resource block, assuming that resource reuse between D2D and cellular users is
allowed. Further, D2D communication (from UE2 to UE3 and from UE4 to UE5) may
take place at the same time as a UL transmission in Cell1 (from UE1 to BS1) and DL
transmission in Cell2 (from BS2 to UE6). A variety of challenging cross-interference
situations arise such as:

- DL-to-UL interference from BS2 to BS1
- DL-to-D2D interference from BS2 to UE5
- D2D-to-UL interference generated by D2D transmitters like UE2 and UE4 toward
  BS1
- D2D-to-D2D interference from D2D transmitters like UE4 to D2D receivers like UE3

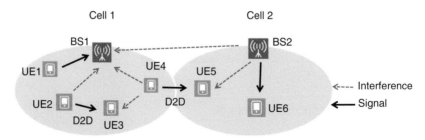

Figure 5.3   Multi-cell D2D in the context of a flexible UL/DL/D2D air interface.

The management of the rapidly changing interference conditions created by flexible TDD and multi-cell D2D is challenging from the scheduler perspective, but it also creates the opportunity for joint fast mode selection and resource allocation based on instantaneous channel conditions, i.e. the scheduling of direct D2D or DID communication depending on the current signal and interference conditions and network load.

5.2.3.2 Decentralized and centralized schedulers

Centralized (coordinated) or decentralized (uncoordinated) resource allocation approaches are considered, leading to two different architecture alternatives (cf. Chapter 11). In the decentralized case, each cell (which could be a small cell) performs its own resource scheduling decisions. In the centralized case, the channel quality information from the users is further forwarded by their respective small cells to a centralized entity in the network, e.g. a macro cell, which performs coordinated scheduling decisions.

The optimization metric is delay-weighted sum rate maximization for each resource block in a cell (in the decentralized case) or group of cells (in the centralized case), considering all cellular (UL and DL) and D2D links in that cell or group of cells, respectively. The scheduling potential of each link, either cellular or D2D, depends on the achievable data rate on that link (based on SINR estimation from interference conditions in the previous scheduling slot) and the packet buffer delay (to provide user fairness in terms of delay) [23].

The scheduler decides for each available resource block which link should make use of it, either an UL, DL or D2D link(s) (possibly with resource reuse between cellular and D2D communications), based on a brute force search of the configuration that provides the highest delay-weighted sum rate out of all the possible combinations. In the decentralized case, the scheduling decisions are made for each cell independently, whereas in the centralized case, they are made jointly for a group of cells including all the possible schedulable links in the cluster. It should be noted that inter-cell D2D is also supported in the decentralized case by means of a simple scheduling conflict resolution mechanism that ensures the fulfillment of the half-duplex constraint in the system [22].

It is worth mentioning that the performance of the brute force scheme is to be seen as an upper bound on the performance of any practical scheduling algorithm, and that the scheduler assumes instantaneous knowledge of all channel gains between cellular and D2D users.

5.2.3.3 Mode selection

Mode selection is especially relevant when the separation distance between users (with traffic to be exchanged) increases. In that case, the routing of the D2D traffic through the infrastructure may be more efficient than making use of a direct link between the devices. Hence, it is important to investigate the adequate time scale to perform MoS between D2D and DID communication. Here the choice is between fast (i.e. based on instantaneous SINR information) and slow time scale (i.e. based on large-scale channel conditions). Clearly, conducting fast MoS would imply that the decision is executed at the MAC layer, whereas in the slow MoS case the decision would be performed at the PDCP or RRC layer. In fact, the following forms of mode selection are considered:

- **Direct D2D only:** All D2D traffic is served through direct links between devices. Reuse of resource blocks is allowed between cellular and D2D users.
- **Indirect D2D only (Device-Infrastructure-Device, DID):** All D2D traffic is routed through the infrastructure. A D2D communication involves two hops, i.e. a UL transmission and a subsequent DL transmission. No direct D2D is allowed.
- **Path loss-based, slow mode selection:** D2D traffic is routed through the infrastructure when the path loss toward the serving base station and a bias is lower than the path loss of the corresponding direct D2D link. The bias favors direct D2D communication over DID due to the inherent advantages of direct D2D. MoS is done before resource allocation.
- **Fast mode selection:** D2D traffic is routed through the infrastructure or through the corresponding direct D2D link depending on the comparison of estimated SINR conditions between the link that connects the D2D UE to the infrastructure and the direct D2D link. This calculation is done per scheduling slot based on the interference conditions in the previous slot. The SINR of direct links is increased by a certain bias in dBs to favor direct D2D decisions. MoS is made jointly with resource allocation. More details can be found in [24], which is an extension and more rigorous implementation of the scheme introduced in [7].

5.2.3.4   Performance analysis

Results are shown for an ultra-dense multi-cell indoor scenario (25 cells, 10 m × 10 m cells, cell-center BSs), with D2D link range up to 4 m. A scheduling slot, e.g. 2 ms, consists of several time slots, with each time slot being 0.25 ms long. The system bandwidth is 200 MHz composed of 100 resource blocks. A bursty traffic model is assumed, with file size ratios of 4:1:1 for DL/UL/D2D traffic, respectively. A file is transmitted as multiple packet segments during the course of a simulation, with packet segment sizes related to the link data rate in a scheduling slot [23].

The cumulative distribution function (CDF) of packet segment serving delay, defined as the difference between the arrival time and the serving time of a packet segment, is depicted in Figure 5.4. The packet segment delay shall not be confused with the MAC latency defined in Chapters 1 and 2. The figure shows the system performance improvement from flexible TDD and centralized scheduling in terms of overall packet segment delay, focusing on the worst-case performance between D2D and cellular links, which is captured by the 99th percentile delay value. No mode selection is carried out in this case, forcing all D2D traffic to be served through direct D2D links. In the decentralized fixed TDD scheme, the first four out of the five scheduling slots are assigned to DL, whereas one slot is used for both UL and D2D. In the flexible TDD case, there is full flexibility to schedule UL, DL or D2D (with or without resource reuse) in every scheduling slot depending on short-term traffic requirements. Decentralized flexible TDD reduces the worst-case delay by 36% in comparison with decentralized fixed TDD. With centralized flexible TDD, the overall delay is further reduced by 24% from 245 ms to 185 ms. In fact, the centralized scheduler allows to balance the delays of the different users and traffic types by means

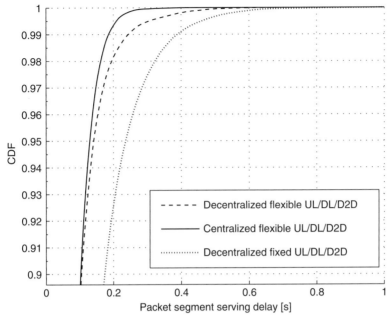

Figure 5.4 Overall (including UL, DL and D2D traffic) packet delays for decentralized fixed and flexible TDD and centralized flexible TDD.

of its global knowledge and coordinated decisions, improving fairness and worst-case user experience.

The maximum allowed range of D2D links is now extended from 4 m to 8 m (with 10 m × 10 m cells) and mode selection is enabled. The results are shown in Figure 5.5, that presents an overview of the compromise reached between cellular and D2D delay performance for the different MoS variants described in Section 5.2.3.3. The vertical axis averages the values of UL and DL packet segment serving delay at both 95th and 50th percentile. Proximity to the origin of coordinates means overall improved latency experience, with the possibility to balance out cellular and D2D delays or to give priority to one specific kind of traffic by applying different biases. The decentralized variants (in grey) perform better for median delay values whereas the centralized ones (in black) improve the delay experience at the 95th percentile. In general, fast MoS is able to reduce the D2D delay (by around 20%), while keeping similar cellular delay values as for the path loss-based MoS. The results in Figure 5.5 and in [24] show that fast MoS can indeed bring gains in the form of a reduced 95th percentile packet delay for D2D transmissions, without sacrificing cellular performance, but it should ideally be done in conjunction with coordinated RRM across cells. Furthermore, the aforementioned gains are on an order that requires careful consideration whether performing D2D MoS on MAC layer is justified, with the associated likely larger burden in terms of signaling overhead and complexity.

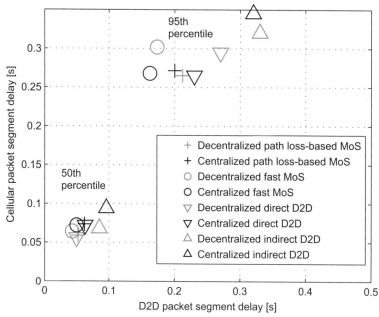

Figure 5.5   Cellular vs. D2D packet segment delay for different MoS variants.

5.3   Multi-hop D2D communications for proximity and emergency services

While cellular-network-assisted D2D communications can capitalize on proximity, reuse and hop gains [13], so far the main driver for standardizing D2D protocols was initially the requirement to support Public Protection and Disaster Relief (PPDR) and National Security and Public Safety (NSPS) services [25]. More precisely, from a PPDR and NSPS perspective, it is important that, as long as a cellular infrastructure is operable, communicating devices should have access to broadband services and local communication should be maintained when cellular coverage becomes unavailable due to a disaster or emergency situation [26]. Along a related line of technology development, the use of fixed and mobile relays provides a cost-efficient way of extending the coverage of cellular networks and can help maintain access to cellular services when some of the infrastructure nodes become dysfunctional, for example, in a PDPR or NSPS scenario. In the remainder of this section, some of the key requirements for NSPS services are highlighted. Afterward, two technology components that play a key role in meeting these requirements are discussed. Both D2D discovery and radio resource management for multi-hop connections should benefit from network assistance when the cellular infrastructure is intact and should remain operational, through a graceful degradation, when parts of the network become dysfunctional.

5.3.1   National security and public safety requirements in 3GPP and METIS

NSPS and PPDR scenarios pose a number of specific requirements that are not typically found in traditional cellular communications. One of the key requirements is robustness and ability to communicate irrespective of the presence or absence of a fixed infrastructure. In many cases, there is at least partial cellular coverage in a geographical area affected by a disaster or emergency situation, which can be exploited for communication. Although some of these scenarios can be addressed by temporary truck-mounted BSs moved into the disaster area, support for proximal or direct D2D communication – to maintain connectivity among rescue personnel or between officers and people in need – remains a critical requirement for NSPS systems [25][26]. Broadband group communication is an example of a requirement typically not supported or deployed in practice in traditional cellular systems; for example, when a dispatcher needs to address multiple officers working in an emergency situation, possibly outside network coverage. Figure 5.6 illustrates some of the use cases that must be supported by the combination and integration of cellular and D2D technologies.

As illustrated in Figure 5.6, in NSPS and PPDR situations, the rescue personnel, including officers with public safety UEs, must be able to communicate in situations in which the cellular BS may provide only partial network coverage. According to the 3GPP requirements [25], such scenarios include proximity services discovery, proximity services traffic initiation, UE with multiple traffic sessions, and proximity services relay. Proximity service discovery is the scenario where a given UE

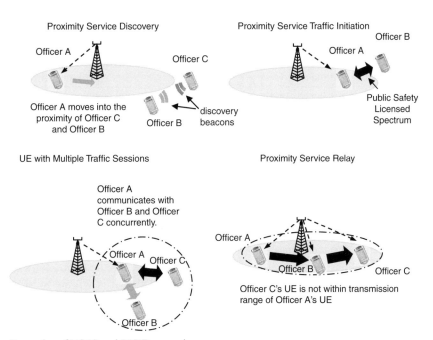

Figure 5.6   Examples of NSPS and PPDR scenarios.

discovers one or more other UEs and the given UE can be with or without network coverage. Proximity services traffic initiation refers to the use case where a public safety UE initiates one-to-one direct user traffic with another UE. UE with multiple traffic sessions means that a given public safety UE can concurrently maintain one-to-one user traffic sessions with several other UEs. With proximity services relay, it is meant that a given UE can act as proximity communication relay for one or more UEs.

An important aspect of these scenarios and requirements is the inherent support for in-coverage, out-of-coverage and partially in-coverage services and specifically the requirement that local (proximal) communication services must be maintained in the absence or partial availability of the cellular infrastructure.

5.3.2   Device discovery without and with network assistance

Peer and service discovery is a key design issue, both in mobile ad-hoc networks operating in unlicensed spectra and in cellular network-assisted D2D communications. The issue stems from the fact that, before the two devices can directly communicate with one another, the devices or a network entity (such as a cellular BS or a core network node) must recognize (discover) that they are near each other. In NSPS and PPDR scenarios, peer discovery is an important service in itself, even without a subsequent communication session. In fact, discovering devices may help rescue personnel take appropriate measures without launching further cellular or D2D communication sessions.

Peer discovery without network support is typically time- and energy-consuming, as it involves beacon signals and sophisticated scanning and security procedures that often include higher layers and/or end users. Therefore, when a cellular network is available, it should assist peer discovery to reduce the discovery time and to increase the energy efficiency of the discovery process. As shown in [12][27][28], peer discovery resources in network-assisted mode can be made available and managed efficiently by the network, which can make such peer discovery and pairing procedures faster, more efficient in terms of energy consumption and more user-friendly. For a deeper analysis of the achievable gains due to various levels of network assistance, see [26].

5.3.3   Network-assisted multi-hop D2D communications

Although multi-hop D2D communication requirements have been primarily defined with NSPS scenarios in mind, it is clear that commercial and traditional broadband Internet services can also benefit from range extension or multi-hop proximity com-munications, as illustrated in Figure 5.7. As shown in the figure, for a UE positioned outside the coverage area, it needs another UE that is willing to provide relaying assistance hence extending the range of a cellular BS. The example in the figure has two single-hop and two two-hop routes (Route 1, Route 2 and Route 3, Route 4, respectively). Resources R-1 and R-3 are reused, while R-2 and R-4 are dedicated.

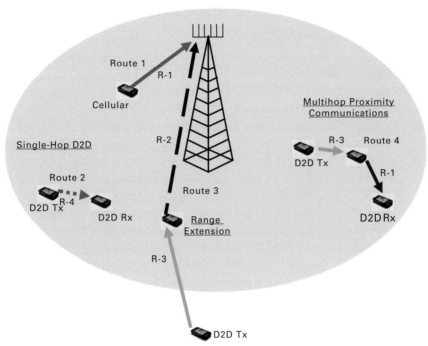

Figure 5.7   Single and multi-hop routes partially under network coverage [29], reproduced with permission (Lic. no. 3664040827123).

Between each Source-Destination (S-D) pair, a route must be defined and resources need to be allocated to each link along the route. In Figure 5.7, different line types indicate different time and frequency resources (Resource Blocks, RBs), while the same line type for different links indicates resource block reuse. Further it is assumed that in the multi-hop case, the incoming and outgoing links of a relay node must use orthogonal resources. A given S-D pair may have the possibility to communicate in cellular mode through the BS or using single- or Multi-Hop (MH) D2D communications.

Recall that for D2D communications in cellular spectrum, MoS and resource allocation (scheduling) and power control are essential. However, extending these key RRM algorithms to MH D2D communication is non-trivial, since

1. Existing single-hop MoS algorithms must be extended to select between the single-hop D2D link, MH D2D paths and cellular communications.
2. Existing single-hop resource allocation algorithms must be further developed in order to not only manage spectrum resources between cellular and D2D layers, but also to comply with resource constraints along MH paths.
3. Available D2D PC algorithms must be made capable of taking into account the rate constraints of MH paths. Specifically, it must be taken into account that, along the multiple links of a given path, only a single rate can be sustained without requiring large buffers or facing buffer underflow situations at intermediate nodes.

5.3.4   Radio resource management for multi-hop D2D

A system model that is appropriate for modeling RRM algorithms in MH D2D networks consists of two parts. The first one is a routing matrix that describes the network topology and associates links with resources. The second one is a utility function associated with a S-D pair that characterizes the utility of supporting some communication rate between the end nodes of the S-D pair.

Recall from Figure 5.7 that MH D2D communications can be used advantageously in two distinct scenarios. In the proximity communication scenario, a D2D relay node helps a D2D pair to communicate. In the coverage or range extension scenario, a D2D relay node assists a coverage-limited D2D Tx node to boost its link budget to a BS, or in NSPS scenarios to a so-called Cluster Head (CH) node that is capable of taking over the core functionalities of a cellular base station [26][28]. In the proximity communication scenario, the mode selection problem consists of deciding whether the D2D Tx node should communicate with the D2D Rx node (1) via a direct D2D (single-hop) link, (2) via a 2-hop path through the D2D relay node, or (3) through a cellular BS or ad-hoc CH node. In the range extension scenario, by contrast, the mode selection problem consists of deciding whether the D2D Tx node should communicate via a direct transmission with its serving BS or via the D2D relay node. In the next sub-section, mode selection algorithms are considered for the proximity communications and range extension scenarios (see Figure 5.7).

5.3.4.1   Mode selection for proximity communications

For the proximity communication scenario, the notion of the equivalent channel from a D2D transmit (Tx) device to a D2D receive (Rx) device through a D2D relay based on the harmonic mean of the composite channels from D2D Tx to D2D relay (G_{TxRe}) and from D2D relay to D2D Rx (G_{ReRx}) has been proposed [29]:

$$\frac{1}{G_{eq}} = \frac{1}{G_{TxRe}} + \frac{1}{G_{ReRx}} \tag{5.1}$$

The intuition of defining the equivalent channel according to the above is that the equivalent channel gain tends to be high only when both composite channels are high; this makes it an appropriate single measure for mode selection purposes. A pseudo-code of a heuristic mode selection algorithm based on the equivalent channel is given by Algorithm 1 below, where the channels are needed from the D2D Tx to the BS (G_{TxBS}) and to the D2D Rx (G_{TxRx}):

ALGORITHM I   Harmonic Mode Selection (HMS) for Proximity Communication
1: **if** $G_{eq} \geq \max\{G_{TxRx}, G_{TxBS}\}$ **then**
2:    Choose D2D two-hop communications
3: **else if** $G_{TxRx} \geq G_{TxBS}$ **then**
4:    Choose D2D single-hop communications
5: **else**
6:    Choose cellular mode, that is D2D Tx and Rx communication through the BS.
7: **end if**

5.3.4.2   Mode selection for range extension

In the range extension scenario, there are only two possible communication modes (direct or relay-assisted) between the D2D Tx device and the BS or CH device. Therefore, in this scenario, the definition of the equivalent channel must be modified such that it includes the path gain between the relay device and the BS (G_{ReBS}):

$$\frac{1}{G_{eq}} = \frac{1}{G_{TxRe}} + \frac{1}{G_{ReBs}} \tag{5.2}$$

This makes it possible to use the following modified version of the Harmonic Mode Selection (HMS) algorithm:

ALGORITHM 2   Harmonic Mode Selection (HMS) for Range Extension
1: **if** $G_{eq} \geq G_{TxBS}$ **then**
2:   Choose D2D relay-assisted communication
3: **else**
4:   Choose cellular mode that is D2D Tx transmits directly to the BS.
5: **end if**

5.3.5   Performance of D2D communications in the proximity communications scenario

An effective way to control the fundamental trade-off between power consumption and system throughput is to employ D2D power control that is not necessarily based on fixed transmit power levels or the well-known LTE path loss compensating Open Loop (OL) method. To this end, several power control algorithms have been proposed, the objective of which is not only to ensure high throughput and energy efficient operation, but also to protect the cellular layer from harmful interference caused by D2D traffic. Specifically, the algorithm proposed in [20] and [29] can tune the power consumption and the throughput of the cellular and D2D layers in single-hop D2D scenarios by setting a parameter that can be seen as the cost of a unit power investment (i.e. a higher cost of unit power implies a higher cost of increasing the system throughput by investing higher transmit power levels). This basic idea has been extended for multi-hop D2D communication scenarios, including the range extension and proximity communication scenarios of Figure 5.7.

The following figures compare the performance achieved by transmitting with some fixed power level ("Fix") or the legacy open loop ("OL") power control algorithm (employed by both the cellular and the D2D layers, using 12 dB of SNR target) with the Utility Maximizing (UM) scheme with some parameter ω ("UM $\omega = 0.1$" and "UM $\omega = 100$"). The parameter ω represents a trade-off between power consumption and utility maximization [10][20][29]. Specifically, Figures 5.8 and 5.9 show the invested power and achieved throughput trade-off in the range extension and proximity communication scenarios, respectively. These results were obtained in a seven-cell system of a

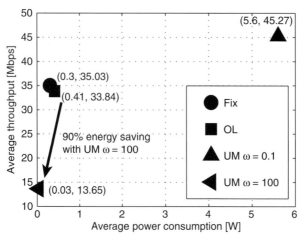

Figure 5.8   The impact of power control on the power consumption–throughput trade-off in the range extension scenario.

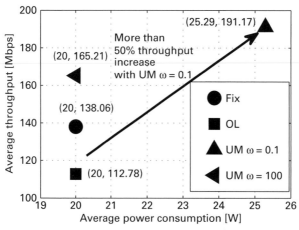

Figure 5.9   The impact of utility maximizing power control on the fundamental trade-off between power consumption and throughput in the Proximity Communications scenario.

cell radius of 500 m. The D2D users are randomly dropped in the coverage area of a cell such that their distance is between 75 m and 125 m. There are 18 uplink physical RBs in each cell. The other parameters of this system are given in [29]. Further, in this system, D2D communications are supported in UL physical resource blocks using the Harmonic Mode Selection algorithm detailed in Section 5.3.4.

Figure 5.8 is a scatter plot for the range extension scenario. The fixed power level of the "Fix" power control scheme is set such that its performance becomes similar to that of the "OL" scheme. Note that (x, y) near each symbol shows the x-axis (power consumption in W) and y-axis (throughput in Mbps) values. Compared with the traditional OL power control, utility maximizing power control (UM with $\omega = 100$) reduces overall power consumption at the expense of reducing system throughput. For

UM $\omega = 0.1$, the utility maximization power control algorithm reaches the highest average throughput, with a gain of approximately 34% over LTE OL power control. However, this gain comes at the expense of transmitting at much higher power levels. In contrast, with $\omega = 100$, utility maximizing power control minimizes power consumption at the expense of reducing the achieved throughput. Clearly, utility maximizing PC can reach high throughput when using low values of ω and can transmit at low power levels with high values of ω.

Figure 5.9 is a scatter plot for the proximity communication scenario. Similarly to Figure 5.8, with UM $\omega = 0.1$ the average throughput gain is large (approximately 69%) over the LTE OL scheme, at the cost of using approximately 26% more power. Notice that in Figure 5.9 the average power consumption includes the power consumption of the BS. However, with UM $\omega = 100$ the average throughput gain is approximately 20% using similar transmit power levels as LTE OL. UM $\omega = 100$ boosts the average throughput at the expense of a small increase in the transmit power level. If the power consumption must be kept at low values with reasonable throughput values, utility maximization with higher ω values or using the LTE OL power control technique is a good design choice.

5.4   Multi-operator D2D communication

The business potential of commercial D2D would be rather limited if direct communication between devices subscribed to different cellular operators is not supported. Inter-operator D2D support is also needed to meet the requirements resulting from D2D-relevant scenarios, e.g. vehicle-to-vehicle communications [7]. In general, D2D support in inter-operator scenarios becomes more complex as compared to single-operator D2D. For instance, operators may not be willing to share operator-specific information, e.g. network loads, utility functions, between each other or with external parties to identify how much spectrum to allocate for inter-operator D2D communication. In this section, inter-operator D2D discovery, mode selection, and spectrum allocation schemes are discussed. Further, single-hop unicast D2D is considered.

5.4.1   Multi-operator D2D discovery

In a multi-operator setting, the D2D discovery cannot be based, for instance, on the time synchronization and distribution of common peer discovery resources unless the operators agree to do so. Further, the D2D discovery should rely on both ends of the D2D pair and on the networks of both operators. In Figure 5.10, an example procedure that enables multi-operator D2D discovery is shown.

In this example procedure, using LTE terminology, the D2D devices send discovery messages only on their home operator's spectrum and hence no change to spectrum regulation or roaming rules is required. Taking the UE#A as an example, after registration of D2D operation and authorization process between UE#A, MME#A and MODS (Multi-operator D2D Server), UE#A can obtain information on discovery resources

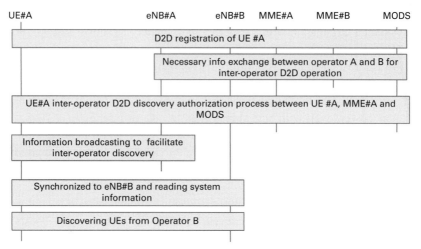

Figure 5.10   Procedure to enable multi-operator D2D discovery.

(both from the home operator and the other operators) based on the broadcasted information from its home operator. MODS is a new logical network entity which could be co-located with certain network elements within an operator's network or running independently, e.g. as a network service provided by a 3rd party. Example functionalities of MODS can include D2D subscription management, network access control, centralized security and radio resource management functions and so on. The broadcasted key parameters from the home operator include, for example, radio resource information related to different operators such as operator identifiers and the corresponding operating frequency bands to facilitate inter-operator discovery. UE#A will listen to both the home and the other operator's resources to detect the presence of discovery messages.

5.4.2   Mode selection for multi-operator D2D

D2D mode selection algorithms developed for single-operator networks may not be directly applicable in a multi-operator system setup. Operators may not want to share information regarding the locations of users or path loss data as in [30], or CSI between the D2D users and their home operator's BS (as required in the mode selection algorithm described in Section 5.2.3). Moreover, operators may not want to cooperate in order to estimate the D2D pairwise distance, see for instance [31], and use it as a criterion for mode selection.

In a single-operator network, either dedicated spectrum can be allocated to the D2D users (also known as D2D overlay), or D2D and cellular users can be allocated to the same resources (also known as D2D underlay). In a multi-operator D2D underlay, the cellular users are exposed to inter-operator interference generated from the D2D users involved in inter-operator communication sessions. The problem of inter-operator interference between cellular users and D2D users needs to be resolved without an excessive information exchange between the operators.

Clearly, at a first stage, an overlay multi-operator D2D scheme would be easier to implement. In the overlay D2D setting, the key design issue is the way to divide spectral resources between cellular and D2D users and the communication mode selection scheme. One method to select the communication mode without incurring excessive communication signaling overhead may rely on the received signal level at the D2D receiver. Such algorithm has been proposed in [32] and it is straightforward to extend it in a multi-operator setting because it does not require proprietary information exchange between the operators.

5.4.2.1   Mode selection algorithm

Given the spectral resources allocated for inter-operator D2D communication, the D2D receiver measures the interference level and communicates a quantized version of the interference to its home BS. The BS compares the measurement report with a decision threshold and chooses the D2D communication mode only if the measured interference is low. The D2D receiver should signal the selected communication mode back to the D2D transmitter, i.e. the source UE subscribed with the other operator, and the session in which it may start.

Note that the mode selection threshold impacts the overall network performance because it determines the amount of inter-operator D2D sessions and also the portion of users in cellular communication mode. The mode selection threshold should be a priori agreed i.e. optimized between the operators.

The mode selection algorithm described above could also be implemented in the following manner: the interference measurements could be carried out at the D2D transmitter instead of the receiver. In that case, the transmitter would be responsible for reporting the measurements to its home BS. While discussing the spectrum allocation algorithm for inter-operator D2D in Section 5.4.3, it is assumed that the mode selection takes place at the transmitter, since the performance can be assessed using analytical means (as long as the D2D pair distance is short).

5.4.3   Spectrum allocation for multi-operator D2D

D2D communication can be enabled either over licensed or unlicensed spectrum. D2D communication in unlicensed bands would suffer from unpredictable interference. Licensed spectrum seems to be the way forward to enable LTE D2D communication, especially considering safety related scenarios such as vehicle-to-vehicle communication (see Chapters 4 and 7 for more information about V2V).

Overlay inter-operator D2D communication takes place over dedicated spectral resources possibly originating from both operators. For FDD operators, the spectral resources may refer to Orthogonal Frequency-Division Multiplexing (OFDM) sub-carriers, while for the TDD operators they may refer to time-frequency resource blocks. In the TDD case, inter-operator D2D support would pose a requirement for time synchronization between the operators, which is more challenging. In Figure 5.11, two FDD operators contribute a part of their cellular spectrum, β_1 and β_2 respectively, for inter-operator D2D communication. Also, each operator $i = (1, 2)$

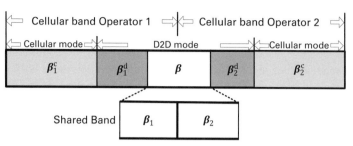

Figure 5.11　Spectrum divisions for two operators supporting inter-operator D2D communication.

allocates fractions β_i^c and β_i^d for cellular and intra-operator D2D communication respectively.

When more than two operators are involved in spectrum sharing, it is possible to realize inter-operator D2D communication based on bilateral agreements between operators, or alternatively, all operators may commit some of their spectral resources in a common spectrum pool, see also Chapter 12 for a detailed description of inter-operator spectrum sharing based on mutual renting and spectrum pooling. The operators should negotiate the amount of resources they want to commit, but they should not be forced to take action. However, once the operators agree to share spectrum for some time and commit certain resources for multi-operator D2D, they are not allowed to break the agreement. The duration of the agreement should be set in advance and may depend on the expected network traffic dynamics.

In general, operators are competitors and they may not want to reveal proprietary information, e.g. utility functions, and network load. Ideally, the negotiations about spectrum allocation for multi-operator D2D should be completed without exchanging proprietary information. One possible way to do that is to model the operators as selfish players, and use a non-cooperative game theoretical approach. For instance, an operator can make a proposal about the amount of spectral resources it is willing to contribute, taking into consideration its own reward and the proposals made by the competitors. All operators can update their proposals based on the proposals submitted by the competitors until consensus is reached. This kind of updating procedure is also known as best response iteration and it is a common method to identify the Nash equilibrium of a one-shot non-cooperative game [33].

In a non-cooperative game, one of the most important aspects is the existence and uniqueness of a Nash equilibrium. A situation where there are multiple equilibrium points may be undesirable because the realized equilibrium will depend on the selection order and the initial proposals of the operators. As a result, it is important to note that operators may be interested to share spectrum only if a unique Nash equilibrium exists. For the time being, the spectrum allocation algorithm does not support coupled constraints between the operators. In that case, there may exist infinite normalized equilibrium points [34]. Hence, some sort of extensive information exchange between the operators might be needed to obtain an efficient equilibrium.

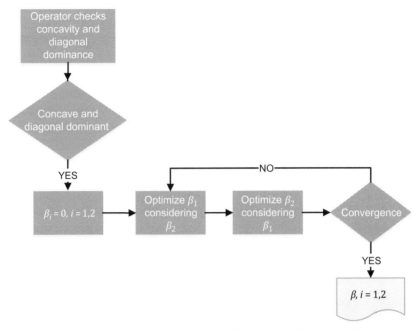

Figure 5.12   Best response iteration algorithm for spectrum allocation in multi-operator D2D communication.

5.4.3.1   Spectrum allocation algorithm

A sequential updating procedure is considered until consensus is reached where each operator strategy consists of responding to the others with the amount of spectrum each operator is willing to contribute for multi-operator D2D support. This strategy is one-dimensional. Further, each operator considers only its individual network utility and performance constraints. It is well known that, for concave utilities and constraints, an equilibrium exists. However, to establish uniqueness, the best response operator should also be a contraction [35]. For one-dimensional strategies, the contraction principle can be degenerated to the dominance solvability condition, which essentially means that an operator can control its own utility more than all other operators can do. Fortunately, each operator can check independently whether its optimization criteria are concave or not and whether the dominance solvability condition holds true. The operators can exchange binary messages regarding these conditions and provided that all indications are positive, the operators become automatically aware about the uniqueness of the equilibrium. Therefore, in that case, the best response iteration can start. Any operator can be ranked first. In case an operator experiences a performance loss as compared to no sharing, it should immediately break the agreement. The best response updating procedure is also summarized in Figure 5.12.

5.4.3.2   Numerical example

Assuming that each operator wants to maximize its average D2D user rate including own operator and inter-operator D2D users subject to transmission rate constraints for cellular communication mode and intra-operator D2D users. With the MoS

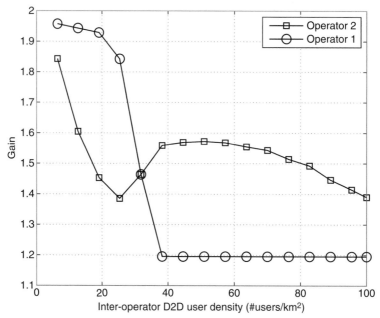

Figure 5.13   Performance gain in terms of average user rate for two operators as compared to the case without multi-operator D2D support.

scheme described in Section 5.4.3.1, it can be shown that the utilities as well as the constraint are concave [36]. Besides, in a spectrum-sharing scenario between two operators, the dominance solvability condition holds always true irrespective of the user densities [36].

The network is modeled for each operator using a Voronoi tesselation[2] with an average inter-site distance of 100 m. Full-buffer traffic model is assumed, with the user density directly related to the network load. The densities of cellular and inter-operator D2D users is 30 users/km^2 (per operator) to model a scenario where the densities of the users are comparable to the densities of BSs. The density of intra-operator D2D users is 30 users/km^2 for Operator 1 and it varies for Operator 2 to model asymmetric network loads between the operators. A 3GPP propagation environment is used with Rayleigh fading [37]. The average D2D link distance is 30 m. The MoS threshold is fixed to -72 dBm both for inter-operator and intra-operator D2D users. The decision threshold impacts the density of users selecting a D2D communication mode. A performance evaluation with other threshold values is available in [36]. The baseline scheme for comparison is not supporting multi-operator D2D communication. In that scheme, all inter-operator D2D traffic is routed toward the cellular infrastructure.

In Figure 5.13, the performance gain is shown in terms of average user rate for both operators. When both operators have an equal network load, they both experience

---

[2] The base stations are distributed uniformly and each point of the plane is associated with the nearest base station.

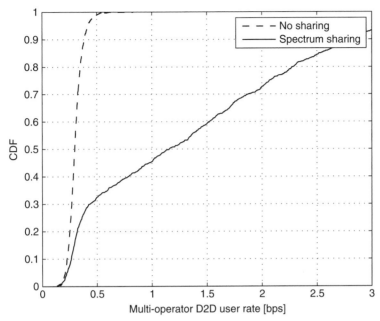

Figure 5.14 Rate distribution for multi-operator D2D users with and without multi-operator D2D support.

around 50% performance gain. The gains for both operators become high when the network load of Operator 2 becomes low. In that case, Operator 2 is able to contribute a high amount of spectral resources for multi-operator D2D support, and both operators can enjoy performance gains close to 100% due to spectrum sharing and D2D proximity.

Figure 5.14 depicts the rate distribution for multi-operator D2D users, where symmetric operators are assumed. Without spectrum sharing, all inter-operator D2D traffic is routed to the cellular infrastructure and the achievable D2D user rate is low. One can see that multi-operator D2D support can boost the median D2D user data rate by up to a factor of 4. Hence, multi-operator D2D support is required in order to harvest the business potential of D2D communications, e.g. in the context of vehicular communication for traffic efficiency and safety.

5.5 Conclusions

It is envisioned that integrated D2D communication will play a more important role in the future 5G system thanks to the promising benefits on both network and end-user sides, contributing to traffic offloading, very high throughput, significantly reduced latency and low power consumption. In addition, D2D has the potential to increase communication availability and reliability, and provide additional diversity. From a service and application perspective, D2D possesses the capability of enabling a number of new applications such as V2V and machine-type communications. Finally, in order to capitalize on the full benefits of a system with native support of D2D operation, it is needed to address in the coming years additional challenges related to for example mobility management (exemplary solutions can be found in Chapter 11) and security.

References

[1] A. Osseiran, F. Boccardi, V. Braun, K. Kusume, P. Marsch, M. Maternia, O. Queseth, M. Schellmann, H. Schotten, H. Taoka, H. Tullberg, M. A. Uusitalo, B. Timus, and M. Fallgren, "Scenarios for 5G mobile and wireless communications: The vision of the METIS Project," *IEEE Communications Magazine*, vol. 52, no. 5, pp. 26–35, May 2014.

[2] NGMN Alliance, 5G White Paper, February 2015, www.ngmn.org/uploads/media/NGMN_5G_White_Paper_V1_0.pdf

[3] Qualcomm, "LTE Device to Device Proximity Services," Work Item RP-140518, 3GPP TSG RAN Meeting #63, March 2014.

[4] 3GPP TS 36.211, "Evolved Universal Terrestrial Radio Access (E-UTRA); Physical channels and modulation," Technical Specification TS 36.211 V11.6.0, Technical Specification Group Radio Access Network, September 2014.

[5] Qualcomm, "Enhanced LTE Device to Device Proximity Services," Work Item RP-150441, 3GPP TSG RAN Meeting #67, March 2015.

[6] ICT-317669 METIS project, "Initial report on horizontal topics, first results and 5G system concept," Deliverable D6.2, March 2014, www.metis2020.com/wp-content/uploads/deliverables/

[7] Z. Li, M. Moisio, M. A. Uusitalo, P. Lundén, C. Wijting, F. S. Moya, A. Yaver, and V. Venkatasubramanian, "Overview on initial METIS D2D concept," in International Conference on 5G for Ubiquitous Connectivity, Levi, November 2014, pp. 203–208.

[8] ICT-317669 METIS project, "Intermediate system evaluation results," Deliverable D6.3, August 2014, www.metis2020.com/wp-content/uploads/deliverables/

[9] T. Peng, Q. Lu, H. Wang, S. Xu, and W. Wang, "Interference avoidance mechanisms in the hybrid cellular and device-to-device systems," in IEEE International Symposium on Personal, Indoor and Mobile Radio Communications, Tokyo, September 2009, pp. 617–621.

[10] G. Fodor, M. Belleschi, D. D. Penda, A. Pradini, M. Johansson, and A. Abrardo, "Benchmarking practical RRM algorithms for D2D communications in LTE advanced," *Wireless Personal Communications*, vol. 82, pp. 883–910, December 2014.

[11] A. Asadi, Q. Wang, and V. Mancuso, "A Survey on device-to-device communication in cellular networks," *IEEE Communications Surveys & Tutorials*, vol. 16, no.4, pp. 1801–1819.

[12] S. Mumtaz and J. Rodriguez (Eds.), *Smart Device to Smart Device Communication*, New York: Springer-Verlag, 2014.

[13] G. Fodor, E. Dahlman, G. Mildh, S. Parkvall, N. Reider, G. Miklós, and Z. Turányi, "Design aspects of network assisted device-to-device communications," *IEEE Communications Magazine*, vol. 50, no. 3, pp. 170–177, March 2012.

[14] N. Reider and G. Fodor, "A distributed power control and mode selection algorithm for D2D communications," *EURASIP Journal on Wireless Communications and Networking*, vol. 2012, no. 1, December 2012.

[15] S. Hakola, Tao Chen, J. Lehtomaki, and T. Koskela, "Device-To-Device (D2D) communication in cellular network: Performance analysis of optimum and practical communication mode selection," in IEEE Wireless Communications and Networking Conference, Sydney, April 2010.

[16] K. Doppler, C.H. Yu, C. Ribeiro, and P. Janis, "Mode selection for device-to-device communication underlaying an LTE-Advanced network," in IEEE Wireless Communications and Networking Conference, Sydney, April 2010.

[17] G. Fodor and N. Reider, "A distributed power control scheme for cellular network assisted D2D communications," in IEEE Global Telecommunications Conference, Houston, December 2011.

[18] C.H. Yu, O. Tirkkonen, K. Doppler, and C. Ribeiro, "Power optimization of device-to-device communication underlaying cellular communication," in IEEE International Conference on Communications, Dresden, June 2009.

[19] H. Xing and S. Hakola, "The investigation of power control schemes for a device-to-device communication integrated into OFDMA cellular system," in IEEE International Symposium on Personal Indoor and Mobile Radio Communications, Istanbul, September 2010, pp. 1775–1780.

[20] G. Fodor, M. Belleschi, D. D. Penda, A. Pradini, M. Johansson, and A. Abrardo, "A comparative study of power control approaches for D2D communications," in IEEE International Conference on Communications, Budapest, June 2013.

[21] P. Mogensen et al., "5G small cell optimized radio design," in IEEE Global Telecommunications Conference Workshops, Atlanta, December 2013, pp. 111–116.

[22] E. Lahetkangas, K. Pajukoski, J. Vihriala, and E. Tiirola, "On the flexible 5G dense deployment air interface for mobile broadband," in International Conference on 5G for Ubiquitous Connectivity, Levi, November 2014, pp. 57–61.

[23] V. Venkatasubramanian, F. Sanchez Moya, and K. Pawlak, "Centralized and decentralized multi-cell D2D resource allocation using flexible UL/DL TDD," in IEEE Wireless Communications and Networking Conference Workshops, New Orleans, March 2015.

[24] F. Sanchez Moya, V. Venkatasubramanian, P. Marsch, and A. Yaver, "D2D mode selection and resource allocation with flexible UL/DL TDD for 5G deployments," in IEEE International Conference on Communications Workshops, London, June 2015.

[25] 3GPP TR 22.803, "Feasibility study for Proximity Services (ProSe)," Technical Report TR 22.803 V12.2.0, Technical Specification Group Radio Access Network, June 2013.

[26] G. Fodor et al., "Device-to-sevice communications for national security and public safety," *IEEE Access*, vol. 2, pp. 1510–1520, January 2015.

[27] Z. Li, "Performance analysis of network assisted neighbor discovery algorithms," School Elect. Eng., Royal Inst. Technol., Stockholm, Sweden, Tech. Rep. XR–EE–RT 2012:026, 2012.

[28] Y. Zhou, "Performance evaluation of a weighted clustering algorithm in NSPS scenarios," School Elect. Eng., Roy. Inst. Technol., Stockholm, Sweden, Tech. Rep. XR-EE-RT 2013:011, January 2014.

[29] J. M. B. da Silva Jr., G. Fodor, and T. Maciel, "Performance analysis of network assisted two-hop device-to-device communications," in IEEE Broadband Wireless Access Workshop, Austin, December 2014, pp. 1–6.

[30] C.-H. Yu, K. Doppler, C. B. Ribeiro, and O. Tirkkonen, "Resource sharing optimization for device-to-device communication underlaying cellular networks," *IEEE Transactions on Wireless Communications*, vol. 10, no. 8, pp. 2752–2763, August 2011.

[31] X. Lin, J. G. Andrews, and A. Ghosh, "Spectrum sharing for device-to-device communication in cellular networks," *IEEE Transactions on Wireless Communications*, vol. 13, no. 12, pp. 6727–6740, December 2014.

[32] B. Cho, K. Koufos, and R. Jäntti, "Spectrum allocation and mode selection for overlay D2D using carrier sensing threshold," in International Conference on Cognitive Radio Oriented Wireless Networks, Oulu, June 2014, pp. 26–31.

[33] M. J. Osborne, *An Introduction to Game Theory*, Oxford: Oxford University Press, 2003.

[34] J. Rosen, "Existence and uniqueness of equilibrium points for concave n-person games," *Econometrica*, vol. 33, pp. 520–534, July 1965.

[35] D. Gabay and H. Moulin, "On the uniqueness and stability of Nash equilibrium in non-cooperative games," in *Applied Stochastic Control in Econometrics and Management Sciences*, A. Bensoussan, P. Kleindorfer, C. S. Tapiero, eds. Amsterdam: North-Holland, 1980.

[36] B. Cho, K. Koufos, R. Jäntti, Z. Li, and M.A. Uusitalo "Spectrum allocation for multi-operator device-to-device communication," in IEEE International Conference on Communications, London, June 2015.

[37] 3GPP TR 30.03U, "Universal mobile telecommunications system (UMTS); Selection procedures for the choice of radio transmission technologies of the UMTS," Technical Report TR 30.03U V3.2.0, ETSI, April 1998.

6 Millimeter wave communications

Robert Baldemair, Kumar Balachandran, Lars Sundström, and Dennis Hui

Certain 5G METIS scenarios [1] such as *Amazingly Fast, Best Experience Follows You*, and *Service in a Crowd* create extreme requirements on data rate, traffic handling capability, and availability of high capacity transport respectively. These scenarios map to corresponding requirements that will entail support of over 10 Gbps, 10–100 times the number of connected devices, 1000 times the traffic, and 5 times lower end-to-end latency than possible through IMT-Advanced. The peak data rate requirements of these scenarios will entail acquisition of several hundreds of MHz of spectrum. These requirements do not encompass 5G, but instead offer one avenue of stressing system capabilities along a limited set of dimensions. Several traffic forecasts [2][3] also predict a tenfold increase in traffic volume from 2015 to 2020.

The 5G requirements of interest to this chapter relate mainly to data rates and traffic volumes and can be met using techniques that are tried and tested in past generations of mobile networks. These are to (1) gain access to new spectrum, (2) improve spectral efficiency, and (3) densify the networks using small cells. In the case of 5G, these techniques are given new life using two means: the use of millimeter Wave (mmW) spectrum for the availability of large blocks of contiguous spectrum, and the subsequent adoption of beamforming as an enabler for high spectrum efficiency. The propagation of millimeter waves is naturally affected by physics to reduce coverage to shorter ranges. Ultra-Dense Network (UDN) deployments are therefore a consequence of the choice of frequency band, and will lead to a tremendous increase in capacity over the covered area. The increase in spectral efficiency arises out of the drastic reduction of interference in relation to signal power due to the high gain beamforming.

6.1 Spectrum and regulations

The primary motivation for using millimeter waves is the promise of abundant spectrum above 30 GHz. While mmW spectrum spans the range from 30 GHz–300 GHz, it is widely believed that the reach of mass market semiconductor technology extends up to around 100 GHz and will inevitably surpass that limit with time. Microwave bands from 3 GHz–30 GHz are just as relevant to meeting extreme requirements for 5G, and much of

5G Mobile and Wireless Communications Technology, ed. A. Osseiran, J. F. Monserrat, and P. Marsch.
Published by Cambridge University Press. © Cambridge University Press 2016.

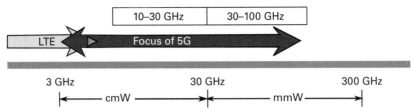

Figure 6.1   The focus of 5G deployment is on frequency bands up to 100 GHz, with the lower end of the range being favored; LTE expands reach into higher frequency bands around 6 GHz.

the discussion in this chapter is relevant to those parts of the centimeter Wave (cmW) band outside of the reach of existing systems as well, namely the region 10 GHz–30 GHz (see Figure 6.1). The technological abilities of the semiconductor industry do not however automatically translate to viable network architectures. In general, lower frequency bands are more appealing for implementation and pose fewer risks for systemization, while higher ranges of frequency are better for access to wide bandwidth, but worse for device and system complexity. Most frequency bands below 60 GHz are already assigned to various services, including mobile services; these assignments are made through treaty arrangements within the three regions addressed by the ITU-R in periodic World Radio Conferences (WRC). Millimeter wave bands in current use are predominantly employed for radar, earth exploration, point-to-point services, and satellite communications, etc. Some of these bands are assigned in co-primary fashion to mobile services; there are of course no terrestrial mobile services in operation above 6 GHz. The 60 GHz ISM band does provide as much as 7 GHz of spectrum for unlicensed use in most parts of the world; the band has been used by IEEE 802.11 in the recent "ad" amendment to create a physical and Medium Access Control (MAC) layer capable of peak rates up to 7 Gbps [4]. The specification is being utilized by WiGig for point-to-point mmW links over 2.16 GHz channels for video and data transfer. In addition, the 802.11ay task group in IEEE 802.11 is examining channel bonding and MIMO as solutions for even higher throughput systems (over 30 Gbps) for video transfer, data center applications, and point-to-point communication [5].

There is considerable interest in industry in expanding mobile services further into microwave spectrum, including mmW spectrum. Inquiries by the Federal Communications Commission (FCC) in the USA [6] and the UK regulator Ofcom [7] have also attempted to gauge industry seriousness about venturing into such spectrum frontiers. These references and subsequent responses from industry partners provide an incomplete but evolving picture of regulatory issues. For a more in-depth discussion on spectrum for 5G systems see Chapter 12.

At the time of writing Electromagnetic Field (EMF) exposure limits determined by the FCC [8] and ICNIRP [9] independently create inconsistencies in allowable power limits above the transition frequencies of 6 GHz and 10 GHz, respectively [10]. The policy guidelines pertaining to EMF will likely need modification if cmW or mmW operation above the transition frequency is permitted for mobile services.

6.2   Channel propagation

Millimeter wave bands pose unique challenges for radio communication. Large-scale losses over line-of-sight paths generally follow free space loss values, and attenuation relative to isotropic radiators increases proportional to the square of operating frequency. It must be pointed out that coupling loss can be kept independent of frequency if the aperture of transmitting or receiving antennas is kept constant over the variation of frequency; the frequency dependence of isotropically referred free space loss can be more than compensated with high-gain antenna designs at the transmitter and receiver. Any mobile radio system at millimeter waves will need beamforming using adaptive antenna arrays or very high order sectorization.

Millimeter wave path losses are affected by a variety of other additional factors, all of which are generally frequency dependent: (a) atmospheric losses due to gases, notably water vapor and oxygen, (b) rain attenuation, (c) foliage loss, (d) diffraction loss. Below 100 GHz, two atmospheric absorption peaks occur at 24 GHz and 60 GHz, due to water and oxygen. The presence of oxygen in the atmosphere contributes an additional 15 dB/km of specific attenuation. For short distances, this additional attenuation is not significant. Obstacles in the signal path typically reflect energy, and the effect of foliage is rapid attenuation along the incident signal path and diffuse scattering from reflectors. Diffraction attenuation increases as the wavelength gets shorter [11].

Small-scale variations can be modeled using site-specific geometric models, statistical models based on general characteristics of propagation, and hybrid approaches. It is expected that narrow beamforming using high-gain antennas will reduce channel dispersion. Ray tracing is a useful tool in modeling propagation, with statistical variations provided by the presence of diffuse scatterers in the environment representing objects and non-smooth characteristics of surfaces, and by the modeling of corner effects due to diffraction. Building materials will differ in the absorption, reflectivity, and transmission characteristics and will be affected by incident angles to the surface. Unshielded windows can provide ingress to signals, while exterior walls are usually opaque. Wall losses within a building can be severe and outdoor to indoor connectivity will often need site planning through placement of antennas on all sides of a building, especially at the higher mmW frequencies. The effect of body loss and attenuation from mobile users or mobile objects is significant.

A detailed discussion on propagation modeling is provided in Chapter 13 of this book. For mmW propagation measurements, see e.g. [12][13].

6.3   Hardware technologies for mmW systems

6.3.1   Device technology

Radio Frequency (RF) building block performance generally degrades with increasing frequency. The power capability of power amplifiers for a given integrated circuit technology roughly degrades by 15 dB per decade, as shown in Figure 6.2. There is

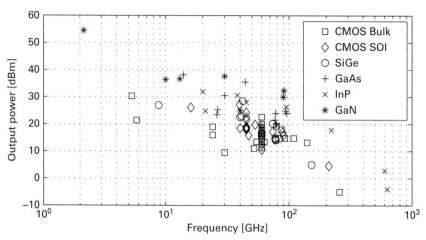

Figure 6.2　Power capability of various power amplifiers in different technologies versus frequency.

a fundamental cause for this degradation; increased power capability and increased frequency capability are conflicting requirements as observed from the so-called Johnson limit [14]. In short, higher operational frequencies require smaller geometries, which subsequently result in lower operational power in order to prevent dielectric breakdown from the increased field strengths. Moore's Law does not favor power capability performance. A remedy is however found in the choice of integrated circuit material. Millimeter wave integrated circuits have traditionally been manufactured using so called III-V materials, i.e. a combination of elements from groups III and V of the periodic table, such as Gallium Arsenide (GaAs) and Gallium Nitride (GaN). Integrated circuit technologies based on III-V materials are substantially more expensive than conventional silicon-based technologies and they cannot handle the integration complexity of e.g. digital circuits or radio modems for cellular handsets. Nevertheless, GaN-based technologies are now maturing rapidly and deliver power levels an order of magnitude higher compared to conventional technologies. Thus, there is a strong interest in new building practices where different technologies can be mixed (heterogeneous integration) in a cost-efficient way to exploit their respective strengths. Such building practices are also well aligned with integration needs expected for cost-efficient implementation of beamforming architectures.

Integrated Local Oscillator (LO) phase noise is another key parameter that worsens with frequency and ultimately limits the attainable Error Vector Magnitude (EVM). The Phase Locked Loop (PLL) used to generate the LO signal has a Voltage Controlled Oscillator (VCO) that dominates power consumption and phase noise. The VCO performance is commonly captured through a Figure-of-Merit (FoM) allowing for a comparison of different VCO implementations and is defined by

$$\text{FoM} = \text{PN}_{\text{VCO}}(df) - 20\log_{10}\left(\frac{f_o}{df}\right) + 10\log_{10}\left(\frac{P_{\text{DC}}}{1\text{mW}}\right). \tag{6.1}$$

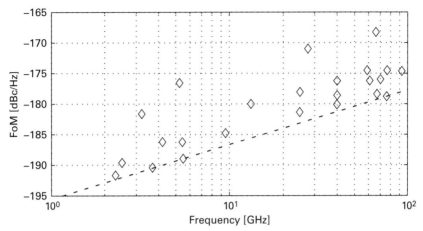

Figure 6.3 VCO FoM versus frequency for recently published VCOs. The dashed line indicates state-of-the-art performance.

Here $\mathrm{PN_{VCO}}(df)$ is the phase noise of the VCO in dBc/Hz at a frequency offset df with oscillation frequency f_o (both in Hz) and power consumption P_{DC} in mW. One noticeable result of this expression is that both phase noise and power consumption in linear power are proportional to f_o^2. While the FoM definition may seem frequency agnostic there is an additional penalty associated with higher frequencies as shown in Figure 6.3 where FoM of recently published VCO designs are compared. Thus, stepping up from the low-GHz regime of today's cellular systems to the mmW regime while preserving the same level of integrated phase noise obviously calls for a re-evaluation of how the LO generation should be implemented. One mean of suppressing VCO phase noise is to increase the bandwidth of the PLL. This effectively yields an LO phase noise characteristic that to a larger extent tracks the phase noise of the crystal oscillator (XO) serving as reference for the PLL and thus push requirements on the XO instead. Also, since the XO phase noise is amplified as $20\log_{10}(f_o/f_{XO})$ when referred to the LO signal the XO frequency f_{XO}, commonly in the low tens of MHz today, would need to be increased to 200 MHz–500 MHz. Although it is possible to implement crystals for this frequency range they do exhibit significantly higher tolerances and drift that in turn may affect terminal-base station synchronization time and complexity of tracking crystal drift in terminals.

The larger bandwidths anticipated with mmW communication will also challenge the data conversion interfaces between analog and digital domains in both receivers and transmitters. This is particularly true for Analog-to-Digital Converters (ADCs). Similar to VCOs there are FoM expressions to capture merits of ADC designs, e.g. the Walden FoM defined by $\mathrm{FoM} = P_{DC}/(2^{\mathrm{ENOB}}f_S)$ with power consumption P_{DC} in W, the Effective Number of Bits for the ADC denoted by ENOB, and sampling frequency f_S in Hz. Figure 6.4 plots the Walden FoM for a large number of published ADCs [15] against the Nyquist sampling frequency f_{snyq} for each design. The Walden FoM can be interpreted as energy per conversion step, and the plot clearly demonstrates a penalty in

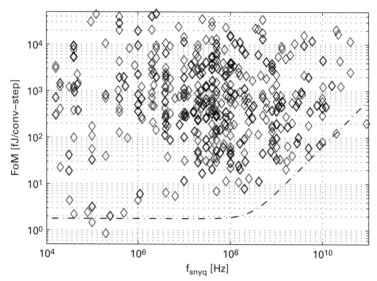

Figure 6.4	The Walden FoM representing the energy per conversion step is plotted against Nyquist sampling frequency for large number of published ADCs. The FoM envelope (dashed) represents the achievable lower limit using technology from around 2015.

conversion beyond a few hundred MHz of sampling rate, roughly amounting to 10 times increase per decade. Although this FoM envelope (dashed line in plot) is expected to be slowly pushed toward higher frequencies by continued development of integrated circuit technology, RF bandwidths in the GHz range are still expected to give poor power efficiency in the analog-to-digital conversion.

Larger signal bandwidths also impact the complexity and power consumption of the digital circuitry. While Moore's Law has enabled a virtually exponential complexity growth for decades, in recent years concerns have gradually been raised on the longevity of this technology evolution. The problem lies in that pure scaling of geometric features as the engine for this progress is quickly approaching its limits. Solutions to this problem considered include the introduction of III-V materials, new device structure (FinFET, nanowire transistors, etc.), and 3D integration. However, none of these will serve as a vehicle for continued exponential improvements. Another problem is that the cost per digital transistor or function has been seen to flatten out or even increase as CMOS technology feature size goes below 28 nm. Nevertheless, a few more technology cycles, with a corresponding reduction of digital power consumption, are expected before 2020.

There are many more building blocks and associated limitations beyond the ones mentioned above, but those treated are viewed as the most challenging, and are worthy of further study.

## 6.3.2	Antennas

The small footprint of antenna elements at mmW frequencies is on par with the size of the RF front-end circuitry driving the antenna element. This enables and possibly requires change in comparison to operation in the low GHz regime. Antennas for milli-meter waves can be integrated together with radio front-ends on the same chip. This eliminates complex low-loss RF interconnects between chip and a separate antenna substrate. On-chip antennas do, however, suffer from very low efficiency due to high permittivity and typically high doping of the substrate. Tailoring the substrate character-istics to mitigate this effect is generally not possible as the substrate properties are already tightly tied with the integrated circuit technology in use. Also, as the cost of chip area (in cents/mm^2) increases with every generation of integrated circuit technology scaling on-chip antennas seems doomed except for low performance low cost solutions implemented in older integrated circuit technologies. In-package antennas, with RF front-end chip and antenna substrate in the same package, offer substantially higher efficiency than on-chip antennas, but they have higher costs due to more complex packaging processes.

Larger antenna arrays help preserve the antenna aperture for increasing frequency. Consequently, the directivity increases and the antenna arrays should be implemented as a phased array that can electrically steer the beam direction, or be able to form beams using a more general beamforming architecture.

6.3.3   Beamforming architecture

A digital beamformer, see Figure 6.5(a), where each antenna element has its own corre-sponding baseband port offers the largest flexibility. However, ADCs and Digital-to-Analog Converters (DACs) operating at multi-GHz sampling rates are very power consuming; a full digital beamformer with several hundred antenna elements might be infeasible or at best feasible but very power hungry and complex. Therefore early mmW communication systems are expected to use analog or hybrid beamforming architectures.

In analog beamforming, see Figure 6.5(b), one baseband port feeds an analog beam-forming network where the beamforming weights are applied either directly on the

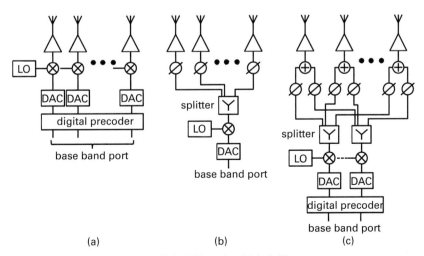

(a)   (b)   (c)

Figure 6.5   Beamforming architectures (a) digital (b) analog (c) hybrid.

analog baseband components, at some intermediate frequency, or at RF. For example, an RF beamforming network may consist of several phase shifters, one per antenna element, and optionally also variable gain amplifiers. In any case, an analog beamforming network typically generates physical beams but cannot generate a complex beam pattern. Especially in a multi-user environment this can lead to interference, if pure beam separation is not sufficient.

Hybrid beamforming, see Figure 6.5(c), is a compromise between those two where a digital beamformer operating on a few baseband ports is followed by an analog beamforming network. This architecture enables a compromise with respect to both complexity and flexibility between analog and full digital beamformer.

A beamforming receiver provides spatial selectivity, i.e. efficient reception of signals in desired directions while suppressing signals in other directions. Each individual antenna element, however, does not provide much spatial selectivity. For a digital beamforming receiver this means that each signal path, extending from respective antenna element all the way to the baseband port, will have to accommodate desired as well as undesired signals. Thus, to handle strong undesired signals, requirements on dynamic range will be high for all blocks in the signal path and that will have a corresponding impact on power consumption. In an analog beamformer, however, the beamforming may be carried out already at RF and thus all subsequent blocks will need less dynamic range compared to the digital beamforming receiver.

While the digital beamformer requires a complete analog RF front-end including ADCs and DACs it does on the other hand not need to distribute RF signals over long distances to a large number of antenna elements as is the case for analog RF beamforming. The power consumption saved by this does not compensate for the power footprint of a digital beamforming architecture.

The ability to create a physical beam in a given direction does not require high resolution and precision of the beamforming weights. An analog beamformer in many cases will be just as good as the digital counterpart with regards to antenna gain. The challenge rather lies in the degree of suppression of side lobes. Accuracy of directed nulls in the radiation pattern is even harder to realize. In these matters, analog beamforming is inferior to digital or hybrid beamforming, particularly for higher frequencies.

The reader is referred to Chapter 8 for more information on beamforming.

6.4   Deployment scenarios

Most outdoor deployments of 5G mmW networks will initially appear in bands above 10 GHz. As the frequency of operation increases, they will be made up of very dense clusters of infrastructure nodes or UDNs in urban areas (cf. Chapters 7 and 11 for more information about UDNs). While the use of highly directive transmissions can provide significant Signal-to-Noise Ratios (SNRs) and even reach long ranges with a good channel realization, area coverage will be limited by the relatively low power levels conducted into antenna ports due to the technological limitations of the hardware.

Typical deployments will use the same approximate site grid as very dense networks in the lower microwave bands that LTE occupies today, with inter-site distances of 40 m–200 m. A deployment would primarily provide coverage using macro-sites above rooftops, with coverage extension being provided by street level pico-base stations. The primary reason to consider cmW technology is the expectation of improved outdoor to indoor performance below 30 GHz.

Spectrum above 30 GHz is useful for near line-of-sight environments; these are typically interior or exterior spaces that are connected in a way that allows propagation of electromagnetic fields. Coverage in such environments is provided by dense deployment of infrastructure nodes within and around hot spots that tend to concentrate traffic.

Millimeter wave bands above 60 GHz are well suited for short range point-to-point links for backhaul. These bands will typically support higher bandwidth communication than access links, and can support high-reliability performance requirements for the data plane as well as provide excess bandwidth for link management and radio system monitoring. Such bands will also find use for short range applications that need very wide bandwidth capability, such as video transmission, virtual office or augmented and virtual reality applications.

One deployment scenario for systems in mmW bands is self-backhauling, which may be formally defined as use of an integrated air interface to provide multiple access as well as transport over one or more hops. Self-backhauling will use the same basic physical layer as the access link and specialized MAC modes. Although not required, access and backhaul could share the same spectrum band. There are several scenarios that are enabled by self-backhauling in indoor and outdoor locations:

- Macro-to-pico deployments, typically from above rooftop to below
- Elevation coverage from ground level
- Outdoor to indoor coverage
- Successive links along a roadway or an open space

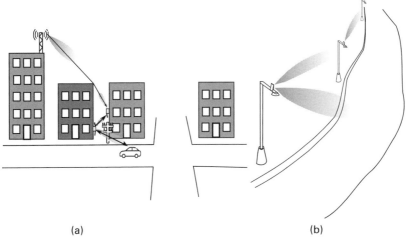

(a)   (b)

Figure 6.6   (a) Outdoor self-backhaul for small-cell connectivity and (b) access and transport using self-backhaul along a road.

The general topology of a self-backhaul scenario is a mesh, although other topologies such as a single path route, a tree, etc. can be superimposed.

Self-backhauling has two major purposes, extension of coverage over short distances without having to provision fiber-access, and the creation of diverse opportunities for connectivity, so that base stations in the infrastructure can share information, and mobility is supported by the quick transfer of information to and from the best access resource. Diverse backhaul paths also offer redundancy, and this may improve per-link reliability requirements without having to operate at unreasonably high Signal-to-Interference-and-Noise-Ratio (SINR). It is expected that the number of hops to fiber infrastructure from any node will not be more than two or three hops.

While self-backhauling is expected to share the same physical and MAC layer as the access, the specific MAC protocol modes that will be suitable for backhaul communication will differ from the MAC procedures used for mobile users. In most cases, infrastructure nodes will be stationary, and self-backhaul links can associate each transmitter to multiple receiving nodes, with a scheduling algorithm to determine the rate of active communication between any two nodes and the roles of the nodes, transmission or reception. The MAC protocol for self-backhauling can change routes and bandwidth assignments dynamically on the basis of changes in traffic distribution, and changes in the interference environment. Self-backhaul can therefore be provisioned for higher degrees of reliability than possible for the multiple access links, where interference statistics are more unpredictable.

Several other deployment scenarios have been identified in the IEEE 802.15.3c and IEEE 802.11ad amendments. A new project within IEEE known as 802.11ay is examining channel-bonding and MIMO for deployment scenarios such as backhaul. Millimeter wave radio is also being considered for inter-chip communication, communication between racks of a data center [16], etc.

6.5   Architecture and mobility

It is important to recognize that the 5G system will not be a complete replacement of LTE with a new air interface. Future releases of LTE in lower spectrum bands will provide a foundation for 5G that ensures wide coverage. It is important that new 5G air interfaces in higher spectrum bands are integrated into mobile networks in a way where they can transform the ability of an operator to handle traffic through the network, and to improve network utility and user experience with great flexibility, by enabling improvements in delay, reliability, data rate or data volume. Some of these improvements will be engineered by revolutionary ways of engineering the core network, e.g. the use of Software Defined Networking (SDN) to configure and isolate network resources, and employing Network Function Virtualization (NFV) to dynamically provision processing or storage. Air interfaces at bands above 10 GHz will also need to be integrated with LTE to improve connectivity and mobility.

While Chapter 3 contains a comprehensive discussion of 5G architecture, this chapter focuses on three aspects of particular relevance for mmW communication systems: dual connectivity, mobility handling using phantom cells, and terminal-specific serving cluster. Additional information on mobility management for phantom cells appears in Chapter 11.

6.5.1   Dual connectivity

When a new mmW air interface is introduced, dual connectivity will be an important feature to prevent loss of coverage. Even for co-located deployments, lower frequency operation will provide better coverage due to lower diffraction loss and improved indoor reach. The primary feature of dual connectivity is to ensure that the terminal avails of lower spectrum bands for control signaling. The primary purpose is fast fall-back to a connectivity option with better coverage if mmW band connectivity is lost. The requirements of 5G to enable dense deployments with wireless self-backhauling, or enabling high reliability or low latency will result in a MAC layer that is quite different from LTE. As it is unlikely that the lower layers of the radio protocol stack will be reused by the 5G air interface, a tight integration of LTE and the 5G mmW radio access technology will occur at an evolved multi-protocol convergence layer at or above Layer 3. The convergence layer would handle data plane and radio resource control flows independently [17]. Dual connectivity is not meant to be a requirement, but a convenience.

6.5.2   Mobility

6.5.2.1   Phantom cell

In a heterogeneous deployment of a covering cell and small cells, it is convenient if mobility between small cells is transparently handled. Transparent tracking of terminal mobility in a small-cell layer covered by an overlay network is known as the phantom cell concept [18][19]. The concept is already well known and proposed for LTE, and can be applied in a similar form for the mmW air interface, although the exact timing between the overlay network and the 5G mmW air interface will not coincide as in a synchronized LTE deployment of small cells.

The phantom cell concept, illustrated in Figure 6.7(b), is a special case of dual connectivity and involves separating the control and user plane of the overlay cell and small cell so that the control and signaling pertaining to the small cell always occurs from the overlay cell. The consequence of this is that radio level mobility is tracked only within the overlay network, while the radio equipment corresponding to the lower layer made of small cells appears indistinguishable from antenna resources belonging to the overlay network. The differences in data rate and latency between traditional networks and the 5G air interface will pose challenges for the specification of the phantom cell concept, but the principles are without doubt useful for 5G.

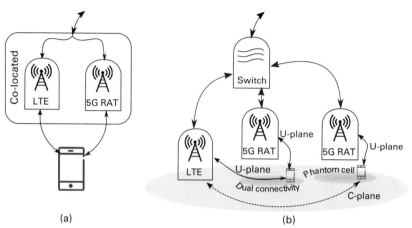

(a)   (b)

Figure 6.7   5G mmW radios can support dual connectivity, while (a) co-located with LTE, (b) deployed as small cells for high capacity, or support phantom cell operation, with LTE assuring coverage.

6.5.2.2   Terminal-specific serving cluster

At mmW frequencies, a terminal can quickly lose connection with its serving base station due to shadowing by other moving objects. It is important that a terminal has a backup connection (either via another path to the same base station or to another base station) already prepared so that it can quickly switch if needed. In this way, the currently serving base station and those base stations with which the terminal has a prepared backup connection form a terminal-specific Serving Cluster (SvC), which typically consists of base stations that are in the vicinity of the terminal. A related concept [20] has also been proposed for communications in traditional cellular bands albeit for a different purpose.

A Principal Serving Access Node (P-SAN) in each SvC is responsible for the connection between the SvC and its associated terminal. Most of the data flow between the terminal and the network passes directly through the P-SAN. Other base stations in the SvC are Assistant Serving Access Nodes (A-SAN) which act to provide diversity when the connection between the P-SAN and the terminal is lost (e.g. due to an obstacle), as illustrated in Figure 6.8. The P-SAN manages membership in the SvC and can proactively wake up sleeping base stations for inclusion in the SvC. A reliable backhaul connection with sufficiently low latency is required between the P-SAN and each A-SAN in the SvC.

A-SANs are typically lightly loaded nodes and must have spare radio resources and processing capacity to assist the SvC. It is additionally beneficial to provision excess storage at the A-SAN to proactively buffer user data for immediate forwarding to the terminal.

SvCs of different terminals may overlap. Thus, a base station can simultaneously be an A-SAN and a P-SAN for two different SvCs. In the role of A-SAN, a base station temporally assists a P-SAN to communicate with its terminal using spare resources. When a significant amount of data flows through an A-SAN for a prolonged period of

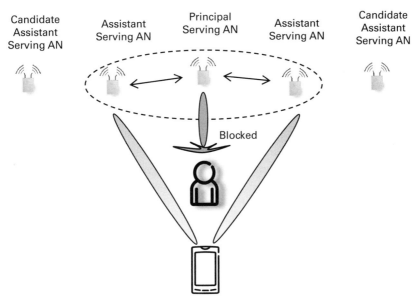

Figure 6.8   Illustration of a terminal-specific serving cluster with a serving beam from P-SAN blocked.

time, the A-SAN will take on the role of P-SAN after the completion of network procedures for transfer of context from the original P-SAN to the new P-SAN, which may subsequently recruit new A-SANs or remove old A-SANs from the SvC. The resulting migration of SvC provides a means for slowly tracking the long-term movement of a terminal across the network. It yields a desirable hysteresis effect in handling the potential rapid changes of connection points as terminals move in a densely, and possibly irregularly, deployed network. It is also well suited for distributed mobility management where no centralized mobility management entity is needed, which is particularly attractive for user-deployed, self-organizing networks.

6.6   Beamforming

6.6.1   Beamforming techniques

As outlined in Section 6.3.3, beamforming functionality can be implemented in several ways. Figure 6.5(b) shows an analog beamformer where the desired beam direction is steered via the phase of the analog phase shifters. Created beams are wideband, i.e. the beam direction is the same across the system bandwidth. Data rate can be increased with two analog beamformers implementing spatial multiplexing using polarization diversity.

   In principle, if the propagation environment is rich enough and multiple strong paths exist between source and destination, each beam can convey up to two layers [21]. One analog beamformer is required per layer. However, analog beamforming allows only for simple beam shapes but does not enable advanced techniques such as creating nulls in certain transmit and/or receive directions. Interference between beams can thus be substantial.

This problem can be mitigated by using hybrid beamforming [22]. A hybrid beamformer is shown in Figure 6.5(c). The analog beamformer creates beams directed to the desired users while the digital beamformer has full flexibility and can even apply frequency selective beamforming weights. The flexibility introduced by the digital beamformer can be used to generate nulls in desired directions to suppress interference or to realize more complex precoders. Even though outlined here on the transmitter side, the same principle can be applied to the combining network at the receiver. The possibility to suppress interference makes this structure suitable for transmissions using multiple beams or even multi-user communications. The rank of the channel is typically rather small; if the number of radio chains matches the channel rank, then the performance of hybrid beamforming approaches that one of digital beamforming [22].

In the digital beamformer shown in Figure 6.5(a) the beamforming weights can be continuously (down to baseband accuracy) and even frequency selectively adjusted, allowing the greatest flexibility of possible beam shapes. However, as stated in Section 6.3.3, the price is high complexity and power consumption. Therefore, early mmW communication systems are expected to use analog or hybrid beamforming architectures.

6.6.2   Beam finding

In order to achieve the array gain required for adequate coverage at mmW frequencies, transmit and receive beam directions at a transmitter and a receiver must be properly aligned. Due to the high spatial selectivity resulting from narrow beamforming, a slight error in the choice of beam direction can lead to a drastic decrease in SNR. An effective beam finding mechanism is therefore of fundamental importance in mmW communications.

Since different antenna arrays may be used for transmission and for reception, directional reciprocity does not always hold. As a result, one may have to rely on a feedback-based beam finding mechanism where synchronization pilot signals are periodically scanned by the transmitter at different beam directions for any receiver to identify the beam direction that yields the best reception quality and send the beam index back to the transmitter. Since each base station potentially has to support multiple terminals, such beam scanning process is preferably performed in a receiver nonspecific manner. In what follows, three different types of beam scanning procedures are briefly discussed, along with their advantages and disadvantages.

6.6.2.1   Linear beam scan

The simplest and most commonly used beam scanning method, cf. [23], is to have the transmitter periodically select one beam at a time from a beam codebook $\mathcal{P}_T$ in a round-robin fashion and transmit the pilot signal at the associated beam direction in a corresponding (time or frequency) resource slot. The receiver observes the pilot signal quality at each resource slot and reports back to the transmitter the best resource slot or beam identification that yields the most preferable beam in $\wp_T$. This approach is called a linear beam scan. For a total of $N \equiv |\wp_T|$ beams, this approach requires a total N resource slots. The amount of feedback needed for linear beam scan is therefore $\log_2 N$ bits.

6.6.2.2   Tree scan

A more efficient approach is to divide the beam scanning process into multiple stages of linear beam scans and use a different set of beam patterns in each stage. Each set partitions the coverage area in a different way. By feeding back the time or frequency slot index that attains the maximum received signal power at the receiver in each stage, the transmitter can determine the best transmit beam direction. Such feedback needs not be done immediately after each stage but instead can be done collectively after observing all stages of beam scans. This beam scanning method is known as a tree scan.

A simple and practical method of forming a unique set of beam patterns for each stage of the tree scan for a uniform antenna array is to subsample and activate a subset of antenna elements using a different spacing at each stage, as illustrated in Figure 6.9 for the case of a uniform linear array of eight antenna elements. At the first stage, these antenna elements may be next to each other, and thus forming wide beam patterns. At subsequent stages, the spacing among adjacent antenna elements in each sub-array increases, and thus forming intentional grating lobes of reduced widths. At low SNR where the receiver is power-limited, both linear scan and tree scan require the same number of resource slots for the receiver to accumulate adequate amount of energy to identify the best beam direction. At high SNR, however, this approach only requires a total of $2\log_2 N$ resource slots and is exponentially more efficient than linear beam scan. The amount of feedback needed for tree scan is roughly $\log_2 N$ bits, about the same as that for the linear beam scan.

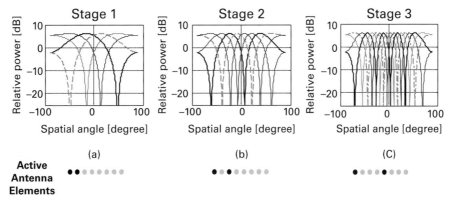

Figure 6.9   Angular power distribution of the beam pattern used in (a) Stage 1, (b) Stage 2, and (c) Stage 3; and the activated antenna elements (black dots) in a uniform linear array of eight antenna elements.

6.6.2.3   Random excitation

Another efficient beam scanning method [24] is to transmit a pilot signal at a pseudo-random direction in each resource slot by applying beamforming weights that are chosen according to a pseudo-random sequence. For example, the beamforming weights may consist of pseudo-random phase shifts. If the receiver has the knowledge of the set of possible pseudo-random beamforming weights used over all resource slots by the transmitter (e.g. through shared random seeds), the receiver can determine with high probability the complex gain of each channel path by exploiting the sparseness of the underlying scattering environment using compressive sensing techniques. From these complex gains, the receiver can then derive the best beam in the beam codebook $\wp_T$, and feeds back its index. The amount of feedback is again $\log_2 N$ bits. Similar to tree scan, at low SNR where the receiver is power-limited, both random excitation and linear beam scan require the same number of resource slots to identify the best beam direction. At high SNR, the number of resource slots needed for random excitation is directly proportional to the number of significant scatterers that exist in the environment, which can be substantially smaller than the number of antennas at the transmitter. However, receivers for this method require more complex processing than those for linear beam scan and tree scan.

6.7   Physical layer techniques

6.7.1   Duplex scheme

Millimeter wave communication systems are expected to be deployed in small cells with infrastructure densities ranging from indoor deployments to densities corresponding to very dense macro cells of today. The number of users within the coverage area of a cell is thus small leading to strong variations in uplink and downlink traffic demands depending on the services currently used by the few covered users. This speaks in favor of dynamic resource partitioning, where resources can be assigned to the two transmission directions on demand.

Millimeter wave communication systems are likely to operate over large bandwidths ranging from several hundreds of MHz up to one GHz and beyond. Such large amounts of spectrum are likely to be unpaired for reasons of convenience and the unavailability of practical duplex filter technology in those bands.

Flexible duplex assigns transmission resources for data dynamically to either transmission direction, allowing more efficient use of bandwidth for communication. To enable power efficient terminal operation control resources will in most cases still follow a fixed structure.

6.7.2   Transmission schemes

Mobile systems operating above 10 GHz will operate over vastly larger bandwidths than cellular systems of today do. Depending on terminal capabilities not all terminals may implement support for the complete system bandwidth. Terminals may not always require the full bandwidth for its current transmission, either because of too few data or power limitations [25].

Based on these observations a transmission scheme supporting operation over a fraction of the system bandwidth seems favorable. Orthogonal Frequency Division Multiplexing (OFDM) and Discrete Fourier Transform Spread OFDM (DFTS-OFDM) which are already used in LTE remain therefore good choices for 5G. Transmission schemes for 5G are analyzed in Chapter 7.

Once the system supports transmission and reception over fractions of the system bandwidth, one can even consider to adopt Frequency Division Multiplex (FDM) and Frequency Division Multiple Access (FDMA), i.e. using remaining parts of the system bandwidth to serve other terminals. When introducing FDM(A) one needs to keep in mind that many mmW systems will require high gain beamforming to close the link at the challenging propagation conditions prevailing at these frequencies. Typically, only a single user will be in the coverage area of a beam. Support of FDM(A) requires a beamforming hardware enabling either frequency selective beam directions or multiple wideband beams. The first option requires digital beamforming while the latter can be implemented using analog or hybrid beamforming.

In [26], a mmW communication system is described for the 72 GHz band based on a single-carrier scheme that is based on Time Division Multiple Access (TDMA) and Time Division Multiplex (TDM), quite unrelated to DFTS-OFDM. The design is straightforward and utilizes very short transmission intervals for multiplexing users. A guard interval, also known as a null cyclic prefix, allows easy frequency domain equalization. The Peak to Average Power Ratio (PAPR) of the scheme is lower than in OFDM and it has lower out-of-band emissions. Complexity is higher than in OFDM.

One attempt to harmonize these contradicting views can be based on operating frequency. The discussion in this book tries to cover the range from around 10 GHz–100 GHz while [26] discusses frequencies around 70 GHz. Systems operating at 70 GHz are expected to cover fewer users and to use shorter transmission times than systems operating closer to 10 GHz, both facts that make pure TDM(A) more attractive.

The large operating range of mmW communication systems (10 GHz–100 GHz) and the ability to support much higher bandwidths at the upper ranges of spectrum suggest changing numerology for the signal waveform. It might well be desirable for 5G systems to be introduced in bands lower than 1 GHz. Figure 6.10 illustrates three different choices of OFDM (or DFTS-OFDM) parameters spanning the range from 1 GHz–100 GHz [27]. Wider subcarrier bandwidth is specified for higher bands, thus providing increased robustness to Doppler and phase noise.

In the following, a short qualitative comparison between single-carrier modulation (incl. DFTS-OFDM) and OFDM is provided, and a more detailed comparison is provided in [28].

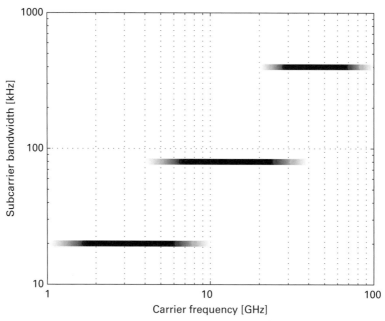

Figure 6.10   Three different OFDM numerologies are proposed to cover the range from 1 GHz–100 GHz.

OFDM has a higher PAPR than single-carrier modulation and thus requires larger power backoff at higher output powers. This does not only limit peak transmit power but also biases the power amplifier into a less power efficient operating region. Furthermore, dynamic range increases with PAPR, which requires higher-resolution ADCs for the same quantization noise. In Section 6.3.1, it is shown that both the power capability of power amplifiers and the resolution of ADCs are important parameters for mmW hardware design.

Link performance comparisons between OFDM and single-carrier modulation not considering hardware impairments often end up in favor of OFDM. OFDM typically outperforms single-carrier modulation over frequency selective channels while over flat fading channels the difference is much smaller. Frequency selective channels correspond to Non-Line-Of-Sight (NLOS) conditions whereas in flat fading channels one path – typically the LOS path – dominates. When adding impairments the comparison becomes less clear [28].

Single-carrier modulation poses more restrictions on the system design than OFDM does since it restricts multiplexing of signals to the time domain. Single-carrier modulation provides lower PAPR than OFDM, a key advantage, provided multiple signals (e.g. data and reference signals) are not transmitted simultaneously from the same power amplifier.

No clear consensus on transmission schemes for 5G mmW communication systems exists. Given the large considered spectrum range from 10 GHz–100 GHz, this is not surprising and the answer can very well depend on the operating frequency and application. This is also reflected in the current landscape of already standardized mmW communication systems where no single trend is visible: IEEE 802.11ad, the wireless LAN standard operating in the 60 GHz band, specifies several physical layers based on both OFDM and single-carrier modulation. Wireless HD, a proprietary standard aiming for wireless transmission of high-definition videos in the 60 GHz band, defines multiple physical layers based on OFDM. The personal area networking standard IEEE 802.15.3c, also operating in the 60 GHz band, specifies physical layers based on both OFDM and single-carrier modulation.

6.8   Conclusions

This chapter has ventured into mmW communications, in fact also covering a large extent of cmW frequencies, offering the potential to unlock a huge amount of spectrum in the 10 GHz–100 GHz frequency range for both wireless backhaul and access. However, it is clear that the usage of millimeter waves, in particular toward the upper end of the stated frequency range, is subject to various challenges that have to be mastered.

One of these is the propagation condition that millimeter waves experience, in particular the effects of foliage, diffraction, and body loss, rendering a large extent of beamforming necessary. These radio conditions imply that for higher frequencies also control signals have to be beamformed, and beam finding and beam scanning techniques have to be applied.

Furthermore, there are various challenges on the hardware side that have to be overcome, namely the decreasing power efficiency and increasing phase noise toward higher carrier frequencies, and also the decreasing A/D conversion power efficiency and increasing device complexity toward larger system bandwidths. These aspects require further innovation on the hardware side, and/or a 5G radio design that is able to compensate or alleviate their impact. On the positive side, shorter signal wavelengths enable a high antenna integration and hence the usage of on-chip or in-package antennas.

Clearly, the more hostile radio conditions for higher frequencies, such as the possibility of sudden signal blockage, require that mmW communication solutions allow for a quick backup connection, for instance through another mmW node or a lower frequency radio interface, as for instance evolved LTE. In this respect, the chapter has pointed out different dual connectivity options, as for instance a control and user plane split between a lower and higher frequency layer as foreseen in the so-called phantom cell concept.

Ultimately, the chapter has compared different physical layer techniques for mmW communication, where there is no consensus yet, but already an indication that OFDM-based solutions are likely most relevant for a large portion of the considered frequency range. For the upper spectrum range, single-carrier approaches may be suitable due to the improved power efficiency.

References

[1] ICT-317669 METIS project, "Scenarios, requirements and KPIs for 5G mobile and wireless system," Deliverable D1.1, April 2013, www.metis2020.com/documents/deliverables/

[2] Ericsson, Ericsson Mobility Report, Report No. EAB-15:037849, November 2015, www.ericsson.com/res/docs/2015/mobility-report/ericsson-mobility-report-nov-2015.pdf

[3] Cisco, Cisco Visual Networking Index: Global Mobile Data Traffic Forecast Update 2014–2019, White Paper, February 2015, www.cisco.com/c/en/us/solutions/collateral/service-provider/visual-networking-index-vni/white_paper_c11-520862.html

[4] IEEE 802.11ad, "IEEE Wireless LAN Medium Access Control (MAC) and Physical Layer (PHY) Specifications Amendment 3: Enhancements for Very High Throughput in the 60 GHz Band," IEEE Standard 802.11ad-2012 Part 11, 2012.

[5] IEEE, "Next Generation 802.11ad: 30+ Gbps WLAN," Document IEEE 11–14/0606r0, May 2014, https://mentor.ieee.org/802.11/dcn/14/11-14-0606-00-0wng-next-generation-802-11ad.pptx

[6] FCC, "NOI to examine use of bands above 24 GHz for mobile broadband," FCC 14–154, October 2014, www.fcc.gov/document/noi-examine-use-bands-above-24-ghz-mobile-broadband

[7] Ofcom, "Call for Input: Spectrum above 6 GHz for future mobile communications," January 2015.

[8] FCC, "47 CFR 2.1093 – Radiofrequency radiation exposure evaluation: portable devices," Code of Federal Regulations (CFR), title 47, vol. 1, section 2.1093, 2010, 47 CFR 2.1093 – Radiofrequency radiation exposure evaluation: portable devices

[9] International Commission on Non-Ionizing Radiation Protection, "Guidelines for limiting exposure to time-varying electric, magnetic, and electromagnetic fields (up to 300 GHz)," *Health Physics*, vol. 74, no. 4, pp. 494–522, October 1998.

[10] D. Colombi, B. Thors, and C. Törnevik, "Implications of EMF exposure limits on output power levels for 5G devices above 6 GHz," *IEEE Antennas and Wireless Propagation Letters*, vol. 14, pp. 1247–1249, 2015.

[11] FCC, "Millimeter Wave Propagation: Spectrum Management Implications," Bulletin no. 70, July 1997.

[12] T. S. Rappaport, S. Sun, R. Mayzus, H. Zhao, Y. Azar, K. Wang, G. N. Wong, J. K. Schulz, M. Samimi, and F. Gutierrez, "Millimeter wave mobile communications for 5G cellular: It will work!," *IEEE Access*, vol. 1, pp. 335–349, 2013.

[13] G. R. MacCartney and T. S Rappaport, "73 GHz millimeter wave propagation measurements for outdoor urban mobile and backhaul communications in New York City," in IEEE International Conf. on Communications, Sydney, June 2014, pp. 4862–4867.

[14] E. Johnson, "Physical limitations on frequency and power parameters of transistors," in *1958 IRE International Convention Record*, vol. 13, pp. 27–34, 1966.

[15] B. Murmann, "ADC Performance Survey 1997–2015," [Online] http://web .stanford.edu/~murmann/adcsurvey.html

[16] H. Vardhan, N. Thomas, S.-R. Ryu, B. Banerjee, and R. Prakash, "Wireless data center with millimeter wave network," in IEEE Global Telecommunications Conference, Miami, December 2010, pp. 1–6.

[17] I. Da Silva, G. Mildh, J. Rune, P. Wallentin, Rui Fan, J. Vikberg, and P. Schliwa-Bertling, "Tight integration of new 5G Air Interface and LTE to fulfil 5G requirements," in IEEE Vehicular Technology Conference, Glasgow, May 2015.

[18] H. Ishii, Y. Kishiyama, and H. Takahashi, "A novel architecture for LTE-B: C-plane, U-plane split and the phantom cell concept," in IEEE International workshop on emerging technologies for LTE Advanced and Beyond-4G, Anaheim, December 2012.

[19] T. Nakamura, S. Nagata, A. Benjebbour, Y. Kishiyama, Tang Hai, Shen Xiaodong, Yang Ning, and Li Nan, "Trends in small cell enhancements in LTE Advanced," *IEEE Communications Magazine*, vol. 51, no. 2, pp. 98–105, February 2013.

[20] D. Hui, "Distributed precoding with local power negotiation for coordinated multi-point transmission," in IEEE Vehicular Technology Conference, Yokohama, May 2011, pp. 1–5.

[21] S. Sun, T. S. Rappaport, R. W. Heath Jr., A Nix, and S. Rangan, "MIMO for millimeter-wave wireless communications: Beamforming, spatial multiplexing, or both?," *IEEE Communications Magazine*, vol. 52, no. 12, pp. 32–33, December 2014.

[22] A. Alkhateeb, J. Mo, N. González-Prelcic, and R. W. Heath Jr., "MIMO precoding and combining solutions for millimeter-wave systems," *IEEE Communications Magazine*, vol. 52, no. 12, pp. 122–131, December 2014.

[23] L. Zhou and Y. Ohashi, "Efficient codebook-based MIMO beamforming for millimeter-wave WLANs," in IEEE International Symposium on Personal, Indoor and Mobile Radio Communications, Sydney, 2012, pp. 1885–1889.

[24] W. U. Bajwa, J. Haupt, A. M. Sayeed, and R. Nowak, "Compressed channel sensing: A new approach to estimating sparse multipath channels," *Proceedings of the IEEE*, vol. 98, no. 6, pp. 1058–1076, 2010.

[25] R. Baldemair, T. Irnich, K. Balachandran, E. Dahlman, G. Mildh, Y. Selén, S. Parkvall, M. Meyer, and A. Osseiran, "Ultra-dense networks in millimeter-wave frequencies," *IEEE Communications Magazine*, vol. 53, no. 1, pp. 202–208, January 2015.

[26] A. Ghosh, T. A. Thomas, M. C. Cudak, R. Ratasuk, P. Moorut, F. W. Vook, T. S. Rappaport, G. R. MacCartney Jr., S. Sun, and S. Nie, "Millimeter-wave enhanced local area systems: A high-data-rate approach for future wireless networks," *IEEE Journal on Selected Areas in Communications*, vol. 32, no. 6, pp. 1152–1163, June 2014.

[27] ICT-317669 METIS project, "Proposed solutions for new radio access," Deliverable D2.4, February 2015, https://www.metis2020.com/documents/ deliverables/

[28] T. S. Rappaport, R. W. Heath Jr., R. C. Daniels, and J. N. Murdock, *Millimeter Wave Wireless Communications*, New Jersey: Prentice Hall, 2014.

7   The 5G radio-access technologies

Malte Schellmann, Petra Weitkemper, Eeva Lähetkangas, Erik Ström,
Carsten Bockelmann, and Slimane Ben Slimane

The radio access for 5G will have to respond to a number of diverse requirements raised
by a large variety of different new services, such as those from the context of massive
Machine-Type Communication (mMTC) and ultra-reliable MTC (uMTC), as discussed
in Chapter 2. Consequently, a "one-size-fits-all" solution for the air interface as pre-
valent in today's radio systems may no longer be the adequate choice in the future, as it
can merely provide an inadequate compromise. Instead, the system should provide more
flexibility and scalability to enable tailoring the system configurations to the service
types and their demands. Moreover, as the data rates to be provided by mobile radio
systems are ever increasing, technologies need to be devised to squeeze out the last bit
from the scarce spectrum resources. This chapter elaborates on novel radio-access
technologies addressing the aforementioned issues, which can be considered promising
candidates for the 5G system. It is noteworthy that there has been flourishing work on
potential radio-access technologies for 5G in recent time; refer to [1][2] for prominent
research activities in the field.

The chapter starts with a general introduction to the access design principles
for multi-user communications in Section 7.1, which build the fundamentals for the
novel access technologies presented in this chapter. Section 7.2 then presents novel
multi-carrier waveforms based on filtering, which offer additional degrees of freedom
in the system design to enable flexible system configurations. Novel non-orthogonal
multiple-access schemes yielding an increased spectral efficiency are presented in
Section 7.3. The following three sections then elaborate on radio access technologies
and scalable solutions tailored for specific use cases, which are considered key drivers
for 5G radio systems. Section 7.4 focuses on Ultra-Dense Networks (UDN), where
also higher frequencies beyond 6 GHz are expected to be used. Section 7.5 presents
an ad-hoc radio-access solution for the Vehicle-to-Anything (V2X) context, and
finally Section 7.6 proposes schemes for the massive access of Machine-Type
Communication (MTC) devices, characterized by a low amount of overhead and
thus enabling an energy efficient transmission.

Table 7.1 gives a brief overview on the radio-access technologies presented in this
chapter, highlighting some of their characteristics and properties. It should be noted that
the gathered information is not exhaustive and only the most important aspects are listed.

5G Mobile and Wireless Communications Technology, ed. A. Osseiran, J. F. Monserrat, and P. Marsch.
Published by Cambridge University Press. © Cambridge University Press 2016.

Table 7.1 5G Multiple and medium access schemes.

Name	Type[1], direction[2]	Separation of resources	Advantage	Disadvantage
OFDM	Multiple, UL & DL	Time, frequency	Simple implementation, simple equalization	Large side lobes require tight sync and large guard bands
FBMC-OQAM	Multiple, UL & DL	Time, frequency	Small side lobes enable coexistence and relaxed sync	Orthogonality in the real field requires redesign of selected algorithms
UF-OFDM	Multiple, UL & DL	Time, frequency	Reduced side lobes, compatible with OFDM, relaxed sync	Vulnerable to large delay spread
SCMA	Multiple, UL & DL	Code & power	Limited CSIT	Complex receiver (MPA)
NOMA	Multiple, UL & DL	Power	Limited CSIT	SIC receiver
IDMA	Multiple, UL	Code	Limited or no CSIT	Iterative receiver
Coded Slotted Aloha	Medium	Not applicable	High reliability with minimal coordination	Complex receiver
Coded Access Reservation	Medium, UL	Not applicable	compatible with LTE	High overhead for small packets
Coded Random Access	Medium, UL	Not applicable	Suitable for small packets	MUD

For comprehensive information, the reader is referred to the details presented in this chapter.

7.1 Access design principles for multi-user communications

Allowing multiple communication links to share the same frequency bandwidth requires a proper signal design so that they will not affect each other. This is usually done by assigning a different waveform to each user. Let us assume that, for a given bandwidth W in Hz, one would like to serve as many users as possible. In the absence of external interference, one can obtain up to $2WT$ orthogonal waveforms when coherent detection is employed (i.e. based on a common phase reference), and up to WT orthogonal waveforms when non-coherent detection is employed (i.e. waveforms can be distinguished without a phase reference), where T is the waveform duration [3]. The basic principle of orthogonal waveforms is to divide the radio resources (in time or in frequency) between the different users.

---

[1] There are two types of radio access: multiple access and medium access.
[2] The transmission direction can be either uplink or downlink.

Spread spectrum signaling is based on one type of waveform that allows multiple users to share the same spectrum in a manner that maximizes the number of users while minimizing the interference between users. The philosophy behind these waveforms is to divide the radio resources between the different users without physical separation in time or in frequency, but rather through the usage of different codes that are orthogonal (or at least moderately orthogonal) to each other. Spread spectrum waveforms allow some controlled internal interference but have the capability to reject external interference and are quite robust in frequency selective fading channels.

The choice of the proper waveforms will provide a certain number of channels where multiple nodes can share a communication medium for transmitting their data packets. However, these waveforms do not specify how they will be shared. The Medium Access Control (MAC) protocol is the one primarily responsible for regulating access to this shared medium. The choice of the MAC protocol has a direct bearing on the reliability and efficiency of network transmissions due to errors and interferences in wireless communications and due to other challenges. The design of MAC protocols should take into account the radio channel and trade-off between energy efficiency and latency, throughput, and/or fairness. MAC schemes can be divided into two categories: contention-free and contention-based protocols. Contention-free medium access can avoid collisions by ensuring that each node can use its allocated resources exclusively. In contention-free protocols, fixed assignments can be found such as Frequency Division Multiple Access (FDMA), Time Division Multiple Access (TDMA) and Code Division Multiple Access (CDMA); and dynamic assignment such as polling, token passing, and reservation-based protocols. Contention-based protocols allow some form of competition, where nodes may initiate transmissions at the same time. This competition will require some mechanisms to reduce the number of collisions and to recover from collisions when they occur. For contention-based MAC, the most common protocols are ALOHA, slotted ALOHA, and Carrier Sense Multiple Access with Collision Avoidance (CSMA/CA) [4].

7.1.1   Orthogonal multiple-access systems

Orthogonal multiple-access methods are based on dividing the radio resources (in time or frequency) between the different users. The corresponding multiple-access schemes are respectively FDMA, TDMA and Orthogonal Frequency Division Multiple Access (OFDMA). FDMA and OFDMA are quite similar but FDMA has non-overlapping frequency sub-bands, while OFDMA has overlapping frequency sub-bands. In a single cell environment with AWGN channels, all orthogonal multiple-access schemes are almost equivalent with respect to capacity [5]. The differences between multiple-access schemes become visible when transmission channels exhibit frequency selectivity and time variability.

7.1.1.1   Frequency division multiple-access systems

In FDMA, the total bandwidth is divided into a set of frequency sub-bands. Individual frequency sub-bands are assigned to individual users as illustrated in Figure 7.1. As the sub-bands do not overlap in frequency, user signals are easily detected by band-pass

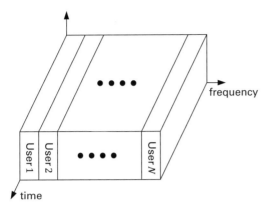

Figure 7.1   Time/frequency diagram of FDMA systems.

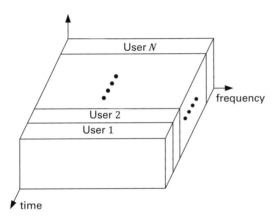

Figure 7.2   Time/frequency diagram of TDMA systems.

filtering, which eliminates all adjacent channel interference. FDMA is, by far, the most popular multiplexing principle for its simplicity and suitability for analog circuit technology. Orthogonality of user signals is maintained between all the channels and multi-user communication occurs independently of each other. To avoid adjacent channel interference, a frequency guard band is inserted between neighboring sub-bands. With narrow-band signaling, FDMA systems suffer from flat fading channels and external interference. FDMA was for instance used in the analog Advanced Mobile Phone System (AMPS) and Total Access Communication System (TACS). The exclusive choice of FDMA as a multiple-access scheme results in the necessity of reduced frequency re-use to cope with inter-cell interference. It also requires careful radio network design taking frequency planning into account.

7.1.1.2   Time division multiple-access systems
A multiplexing scheme that is highly suited for digital transmission is time division multiplexing. Instead of dividing the available bandwidth to each user, users share the entire signal bandwidth but are separated in time as shown in Figure 7.2. The users are

assigned short time slots which are repeated in a cyclic fashion. The number of time slots in TDMA corresponds to the number of channels in an FDMA system.

Coordination between users is needed in TDMA systems so that users do not interfere with each other. In general, a time guard interval is required between consecutive slots to ensure that interference between different users is avoided despite the propagation delays and delay spreads of the different users' signals. As each user occupies the whole bandwidth when active, transmission will suffer from frequency selective fading channels and time (or frequency) equalization is needed. Time equalization has a complexity that increases with the user data rate, whereas the complexity of frequency equalization, although increasing, is maintained in the same order of magnitude. This makes frequency equalization more suitable for high data rate systems such as 4G and 5G systems. Combining TDMA with FDMA is quite efficient for radio systems. Such a combination offers the possibility to average out external interference and avoid deep fading situations. For instance, GSM is based on a hybrid TDMA/FDMA scheme with Frequency Division Duplex (FDD). In fact, most 2G cellular systems are based on the TDMA principle.

7.1.1.3   Orthogonal frequency division multiple-access systems

An efficient way of solving the problem of inter-symbol interference caused by frequency selective fading channels is by using Orthogonal Frequency Division Multiplexing (OFDM). The idea behind OFDM is to transform a very high data rate stream into a set of low data rate streams that are then transmitted in parallel with each stream transmitted over a different sub-carrier frequency. With this structure, a frequency selective fading channel can be transformed into a set of frequency-flat fading channels. More precisely, OFDM divides the available bandwidth into a number of equally spaced sub-carriers and carries a portion of a user's information on each sub-carrier. OFDM can be viewed as a form of FDMA; however, OFDM has an important special property that each subcarrier is orthogonal to every other subcarrier. OFDM allows the spectrum of each subcarrier to overlap, and because they are orthogonal, they do not interfere with each other. By allowing the sub-carriers to overlap, the overall amount of spectrum required is reduced and the obtained access scheme is more bandwidth efficient as compared to FDMA. This is illustrated in Figure 7.3.

Figure 7.4 shows a discrete representation of an OFDM system. The input binary data X is mapped into baseband-modulated constellations in the frequency domain and is then converted to a time domain signal using an IFFT. The last part of the signal is then appended to the beginning of the signal, known as the Cyclic Prefix (CP), in part to fight Inter-Symbol Interference (ISI) and preserve orthogonality between the sub-carriers. One important feature of OFDM is its ability to enable equalization in frequency domain with a simple one-tap equalizer per sub-carrier. The CP duration should be larger than the maximum delay spread of the environment channel where the system is intended to operate, and compensate the impact of transmitter and receiver filters. In addition, CP design is also intended to cope with synchronization mismatches and timing errors e.g. in a cell. In general, OFDM signals are quite sensitive to time/frequency synchronization errors. For instance, Phase Noise (PN) caused by implementation technology and

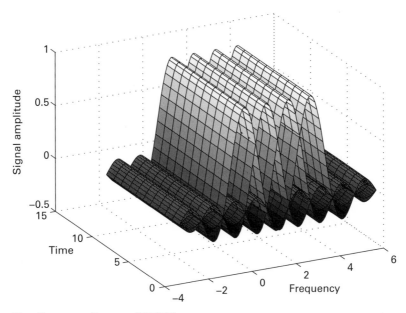

Figure 7.3   Time/frequency diagram of OFDM.

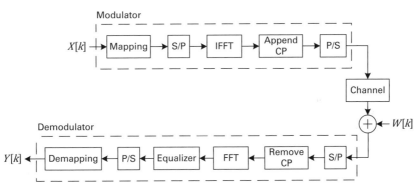

Figure 7.4   Discrete representation of an OFDM system.

hardware design causes Common Phase Errors (CPEs) and inter-carrier interference in OFDM systems.

As a multiple-access technique, an individual subcarrier or groups of sub-carriers can be assigned to different users. Multiple users then share a given bandwidth in this manner, yielding a system called OFDMA. Since OFDMA is based on OFDM, it inherits its avoidance of inter-symbol interference and enables simple equalization for frequency-selective fading channels. In OFDMA, each user can be assigned a predetermined number of sub-carriers when they have information to send, or alternatively, a user can be assigned a variable number of sub-carriers based on the amount of information that they have to send.

OFDM-based waveforms are currently used in several contemporary systems, such as LTE and WLAN.

7.1.2   Spread spectrum multiple-access systems

These methods are usually referred to as Code Division Multiple Access (CDMA) schemes and are characterized by signals with a bandwidth much larger than the user data rate. The two most popular schemes in this class are Frequency Hopping (FH) systems and Direct Sequence (DS) systems.

7.1.2.1   Frequency hop-code division multiple-access systems

A Frequency Hop-Code Division Multiple Access (FH-CDMA) system is a combined FDMA and TDMA scheme, where the available bandwidth is divided into a number of narrow-band channels and the time is also divided into time slots. The user transmits on one frequency during a time slot and on a different frequency during the next time slot. Hence, the user hops from frequency to another according to a hopping sequence. Every user has a unique code known at both the transmitter and the receiver. The receiver tracks the transmitter in every time slot to recover the information. In general, one can distinguish between fast frequency hopping and slow frequency hopping. A fast hopping FH-CDMA system has a hopping rate greater or equal than the user data rate, while a slow hopping FH-CDMA system has a hopping rate less than the user data rate. An example of the latter is GSM, where an entire burst is transmitted on each frequency.

7.1.2.2   Direct sequence-code division multiple-access systems

In Direct Sequence-Code Division Multiple Access (DS-CDMA), users use different spreading waveforms allowing them to share the same carrier frequency and transmit the spread signals simultaneously. There is no physical separation in time or in frequency between signals from different users. Different from TDMA and FDMA, spread signals from different users do interfere with each other unless the users are perfectly synchronized, orthogonal spreading codes are employed and propagation channels are frequency-flat.

DS-CDMA cellular systems employ two-layered spreading codes. This spreading code allocation provides flexible system deployment and operation. In fact, multiple spreading codes make it possible to provide near waveform orthogonality among all users of the same cell while maintaining mutual randomness between users of different cells. Orthogonality can be achieved through the channelization code layer, a set of orthogonal short spreading codes such as the variable-length Walsh orthogonal sequence set [6], where each cell uses the same set of orthogonal codes. A long scrambling code is employed as a second layer to reduce the impact of external interference (inter-cell interference). A cell-specific scrambling code (common to all users in that cell) is employed in the downlink and a user-specific code in the uplink. Hence, each transmission is characterized by the combination of a channelization code and a scrambling code. The IS-95 and WCDMA standards employ DS-CDMA.

7.1.3   Capacity limits of multiple-access methods

Among a vast list of issues to be considered, there are two very basic aspects that influence the suitability of a multiple-access scheme for a specific system. One is the possibility of fast adaptation to the fading channel, the other one is the complexity involved at the transmitter and receiver side. These two aspects are the main drivers for the different options used in the past to design the access schemes described in the previous sections such as FDMA, TDMA, CDMA and OFDMA.

In this section, the fundamental capacity limits of multiple-access methods are described. The focus is initially on the information theoretic notion of the Multiple Access Channel (MAC), typically used to described the uplink of a mobile communications system, where multiple users transmit toward one common receiver, and then on the Broadcast Channel (BC), corresponding to a downlink transmission from one transmitter to multiple users.

7.1.3.1   The multiple-access channel (uplink)

The capacity region of a two user memoryless multiple-access channel is illustrated in Figure 7.5. As one can see, the capacity region is a pentagon with

$$C_1 = \log_2\left(1 + \frac{P_1}{WN_0}\right), \tag{7.1}$$

$$C_2 = \log_2\left(1 + \frac{P_2}{WN_0}\right), \tag{7.2}$$

and

$$C_1 + C_2 = \log_2\left(1 + \frac{P_1 + P_2}{WN_0}\right), \tag{7.3}$$

where W is the total available bandwidth, N_0 is the additive noise power spectral density, and P_i is the power of user i. In general, to achieve the capacity for a multiple-access channel, joint decoding is needed at the receiver. In the figure, the corner segments of the

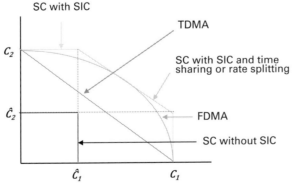

Figure 7.5   Capacity regions of different multiple-access schemes in the uplink.

capacity region are achieved when both users transmit at the same time and in the same frequency band (superposition coding). The combined signals are then separated at the receiver using a successive interference cancellation technique, i.e. one user is decoded first, and then its effect is subtracted before the second user is decoded. Such Superposition Coding (SC) with Successive Interference Cancellation (SIC) is known to be optimal for the multiple-access channel, when the rates and powers are chosen properly according to the channel condition. Consequently, the transmitters have to know the channel condition in advance to determine the optimal rate and power allocation. If this is not possible, e.g. due to missing or outdated channel knowledge, iterative receivers or Multi-User Detection (MUD) can be used instead of SIC, but in this case a lower target rate than for optimal SC with SIC may need to be used to increase the probability of successful decoding. This first basic consideration holds for a single static and frequency-flat channel. In a time- and/or frequency-selective channel, additional multi-user diversity can be achieved by not using the whole bandwidth for each user all the time, but to do some time and frequency-selective scheduling. Also for this approach, reliable channel knowledge at the transmitter side is needed.

The optimal transmission strategy for the multiple-access channel, assuming a frequency-flat and static channel, requires that the entire frequency band is used by both users simultaneously. Hence, the orthogonal access methods are not optimal except in some special cases. For instance, the capacity region corresponding to FDMA is given by

$$C_1 = \alpha \log_2\left(1 + \frac{P_1}{\alpha W N_0}\right), \tag{7.4}$$

$$C_2 = (1 - \alpha)\log_2\left(1 + \frac{P_2}{(1 - \alpha)W N_0}\right), \tag{7.5}$$

where α is the fraction of the total bandwidth used by the first user. As illustrated in Figure 7.5, the capacity region of FDMA is strictly smaller than the optimal capacity region, and FDMA is optimal only at a single point.

For TDMA systems, each user occupies the whole bandwidth when active. Hence, the capacity region of TDMA for a memoryless multiple-access channel is given by

$$C_1 = \alpha \log_2\left(1 + \frac{P_1}{W N_0}\right), \tag{7.6}$$

$$C_2 = (1 - \alpha)\log_2\left(1 + \frac{P_2}{W N_0}\right), \tag{7.7}$$

and is also plotted in Figure 7.5. Hence, TDMA can only achieve the corner points of this region and all points on a straight line between these edges, depending on the choice of α above.

The early mobile communication systems applied orthogonal schemes like TDMA or FDMA as they are simple and fair, but suboptimal as described above. In 3G, CDMA employs a combination of orthogonal and random spreading codes with an objective to reduce external interference and to exploit frequency diversity by spreading over the full bandwidth. The rate and power adaptation was rather slow due to the system design (at least before the introduction of HSDPA), so although applying some kind of superposition of the users, it was far from capacity achieving. SIC was not suitable due to the suboptimal power and rate allocation, so simple detectors were used like rake or MMSE receivers. Furthermore, the code was rather weak, being a combination of a repetition-like code due to the spreading, and a turbo or convolutional code for Forward Error Correction (FEC).

OFDMA in LTE uses fast rate and power adaptation and fast scheduling. Significant gain over non-adaptive schemes can be achieved, but in practical systems no super-position of users was applied except for spatial superposition in case of Multi-User MIMO (MU-MIMO). From information theory perspective, OFDMA can achieve the FDMA capacity region, but not the full multiple-access capacity region. In Section 7.3 it is shown how Non-Orthogonal Multiple Access (NOMA) techniques may be able to overcome this limitation.

7.1.3.2   The broadcast channel (downlink)

The capacity region of a two user Gaussian broadcast channel is depicted in Figure 7.6. It is known that capacity is achieved if the transmitter superimposes the transmissions to the two users and uses a non-linear transmit strategy known as Dirty Paper Coding (DPC) [7]. This coding strategy ensures that the transmission to one user is not impacted by the interference from the transmission to the other user. In fact, DPC can be seen as a reciprocal approach to SIC in the uplink. For the multiple-access channel, it was shown that a capacity-achieving approach is a superimposed transmission of both users with SIC decoding at the receiver, such that one user can be decoded (in theory) free of interference from the other user, as the interference from this is cancelled. For the broadcast channel, a capacity-achieving approach is to apply interference cancelation already at the transmitter side through DPC, again with the result that one of the transmissions is (in theory) free of the interference from the other transmission.

In principle, duality between the uplink multiple-access channel and the downlink broadcast channel applies, i.e. given the same overall transmit power constraint, the

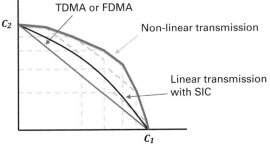

Figure 7.6   Capacity regions of different multiple-access schemes in the downlink.

uplink and downlink capacity regions for the same channel are identical. In practice, a key difference between uplink and downlink is that the uplink is typically subject to a per-user power constraint, while in the downlink an overall transmit power budget is available that can be allocated freely to either of the users. In fact, one can see the capacity region of the downlink broadcast channel as the convex hull around all capacity regions of the corresponding uplink multiple-access channels for different power settings in the uplink (assuming that one could distribute the joint power budget freely among the users), as illustrated by the pentagons with dashed lines in Figure 7.6.

In the downlink, the total transmitted power is split between the users. Denoting by β the fraction of power allocated to user 1 for the two users case, the capacity region corresponding to FDMA can be written as follows:

$$C_1 = \alpha\log_2\left(1 + \frac{\beta P_t}{\alpha W N_0}\right),\tag{7.8}$$

$$C_2 = (1 - \alpha)\log_2\left(1 + \frac{(1 - \beta)P_t}{(1 - \alpha)W N_0}\right),\tag{7.9}$$

where α is the fraction of the total bandwidth used by the first user and P_t is the total transmitted power. Assuming that the channel gains toward the two receivers are identical, it is easy to verify that (i) the sum capacity is maximized when $\alpha = \beta$, and (ii) the capacity region of FDMA reduces to that of the uplink TDMA.

From the above expressions, it can be observed that the performance for time-division or frequency-division transmission to two users is the same straight line in the downlink, as the same overall power constraint applies no matter whether the users are orthogonalized in time or frequency. However, in the uplink, orthogonalizing the transmissions in frequency allows each user to invest its transmit power into a smaller portion of spectrum, increasing its power spectral density, which leads to the curve in the case of frequency division depicted in Figure 7.5.

While SIC, or other joint detection strategies, can be implemented in practice at reasonable complexity for the uplink multiple-access channel, the problem is that DPC (or suboptimal variants thereof, such as Tomlinson-Harashima Precoding (THP) [8]) is typically considered too complex for its practical implementation. Further, it relies on very accurate channel state information at the transmitter side. For this reason, one typically rather considers linear precoding at the transmitter side, as is for instance done in the context of MIMO and Coordinated Multi-Point (CoMP) as described in Chapters 8 and 9, respectively. In cases where there is a very large difference in path loss between the two users, it may also be interesting to apply SIC at the receiver side, as investigated in Section 7.3.1. In fact, for so-called degenerated broadcast channels, where the channel toward one user is a degenerated form of the channel toward the other (for instance with scaled down channel coefficients or an increased noise level), it is known that linear transmission with a SIC strategy at the receiver side is capacity-achieving.

To conclude this section, it is also clear for the downlink broadcast channel that orthogonal transmission to multiple users is in fact suboptimal, and that a superposition of transmissions to multiple users should be considered instead.

7.2   Multi-carrier with filtering: a new waveform

The classical multi-carrier waveform OFDM can be extended by an integrated filtering component, providing good spectral containment properties of the transmit signals. This novel waveform property enables partitioning the spectrum available for mobile radio transmission into independent sub-bands that can be individually configured to optimally adapt to signal conditions of individual user links or to requirements of a particular radio service. This is opposed to the paradigm followed in today's system design, where the selection of the waveform parameters is always made as a "best compromise", meeting the overall needs of the services as a whole and matching all link signal conditions expected to show up during system operation. For OFDM-based systems, for example, this best compromise typically translates to a fixed subcarrier spacing and CP length. Thanks to the good spectral containment of the filtered multi-carrier signals, interference between individually configured sub-bands can be kept to a minimum even if those signals are only loosely synchronized. Hence, independent and uncoordinated operation of different services in the transmission band can be facilitated, and asynchronous system design can be enabled. Thus, multi-carrier waveforms with filtering can be considered key enablers for a flexible air interface design, which has been identified as one of the key components for the future 5G systems. Two promising candidates are under investigation, namely Filter-Bank based Multi-Carrier (FBMC) and Universal Filtered OFDM (UF-OFDM); the latter is also known under the term Universal Filtered Multi-Carrier (UFMC). While both candidates target the same scenarios and can realize gains from the flexible configuration of the spectrum similarly, they use different means to achieve these. Thus, they also differ in their system requirements and implementation aspects. In UF-OFDM, sub-bands being constituted of a minimum number of subcarriers are filtered, which is done to maintain the conventional OFDM signal structure for compatibility reasons. In contrast to that, FBMC offers enlarged degrees of freedom for the system design due to individual filtering of the single subcarriers, which comes with some changes in the signal structure, requiring the redesign of some signal processing procedures. In the following subsections, both waveform candidates are briefly described and the major advances achieved in recent research are summarized, paving the way for a potential application in 5G systems.

7.2.1   Filter-bank based multi-carrier

FBMC represents a multi-carrier system where the single subcarrier signals are individually filtered with a prototype pulse. Spectral containment of the multi-carrier signals is achieved by choosing prototype pulses with steep power roll-off in the frequency

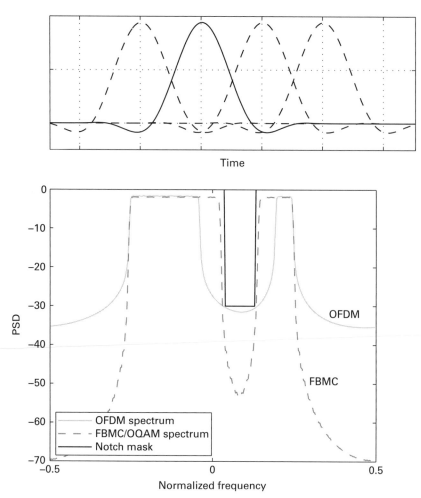

Figure 7.7   Impulse response of the prototype filter (top) and power spectral density of notched FBMC signal in comparison to OFDM (bottom).

domain. These can be realized by allowing the time domain representation of the pulse to expand over the size of the FFT window, resulting in overlapping pulses if several FBMC symbols are transmitted successively in time. An orthogonal pulse design ensures that the overlapping pulses can be (near to) perfectly reconstructed without creating any mutual interference. Practical values for the overlapping factor K reach from 1 up to 4, where K specifies the number of FFT blocks that the time domain pulse spans. For an illustration of overlapping pulses with $K = 4$, see Figure 7.7 (top), where the selected division of the x-axis represents the FFT window size. Since each of the FBMC subcarrier signals are filtered with the prototype pulse, a sub-band can be constituted by the aggregation of adjacent subcarriers of any number, exhibiting the desired spectral containment. Hence, FBMC offers the maximum number of degrees of freedom for the system design. The Power Spectral Density (PSD) of an FBMC multi-carrier signal is shown in Figure 7.7 (bottom).

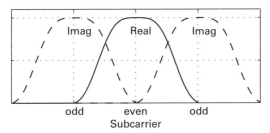

Figure 7.8   Overlapping subcarrier signals in frequency domain and complex modulation pattern for OQAM signaling.

For achieving maximum spectral efficiency in FBMC systems, the same subcarrier spacing as in an equivalent OFDM system (determined by the selected FFT window size) should be chosen. However, the spectral containment of the pulse will then let adjacent subcarrier signals overlap in frequency domain, as shown in Figure 7.8, which will result in inter-carrier interference if complex-valued signals are used on the adjacent subcarriers. This problem can be overcome by introducing a special modulation format, called Offset QAM (OQAM) signaling, where the orthogonality is constrained to the real-valued signal space (also referred to as "real-field orthogonality"). With OQAM, FBMC symbols carry only real-valued data on the subcarriers, and the subcarrier signals are modulated with a complex-valued pattern to make a single (real-valued) subcarrier signal be surrounded by signals carrying (real-valued) data in the orthogonal complex dimension. This is illustrated in Figure 7.8, where the subcarrier with even index carries data in the real signal space, while adjacent subcarriers with odd indices carry data in the imaginary signal space. Through this modulation scheme, it is ensured that interference caused from one subcarrier to its two adjacent subcarriers is orthogonal to the modulated signals on these subcarriers. To compensate for the loss in data rate due to reducing the signal space for the data to the real dimension, successive FBMC symbols are transmitted at double the symbol rate $2/T$. For the additional FBMC symbols transmitted in between the time slots defined by the grid $1/T$, the complex-valued pattern used for modulating the subcarrier signals is inverted, i.e. now subcarriers with even indices carry data in the imaginary signal space and those with odd indices carry data in the real signal space. Similar as for the inter-subcarrier interference, this approach again ensures that inter-symbol-interference always falls into the signal space which is orthogonal to the signals of interest. Thus, the real-field orthogonality of the data signals can be maintained.

Since orthogonality in OQAM exists only in the real field and no longer in the complex field as in OFDM, several schemes designed for OFDM cannot be directly transferred to FBMC, but require some redesign of selected signal processing procedures.

By designing the pulse shapes appropriately, FBMC can significantly increase the robustness against Doppler distortions as well as time and frequency synchronization impairments. Compared to OFDM, FBMC further offers a higher spectral efficiency,

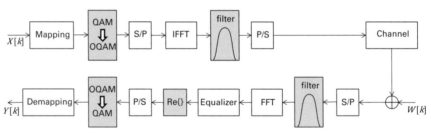

Figure 7.9   FBMC transceiver.

since it requires less guard bands at the band edges thanks to the good spectral containment of the pulse power, and it does not need any cyclic prefix. Compared to OFDM as used in LTE, the spectral efficiency improvement amounts to 13% (see TC6 evaluation in [9]).

The schematic of an FBMC transceiver is shown in Figure 7.9, where the signal blocks that differ from an equivalent OFDM transceiver are given in grey shaded color. As seen in the figure, FBMC requires additional filter banks. The overall framework of FBMC modulation/demodulation can be efficiently realized with Fast Fourier Transforms (FFT) and polyphase filtering [10]. Comparing the transceiver complexity of FBMC with OFDM-based solutions, latest research results have shown that the additional complexity required for implementing the subcarrier filtering is only moderate, amounting to a 30% increase at the transmitter and to a factor of two at the receiver [11].

FBMC has been extensively studied in the past, but its practical application as an enabling waveform for mobile radio has been less in the focus. However, recent research has highlighted the most important aspects of FBMC as an enabler for a flexible air interface design and has focused on solutions for practical challenges arising when applying FBMC as the waveform for the future mobile radio system. The most important findings and achievements are summarized in the following.

7.2.1.1   FBMC: An enabler for a flexible air interface design

Channel adaptive pulse shaping. To match the system configuration to given channel conditions, FBMC allows adapting the pulse shape or the subcarrier spacing of a subband assigned to a user accordingly. The achievable SIR gains measured after reconstructing the signal at the receiver after transmission over a doubly dispersive (2D) channel (considering both delay and Doppler spread) have been evaluated for the most prominent pulse shape candidates available for FBMC in the literature as well as for dynamic subcarrier spacing. Considered pulse shape candidates are the Phydyas pulse [12] and the Enhanced Gaussian Function (EGF) with variable power distribution, indicated by the α factor. Results are illustrated in Figure 7.10 for a 2D Rayleigh fading channel with a constant Doppler/delay spread product $f_D \tau_{rms} > 0$. The Doppler spectrum is modeled according to Jakes, while the delay is modeled as exponentially decaying. With the subcarrier spacing $1/T$, the x-axis shows the normalized delay spread τ/T (bottom) and the corresponding normalized Doppler spread $f_D T$ (top), respectively. Changing the subcarrier spacing implies moving along one curve in the direction of

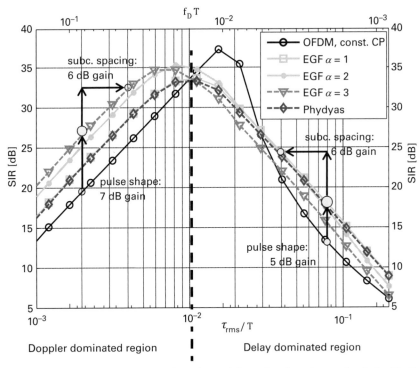

Figure 7.10 SIR gains obtained by pulse shape adaptation and dynamic subcarrier spacing in doubly dispersive (Doppler and delay spread) channels.

the x-axis, while changing the pulse shape implies switching between different curves in the direction of the y-axis. Following that, it can be clearly seen that, by adapting the pulse shape, up to 7 dB gain compared to OFDM can be achieved in the case where the Doppler distortions are the main performance degrading effect (Doppler dominated region: left area of the figure), while changing the subcarrier spacing by a factor of two provides a 6 dB SIR gain. These gains translate to a higher robustness of the transmit signals against distortions from Doppler and delay spread and correspondingly higher throughput. For proper isolation of the signal to adjacent sub-bands with a different configuration, a single subcarrier guard has proven to be sufficient. Details of the research can be found in [13][14].

Synchronization robustness. The choice of the prototype filter has a strong impact on the system's robustness against synchronization errors. Analysis in [15] revealed that filters with good frequency localization are more tolerant to time synchronization errors, while filters with good time localization are more tolerant to Carrier Frequency Offset (CFO) distortions. In that work, evaluation of the gains in terms of SIR after signal reconstruction at the receiver compared to OFDM could be shown to yield 4 up to 10 dB for a set of prototype filters selected from the literature. Moreover, considering the multi-user case, the good spectral containment of pulse shaped FBMC signals enables time asynchronous transmission of different users operating in adjacent frequency bands. Investigations have shown that an interference isolation of more than 60 dB can be achieved, if a single subcarrier is used as guard band between adjacent sub-bands [16]. A timing advance procedure to align multi-user signals, as known from LTE, is then no longer necessary.

Short prototype filters. As detailed above, a steep power roll-off in frequency domain is realized by prototype filters expanding the FFT block size in time domain, i.e. using overlapping factors $K > 1$. However, if a strict time localization of the signal as in OFDM is desired, prototype filters with optimized Time-Frequency Localization (TFL) can be derived also for the case of $K = 1$ (i.e. no symbol overlap), as done in [17]. The key idea is to multiply a single FFT block, obtained as the output of the IFFT operation in Figure 7.9, with a window exhibiting smooth edges, which realizes the subcarrier filtering. These prototype filters may provide means for an efficient transmission of very short messages (e.g. ACK/NACK) in mobile radio systems.

7.2.1.2   Solutions for practical challenges

Filter tails in short package transmission. For overlapping factor $K > 1$, the tails of the prototype filter let the FBMC symbol expand over the FFT block size, as shown for $K = 4$ in Figure 7.7 (bottom), where the prototype filter spans over four FFT block sizes (illustrated by the grid). These tails evoke a signaling overhead in transmissions of data bursts, as the length of a burst is extended by the filter tails which let the last FBMC symbol decay to zero. This overhead has been considered a drawback of FBMC in the context of short package transmission, as its amount relative to the burst length may become significant. Solutions to this problem have been provided by tail cutting methods [12], which, however, slightly degrade the PSD of the FBMC signal. A novel approach realizes the prototype filter by circular convolution of a block of succeeding FBMC symbols, yielding the filter tail to be wrapped into the symbol block. The output signal of the filter is then periodic, so that an additional windowing may be necessary to let the signal smoothly decay to zero to establish the desired time and frequency localization of the signal power. The overhead required for the windowing is in general much smaller than that for the original filter tail. Two variants of this approach have recently been proposed. In [18], the scheme is named Windowed Cyclic Prefix Circular OQAM (WCP-COQAM). Besides solving the tail issue, it also solves other key issues that have been identified for FBMC, namely its sensitivity to large channel delay spreads and its incompatibility with the classical MIMO Alamouti scheme. In [19], another scheme based on circular convolution is introduced under the name "weighted circularly convolved FBMC".

Channel estimation. Channel estimation schemes proposed for FBMC exhibit strong performance degradation in channels with long delay spread. To overcome this problem, a novel pilot design and estimation scheme has been recently developed, enabling improved estimates at the same pilot overhead as in OFDM-based systems with scattered pilots. The novel scheme reuses the idea of auxiliary pilots, introduced in [20], where a pre-calculated symbol is placed next to the channel estimation pilot to cancel the complex interference induced to that pilot from the surrounding data symbols. The key idea of the novel scheme is the use of two instead of one auxiliary pilot, which are placed opposed to each other in the temporal grid (i.e. one to the left and one to the right of the channel estimation pilot), and which are superimposed with the transmitted data. In delay spread channels, both auxiliary pilots suffer from phase shifts. However, due to their opposed placement, phase shifts are in opposite directions and hence compensate each other. It could be shown that with the novel estimation scheme, FBMC can achieve the same coded BER performance as OFDM in typical LTE scenarios [21].

MIMO. It is known in the literature that FBMC-MIMO achieves the same performance as OFDM-MIMO if linear MMSE equalization is used [22].

In multi-user systems, MIMO precoding requires the use of a guard carrier between frequency blocks assigned to different users to avoid inter-block interference caused by the discontinuities in the effective channel transfer function. This issue is attributed to the real-field orthogonality of FBMC and has been considered a major bottleneck so far. A novel approach, to avoid this inter-block interference without sacrificing a subcarrier as guard band, is to establish complex field orthogonality between the adjacent blocks, so that they become immune to the phase differences of the precoders [23]. Different methods have been proposed to implement this approach. One method is to use complex modulated symbols instead of real-valued ones for the boundary subcarrier and cancel the intra-block interference by precoder and receiver design (CQMB – CP-QAM modulated boundary). Another is to use a complex-valued prototype filter (instead of real-valued) on the boundary subcarrier [23].

Radio frequency (RF) imperfections. For multi-carrier transmission, the non-linear amplifiers used in practical systems pose a challenge, as they may cause distortions of the signal and, in particular, may negatively affect the PSD. However, measurements based on an off-the-shelf commercial LTE FDD radio-head unit at 1.8 GHz carrier frequency with output power of 45 dBm have revealed that the linearity of a current practical RF unit is sufficient to sustain the PSD advantages of an FBMC signal with up to 50 dB in-band attenuation [16]. Further results on the effect of RF imperfections in FBMC presented in that reference indicate that the sensitivity to imperfections such as I/Q imbalances, phase noise and power amplifier nonlinearities is close to that of OFDM, and their suppression with appropriate compensation methods results in a small to no performance gap with respect to the state-of-the-art.

7.2.2   Universal filtered OFDM

Universal filtered OFDM, also known as Universal Filtered Multi-Carrier (UFMC), is a modification of the well-known 4G waveform CP-OFDM. Instead of applying a CP, it applies a per sub-band filtering. By doing so, separation of single sub-bands in frequency domain is improved, allowing tuning each sub-band independently according to given link characteristics or service requirements. The PSD for an exemplary setting of UF-OFDM with 12 subcarriers per sub-band, filter length $L = 80$, sidelobe attenuation 60 dB and the corresponding time domain UF-OFDM symbols are illustrated in Figure 7.11. The top figure highlights the significantly improved spectral containment of the UF-OFDM signal compared to classical OFDM. The bottom figure shows the time localization of the UF-OFDM signal, being localized within an FFT window size (depicted by the

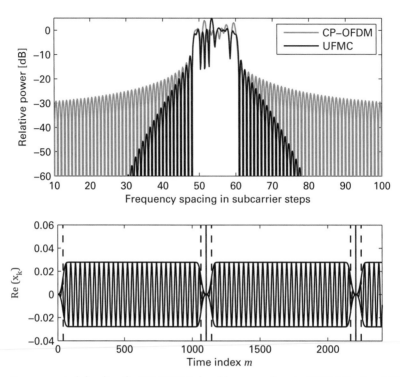

Figure 7.11   Power spectral density of a UF-OFDM signal in comparison to OFDM (top) and UF-OFDM signals in time domain (bottom).

dashed lines) and its filter tails decaying to zero within the duration equivalent to the CP in OFDM.

Figure 7.12 depicts the block diagram of an exemplary UF-OFDM transceiver, explaining the basic principle of the signal generation. Note that alternative transmitter structures performing filtering in frequency domain with significantly reduced complexity are presented in [24]. The time domain transmit vector S_u for a particular multi-carrier symbol of user u is the superposition of the sub-band-wise filtered components, with filter length L and FFT length N (for simplicity, the time index k is dropped):

$$S_u = \sum_{i=1}^{B} F_{iu} V_{iu} X_{iu}. \tag{7.10}$$

For each of the B sub-bands, indexed i, the n_i complex QAM symbols – gathered in X_{iu} – are transformed to time domain by the IDFT-matrix V_{iu}. V_{iu} includes the relevant columns of the inverse Fourier matrix according to the respective sub-band position within the overall available frequency range. F_{iu} is a Toeplitz matrix, composed of the filter impulse response, performing the linear convolution. So far, $n_i = 12$ is applied, leading to reasonable performance measures and being compatible with LTE

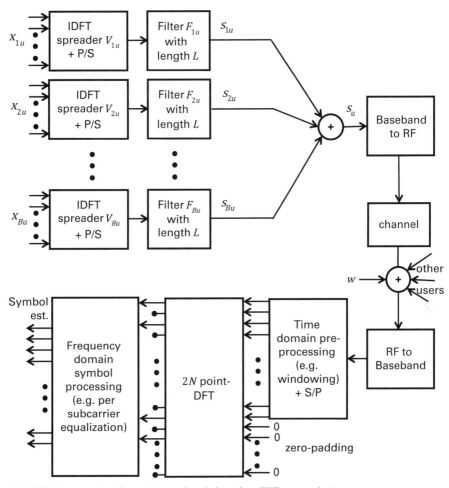

Figure 7.12   UF-OFDM transceiver (here: the receiver is based on FFT processing).

parameters. However, allowing for wider sub-bands (e.g. $n_i = 36$ while keeping the filter settings), even stronger separation between sub-bands is achieved. After superposition, the sum signal is up-converted and RF processed. The detector receives the noisy superposition of all users' transmissions. After conversion to baseband, the received signal vector may optionally be processed in time domain, e.g. by applying windowing, to suppress multi-user interference and thus improve the overall signal quality. After FFT conversion to frequency domain, any procedure known for CP-OFDM can be applied, such as those related to channel estimation and equalization.

A common first impression is to assume UF-OFDM to lose orthogonality between subcarriers in back-to-back mode. However, this holds only for specific receiver types following the matched filter principle. If an FFT-based receiver as presented above is applied, orthogonality between subcarriers is maintained (see proof in [16]).

The key feature of UF-OFDM is its backwards compatibility with OFDM, as the OFDM signal structure is completely maintained. This allows reusing all schemes designed for OFDM without any major modifications. Moreover, UF-OFDM provides the following features and advantages:

- Support of relaxed synchronism both with respect to timing and carrier frequency (originating from e.g. applying open-loop synchronization and cheap oscillators), enabling (1) energy and overhead efficient inclusion of MTC traffic without introducing severe interference [25], and (2) CoMP joint transmission and joint reception [26].
- Support of short burst transmissions without introducing excessive signal overhead thanks to the good time-localization of UF-OFDM (similar to CP-OFDM), while improving frequency-localization on sub-band level [27].
- Support of dynamic subcarrier spacing in the sub-bands, enabling matching the symbol period to the channel coherence time, reducing latency and improving PAPR. A detailed study can be found in [16].
- Enables fragmented spectrum access in a more spectrally efficient way than CP-OFDM since less frequency guards are required [27].

As UF-OFDM in its version being discussed here does not make use of a CP, orthogonality is lost when the signal is transmitted in long delay spread channels. However, with reasonable settings (e.g. 15 kHz spacing) and for relevant channel profiles (e.g. eVEHA), the arising inter-symbol and inter-carrier interference is low enough to be negligible, e.g. against noise for reasonable SNR working points [28]. With channels having an extreme delay spread characteristic, the use of zero-tail DFT spreading is advocated. Alternatively, parameterization of the waveform may be adjusted accordingly.

For an overall system design based on UF-OFDM, including frame structure, synchronization mechanisms, waveform design options and multiple-access schemes, the reader is referred to [29].

7.3   Non-orthogonal schemes for efficient multiple access

Novel multiple-access schemes allow for overloading the spectrum by multiplexing users in the power and the code domain, resulting in non-orthogonal access, where the number of simultaneously served users is no longer bound to the number of orthogonal resources. This approach enables the number of connected devices to be increased by a factor of 2–3 and, at the same time, to obtain gains in user and system throughput of up to 50%. Candidate schemes are Non-Orthogonal Multiple Access (NOMA), Sparse Code Multiple Access (SCMA) and Interleave Division Multiple Access (IDMA). All schemes can be well combined with open- and closed-loop MIMO schemes, so that the MIMO spatial diversity gains could be achieved. If applied in the context of massive MTC, SCMA and IDMA can further reduce the signaling overhead through grant-free access procedures.

As discussed in Section 7.1.3, the information theoretic capacity of the uplink multiple-access channel can in fact only be achieved if transmissions from multiple users occur on the same resource in time and frequency and successive interference cancelation or joint detection is applied at the receiver side. For the downlink broadcast channel, it was shown that superimposed transmission to multiple users with a non-linear precoding strategy is capacity-achieving, but this is typically considered prohibitive in terms of complexity and the need for very accurate channel state information at the transmitter side. In the case where the path losses to multiple users in the downlink are very different, it can be beneficial to use superimposed transmissions in conjunction with successive interference cancelation at the receiver side. All schemes discussed in this section aim at exploiting these potential benefits of non-orthogonal access in both the uplink and downlink of practical mobile communications systems.

More precisely, NOMA directly applies superposition coding with SIC at the receiver but combined with frequency-selective scheduling, relying to some extent on Channel State Information at the Transmitter (CSIT). An alternative option, named SCMA, allows for a flexible usage for systems with different levels of CSIT. The third option called IDMA aims at improving the spreading codes of CDMA to provide further coding gain and apply a receiver of reasonable complexity such as an iterative receiver. This leads to a scheme that can work with or without optimal power and rate allocation and is therefore suited for systems without CSIT. There are also other ways to improve multiple-access schemes, but these three options will be explained in more detail in the sequel.

7.3.1   Non-orthogonal multiple access (NOMA)

The main idea of NOMA [30] is to exploit the power domain in order to multiplex multiple users on the same resources and rely on advanced receivers such as SIC to separate multiplexed users at the receiver side. It is basically the direct implementation of the theoretic idea of superposition coding with SIC, as introduced in Section 7.1.3, combined with advanced scheduling. NOMA assumes OFDM as a basic waveform and can be applied, e.g., on top of LTE, but the basic idea is not limited to a specific waveform. It should be noted that the required extent of CSIT is rather limited, i.e. some knowledge about the SNR is sufficient to do proper rate and power allocation, and no detailed or short-term CSIT as for MIMO precoding is needed. This makes NOMA suitable even for open loop transmission modes such as open loop MIMO under e.g. high user speeds. The main target scenario is classical broadband traffic both in UL and DL, but NOMA can work in any scenario providing some low-rate SNR feedback. In order to illustrate the basic idea, the focus in this section will be on the DL.

Figure 7.13 illustrates the basic principle for DL NOMA on the transmitter and receiver side, respectively, and in Figure 7.14 different achievable capacity regions for two users are shown. Due to the superposition and SIC, the achievable rates R_1 and R_2

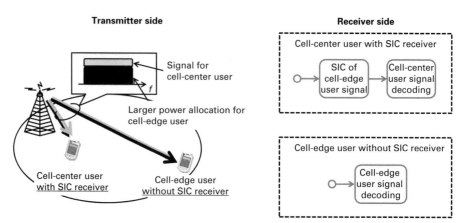

Figure 7.13   Transmitter and receiver side of NOMA in downlink.

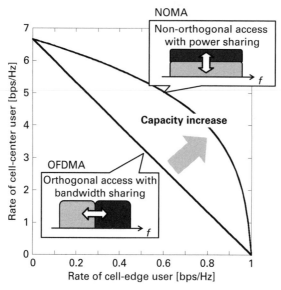

Figure 7.14   Capacity region of NOMA applied in the downlink.

for a two user system with transmit powers P_1, P_2 and channel coefficients h_1, h_2 assuming that $|h_1|^2/N_{0,1} > |h_2|^2/N_{0,2}$ become [31]

$$R_1 = \log_2\left(1 + \frac{P_1|h_1|^2}{N_{0,1}}\right), \quad R_2 = \log_2\left(1 + \frac{P_2|h_2|^2}{P_1|h_2|^2 + N_{0,2}}\right). \tag{7.11}$$

It can be seen that the power allocation for each UE greatly affects the user throughput performance and thus the Modulation and Coding Scheme (MCS) used for data transmission of each UE. By adjusting the power allocation ratio, P_1/P_2, the Base Station (BS) can flexibly control the throughput of each UE.

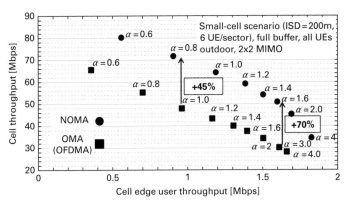

Figure 7.15   Increased cell and cell edge throughput with NOMA compared to OFDMA.

In a practical system with several users in a cell and each having different channel conditions, one challenge is to group users that should be superimposed on a certain time-frequency resource in order to maximize the throughput. When the number of users and possible resources increase, optimal scheduling including grouping of users will become infeasible, so that suboptimal schemes are needed.

In [32], efficient algorithms to do this grouping in a practical setup are presented, and in Figure 7.15 the achieved gain compared with OFDMA can be observed in terms of cell and cell edge throughput. NOMA can improve the cell throughput and the cell-edge throughput at the same time, and the priority of these two measures can be controlled by a fairness parameter α used in the scheduling. As an example, for a target cell edge throughput of ~1 Mbps the overall cell throughput can be increased by 45%, or for a given cell throughput target of ~55 Mbps the cell edge throughput can be roughly doubled.

As MIMO is and will remain a main enabler for high throughput also in future wireless communications generations, the compatibility of new proposals with advanced MIMO schemes is essential. An in-depth analysis of the applicability of the most relevant LTE features has been conducted, showing that NOMA can be combined with single user MIMO (SU-MIMO) [33][34] and even MU-MIMO easily. Further results can be found in [9] and [16].

7.3.2   Sparse code multiple access (SCMA)

Sparse code multiple access is a non-orthogonal code and power domain multiplexing scheme in which data streams or users are multiplexed over the same time-frequency resources in either downlink (DL) or uplink (UL). The channel encoded bits are mapped to sparse multi-dimensional codewords and the signals of different users consisting of one or several so-called layers are superimposed and carried over an OFDMA waveform as illustrated in Figure 7.16. Therefore, the layers or users are overlaid in code and power domain and the system is overloaded if the number of layers is higher than the codeword length. Compared to CDMA, SCMA applies more advanced spreading sequences, providing higher coding gain and additional shaping gain by optimization of the code-book [35]. To achieve a high throughput gain, advanced receivers achieving near-optimal

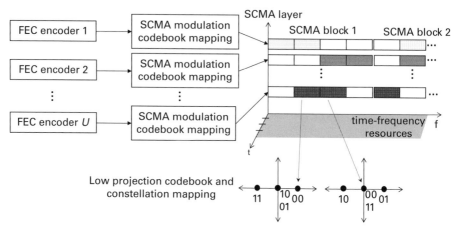

Figure 7.16   Illustration of an SCMA transmitter.

detection are needed, for instance based on a Message Passing Algorithm (MPA), which is usually assumed to be too complex to be implemented. However, thanks to the sparsity of the codewords, the complexity of a MPA can be significantly reduced, for instance inspired by Low Density Parity Check (LDPC) decoders working on sparse graphs [36]. In case CSIT is available, the power and rate can be adapted properly and thus the multiple-access channel capacity region can be achieved as well. Similar to NOMA, the requirements for the CSIT are rather relaxed; instead of full channel knowledge, only the channel quality is needed in order to support SCMA superposition, enabling the application to open loop MIMO as well. Due to the combination of power domain superposition as in NOMA and code domain superposition as in CDMA, SCMA can be applied quite flexibly for scheduled DL [37] as well as unscheduled UL [38][39]. However, SCMA requires a more complex receiver compared to SIC.

In Figure 7.17, simulation results for the scenario of a massive deployment of sensors and actuators are shown for the possible application scenarios of SCMA. For the case of non-delay-sensitive applications, allowing up to three retransmissions for failed packets, an LTE baseline and SCMA are compared. For SCMA, the gains are due to contention-based transmission without LTE's dynamic request and grant procedure together with the SCMA overloading of the physical resources. Depending on the packet size, the gain ranges from around 2x for 125 bytes payload to 10x for 20 bytes payload [9][39]. The LTE baseline is an LTE Release 8 system with 4x2 MIMO assuming 20 MHz bandwidth and operating at 2.6 GHz.

In a macro-cell MIMO setup with full buffer traffic and assuming different user speeds from 3 km/h to 50 km/h, the relative cell average throughput and cell-edge throughput gains of Multi-User SCMA (MU-SCMA) over OFDMA were found to be in the range of 23%–39% for the Spatial Multiplexing (SM) mode and 48%–72% for the transmit diversity mode applying Alamouti's code. These results confirm the capability of MU-SCMA to provide high throughput and high quality of user experience which is independent of the users' mobility status and their speeds. Further results can be found in [9] and [16].

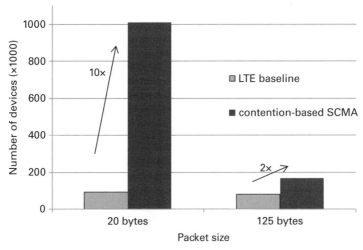

Figure 7.17   Massive MTC capacity in terms of the number of devices per MHz at 1% packet failure rate.

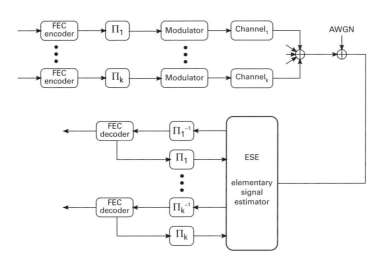

Figure 7.18   Block diagram of an IDMA system.

7.3.3   Interleave division multiple access (IDMA)

IDMA [40] aims at improving the performance of Code Division Multiple Access (CDMA) systems in asynchronous communications [41]. A turbo-type multiuser detector is proposed which includes the simplest receiver, denoted as Elementary Signal Estimator (ESE) or soft rake detector, consisting of a soft demodulator delivering an equivalent performance as the much more complex linear receiver for asynchronous users. A basic block diagram is shown in Figure 7.18.

Similar to CDMA, IDMA applies some kind of spreading to the signal by applying a low-rate channel code, depicted in the figure as "FEC encoder". The main difference with CDMA is that the channel code may not contain a repetition code and may be the same for all users, while the distinction of the users is enabled by different interleavers Π_k, which are anyway usually part of the system to decouple coding and modulation. The spreading can be done either in time or in frequency domain, where frequency domain may be preferred if a multicarrier waveform or a combination with FDMA or OFDMA is assumed.

IDMA is especially suited for the UL including unscheduled communication, as it is robust to asynchronicity and suboptimal rate and power allocation due to the use of iterative receivers instead of SIC. Due to the spreading in frequency domain, frequency diversity can be exploited even without frequency-selective scheduling. As a special case, IDMA can be similar to NOMA if appropriate rate and power allocation is applied. In this case, the iterative receiver degrades to a SIC receiver.

Due to the expected growth of machine-type communication including sensors with very short messages, ensuring fully synchronized users and feeding back CSIT to the BS may not be efficient any more. A solution for mixed traffic of synchronized and adaptive transmission for long packets, and users with relaxed synchronization and without CSIT for short packets, could provide a good compromise. Such a coexistence scenario can be efficiently supported by IDMA due to the robustness to asynchronicity [42]. The results shown here consider a coexistence scenario where the users with relaxed synchronization are superimposed in a certain frequency resource whereas other frequency resources are used for synchronous users at the same time. FDMA is assumed as a baseline where the uplink users are separated in frequency domain. The achieved rate is the same for both schemes. The results shown in Figure 7.19 illustrate the superior performance of IDMA in terms of coded bit error rate compared to FDMA. This confirms the suitability of IDMA for scenarios of relaxed synchronization and suboptimal rate and power allocation, which will play an essential role in 5G system design [43].

7.4   Radio access for dense deployments

Small-cell deployments can be foreseen as a possible solution for realizing ubiquitous 5G extreme mobile broadband (xMBB) with extreme data rate demands of several gigabits per second. Densification enables energy-efficient high data rate transmission due to short and often line-of-sight radio links, lower output powers and access to new spectrum. High-gain beamforming with a large number of antenna elements provides additional energy efficiency and compensates for higher path loss at higher frequencies. Simultaneously, interference from other links using the same physical resources is reduced.

The bursty and variable traffic of small cells leads to the need for efficient spectrum usage and resource management. Time Division Duplexing (TDD) possesses the capability of flexibly and dynamically allocating the available bandwidth to any link direction. In addition, TDD has lower radio component cost, it does not require duplex filters, the amount of available bandwidth is larger compared to FDD, and it enables utilization of

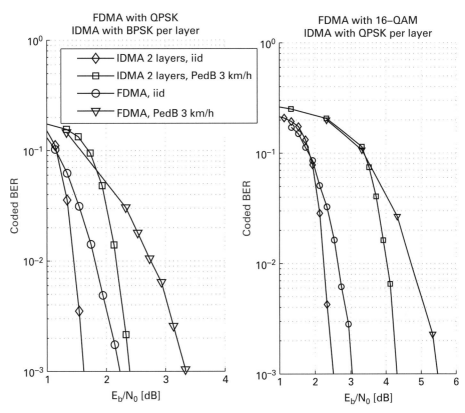

Figure 7.19   Performance comparison of asynchronous FDMA and IDMA for independent and identically distributed (i.i.d.) and Pedestrian B (PedB) channel model, relative delay 0.45, code rate 1/4 with different modulation schemes.

channel reciprocity. For these reasons, TDD is seen as a more attractive duplexing method over FDD for 5G xMBB.

The large increase of 5G data throughput leads to the need of transmitting and processing larger amounts of data, consequently imposing demands on baseband processing. It is possible for a baseband system to cope with the increased throughput demand by decreasing latency. Thus, beside the achievement of high data rates, also latency reductions on the air interface level become vital in order to enable for example long battery lifetime. Low air-interface latency is especially important in scheduled TDD systems where several TDD cycles are required for delivering one control or data round trip transmission. On air interface level, the latency requirements lead to the demand of fast link direction switching and short Transmission Time Interval (TTI) lengths, further leading to the transmission of shorter blocks of data in time and wider blocks in frequency.

This section describes a concept of an air interface targeting xMBB services in dense networks, characterized by high data rate and reduced latency. A harmonized OFDM concept allows using a unified base band design for a broad range of carrier frequencies going up to centimeter (cmW) and millimeter Wave (mmW) bands, enabling a TDD frame structure that is scalable over the range of operating frequencies while facilitating energy efficient solutions and low deployment and hardware cost.

7.4.1   OFDM numerology for small-cell deployments

7.4.1.1   Harmonized OFDM and scalable numerology

In order to fulfill the need for an increased amount of spectrum, the usage of frequencies higher than those currently used for 4G mobile broadband is required. To simplify terminal complexity, the availability of large contiguous spectrum bands is preferable. Alternatively, the user may be required to aggregate multiple frequency bands having differing characteristics. Further, common spectrum usage for different link types, such as access and self-backhauling links, would enable low deployment and HW cost, deployment flexibility, and efficient spectrum usage. Thus, the air interface design needs to be harmonized so that the same framework can be used for different frequency bands and bandwidths together with different link types. To realize this for xMBB, it is therefore proposed to follow the idea of a scalable radio numerology as illustrated in Chapter 6 (Figure 6.10). In other words, when the carrier frequency is increased from conventional cellular spectrum toward cmW and mmW range, the used bandwidth and subcarrier spacing are simultaneously increased while the FFT size is kept within a small set of quantized values. Similar scaling is done in time domain numerology, meaning that for example the CP length, may further be adjusted according to the carrier frequency.

7.4.1.2   OFDM time numerology

Existing 4G standard technologies, such as LTE-A, are not designed for small-cell environments and have legacy restrictions when utilizing evolved component technology. The 5G small-cell environment properties together with evolved component technology aspects, such as reduced TDD switching times and enhanced digital signal processing performance, enable the usage of shorter and 5G-optimized time numerology [16]. In this chapter, the focus is on examining the small-cell numerology especially for cmW frequencies, whereas the mmW-related numerology was already presented in Chapter 6.

A Guard Period (GP) is allocated to the transmission direction switching point in order to obtain a sufficient off power of the transmitter and in order to compensate the cell-size dependent round trip delay. In fact, the time required for the GP in a small-cell operating in a cmW TDD system can be smaller than 0.6 μs (for a cell size of around 50 m) [16].

Since precise timing control procedures, such as timing alignment, are not used due to the limited propagation delay in such small cell scenarios, and since the UE is synchronized to the DL signal, the time uncertainty that needs to be compensated within the CP can be estimated as a two-way maximum propagation delay. The required total 5G CP time can therefore be estimated to be around 1.0 μs. In [16][44], the CP length analysis was further extended by investigating the OFDM spectral efficiency performance for different channel models, such as indoor hotspot and outdoor urban micro channels. It was also concluded here that a CP length of 1 μs seems to be enough for overcoming the channel delay spread in channels varying from indoor hotspots to at least outdoor micro cells in the cmW frequency range. This introduces less than 6% overhead for a 60 kHz SC spacing. Timing alignment has to be used in larger cells to compensate the propagation delay.

7.4.1.3   OFDM frequency numerology

In the physical layer design for dense deployments especially at higher frequency bands, beamforming with a large number of antenna elements is a basic component, which leads to the consideration of in-chip antennas and integrated radio frequency hardware. With such solutions, Phase Noise (PN) will set a cap on the maximum received Signal-to-Noise Ratio (SNR), if currently available technologies are used. This may be improved either by technology progress over time, by the use of a large subcarrier spacing leading to large CP overhead, or by PN estimation and compensation at the receiver. The compensation can be categorized into Common Phase Error (CPE) compensation and full phase-noise compensation. CPE compensation only needs a single phase error estimate for each OFDM symbol in order to de-rotate the entire symbol using that single estimate. Full phase-noise compensation, on the other hand, requires tracking of the PN during each OFDM symbol. For effective full phase-noise compensation, a special phase-noise pilot signal that is present over a certain part of the system bandwidth may be required in every OFDM symbol. The receiver can use this known signal to estimate the PN and then compensate for it. The PN at both the transmitter side and the receiver side can be compensated.

Figure 7.20 illustrates the link performance with and without PN compensation on mmW and with channel measures [45]–[47], including separate curves for both streams in a 2 stream-MIMO simulation. The analysis shows that, with good oscillators with a Figure of Merit (FOM) value of −180 dBc/Hz, the performance loss due to PN is negligible for 16-QAM and 2-stream MIMO even with relatively low subcarrier spacing values, such as 360 kHz. For 64-QAM, the loss is larger, but the performance can be improved with simple PN compensation by using some pilots to estimate the phase errors. The same PN performance improvement gained via PN compensation holds also with lower cost oscillators and more challenging transmission schemes, at least under the assumption of ideal channel estimation. Based on the conducted preliminary analysis on the OFDM numerology, it can be concluded that the estimated subcarrier spacing values even for mmW frequency area are still reasonably low, and, together with a much shorter required CP length, the overhead can be kept in feasible limits. The CPE and different RF technologies can be tackled at the expense of a small increase in the overhead, by increasing slightly the SC spacing, and adding few more pilots for inter-carrier interference estimation. For further details, the reader is referred to [16].

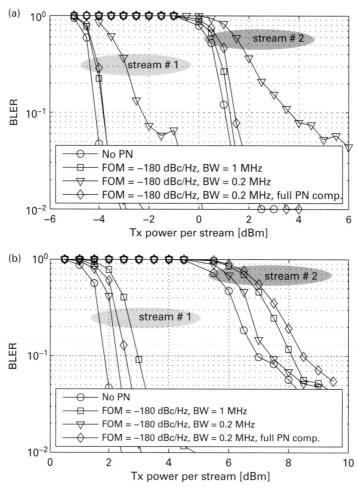

Figure 7.20   PN BLER performance of OFDM NLOS for the two streams of a MIMO simulation (left and right curve set): (a) 16QAM, (b) 64QAM.

7.4.2   Small-cell sub-frame structure

7.4.2.1   Main design principles for small-cell optimized sub-frame structure

The dense deployment optimized TDD numerology introduced earlier enables the usage of a shorter frame length and the design of a small-cell optimized physical frame structure with fast and fully flexible switching between transmission and reception (network-level performance evaluation can be found in Chapter 11). A TDD optimized physical sub-frame structure for a small-cell system is illustrated in Figure 7.21. Here, a bi-directional control part is embedded to each sub-frame. This allows the devices in the network to both receive and send control (ctrl) signals, such as scheduling requests and scheduling grants, in every sub-frame. In addition to the scheduling related control information, the control part may also contain Reference Signals (RS) and

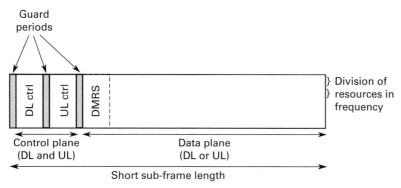

Figure 7.21 Small-cell optimized sub-frame structure.

synchronization signals. The data part in one sub-frame contains data symbols for either transmission or reception in order to realize simple implementation. Demodulation Reference Signal (DMRS) symbols, used to estimate the channel and interference-covariance matrix, are located e.g. in the first OFDM symbol in the dynamic data part and can be pre-coded with the same vector/matrix as data. Short sub-frame lengths, such as for example 0.25 ms on cmW frequencies when assuming 60 kHz SC spacing, have been analyzed to be feasible from the guard and control overhead point of view [16]. By following the principles of a harmonized OFDM concept, the frame numerology is further scaled when moving to mmW, leading to an even shorter frame length, e.g. in the order of 50 µs. In frequency direction, the spectrum can be divided to separate allocable frequency resources.

7.4.2.2   Control part design principles

Multi-cell transmission and massive MIMO, or a combination of these, are essential enablers to reach the 5G capacity targets and, therefore, also need to be supported by control signal and RS structures. Thus, in addition to broadcast/PDCCH[3] type of control, support is also needed for EPDCCH[4] type of control signaling with device specific precoding and support for frequency domain scheduling. Both of these control types require own RS signals. As illustrated in Figure 7.21, control symbols (including both PDCCH and EPDCCH type of control) are located before the data symbols. This is to allow fast and cost-efficient pipeline processing in the receiver and to allow devices to skip the rest of the sub-frame (DRX) if no data is scheduled for that user. Consequently, processing and energy dissipation reductions can be achieved with respect to LTE-A, where EPDCCH is frequency division multiplexed with data and the design is thus significantly breaking the processing pipeline.

---

[3] PDCCH in 3GPP Release 8 uses Cell-specific Reference Signals (CRSs) for demodulation, not allowing e.g. UE specific beamforming.

[4] EPDCCH was introduced in 3GPP Release 11 in order, e.g., to support increased control channel capacity and frequency domain ICIC, to achieve improved spatial reuse of control channel resources, and to support beamforming and/or diversity. Demodulation is based on user-specific DMRS (blind decoding is applied).

7.4.2.3 Sub-frame structure properties and achieved gains

The 5G small-cell optimized physical TDD sub-frame structure enables multiple properties providing gain with respect to the existing 4G technologies. These properties and the provided flexibility fit well not only for access links but also for other communication types in small-cell environments, such as backhaul, D2D and machine-type links.

The possibility to allocate the data part of a sub-frame either to DL or UL direction enables fully flexible UL/DL ratio switching for data transmission. The decreased guard overheads, such as reduced GP and CP, together with the flexibility in UL/DL ratio improve the maximum achievable link spectral efficiency per link direction compared to TDD LTE-A. The simplified bi-directional control plane embedded in each sub-frame enables control signaling that is independent of the allocation of the data part to UL or DL. Moreover, clean TDD Hybrid Automatic Repeat reQuest (HARQ) schemes can be realized with a fixed HARQ timing that does not depend on the UL/DL ratio. Consequently, the proposed scheme reduces TDD HARQ complexity and decreases the related HARQ latency with respect to TDD LTE-A. By using the proposed 5G sub-frame structure, a total HARQ Round Trip Time (RTT) of less or equal to 1 ms can be realized [16]. The reduced HARQ latency leads to the usage of fewer HARQ processes, further reducing the number of needed receiver HARQ buffers and thus leading to lower memory consumption and to lower device cost.

Further, reduced user plane latency can be achieved due to short TTI length, reduced HARQ RTT and shorter BS and UE processing times enabled by pipeline processing. User plane latency [48] is here defined as the one-way transmit time, i.e. the time a Service Data Unit (SDU) packet takes to be sent from the IP layer of the transmitter to the IP layer of the receiver. A characterization of the user plane latency for the presented 5G small-cell concept and the comparison to LTE-A [49] are given in Table 7.2 for cmW. It can be concluded that approximately 5 times user plane latency reduction can be gained with respect to LTE-A.

Table 7.2 User plane latency for TDD with 10% BLER.

	DL		UL		
Delay component	5G small cell (cmW)	LTE-A TDD	5G small cell (cmW)	LTE-A TDD	LTE-A FDD
BS processing	0.25 ms	1 ms	0.25 ms	1 ms	1.5 ms
Frame alignment	0.125 ms	0.6–1.7 ms	0.125 ms	1.1–5 ms*	
TTI duration	0.25 ms	1 ms	0.25 ms	1 ms	1 ms
UE processing	0.375 ms	1.5 ms	0.375 ms	1.5 ms	1.5 ms
HARQ re-transmission (10% × HARQ RTT)	0.1 ms	0.98–1.24 ms	0.1 ms	1.0–1.16 ms*	0.8 ms
Total delay	~1 ms	~5–6 ms*	~1 ms	6–10 ms*	~5 ms

* depends on the TDD configuration

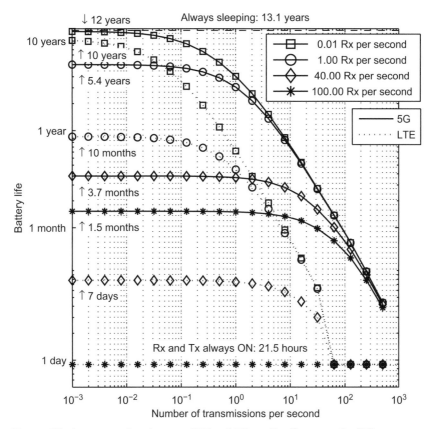

Figure 7.22   Battery life time comparison between LTE and 5G small-cell concept for different reception (Rx) opportunities.

It is possible to enable very fast network synchronization by the network devices, thanks to the frame structure that allows embedding a synchronization signal in the control part of each sub-frame.

By utilizing the fast network synchronization and reduced control and data plane latency, it is possible to enable quick transitions between devices' sleep and active modes, further reducing total energy consumption [44]. A simplified theoretical analysis about battery life time is shown in Figure 7.22 [50]. With the proposed sub-frame structure, it is possible to achieve considerably lower energy consumption compared to LTE-A.

Furthermore, the proposed sub-frame structure allows (UL/DL) symmetric DMRS design. DMRS can be embedded, e.g., to the first symbols in the data part. Since the DMRS symbols between link directions are fully aligned, it is possible to design cross-link orthogonal reference signals, further enabling the usage of advanced receivers [16].

7.4.2.4   Self-backhauling and multi-antenna aspects

In order to support increased user data rates, it may be desirable that the dense BSs are provided with a high capacity backhaul, linking BSs wirelessly over other BSs to one or several aggregation nodes. Multi-hop relaying between UEs/BSs may be required in order to provide improved coverage for the highest bit rates. Furthermore, multi-hop in a mesh of access points will provide robustness against the failure of individual nodes. Relaying aspects are covered more closely in Chapter 10.

For small cells to be deployed in high frequency bands, a considerable number of antenna elements can be implemented even in nodes with small physical dimensions due to smaller radio wavelengths. While MIMO with spatial multiplexing or beam-forming is still expected to be feasible for the lower frequencies, dense deployment networks at mmW frequencies are expected to primarily rely on beamforming to meet the link budget and to reach a certain desired throughput. Using polarization diversity, two layers can be multiplexed even for LOS links, and the minimum assumption for one UE is to receive two spatial layers. Using differently reflected links, higher layer spatial multiplexing can be implemented. In addition, high gain beamforming is an effective mean for the spatial isolation of links to obtain frequency reuse one with simple and low-cost receivers. Simulations performed for a small open office envir-onment with 9 Tx-Rx links in [16] show that a channel capacity of 14–22 Gbps/link (in raw bits) can be reached with 2 GHz bandwidth at 60 GHz carrier frequency, using 64 Tx antenna elements and 16 Rx antenna elements and one layer with Maximal-Ratio Combining (MRC) receivers. Massive antenna solutions are covered in more detail in Chapter 8.

7.5   Radio access for V2X communication

This section will now focus on particular radio-access considerations for Vehicle-to-Anything (V2X) communications. To provide reliable V2X communication, improve-ments of the technology components that are used for V2X are needed in various layers of the OSI stack. In the following, the focus is on the MAC layer, or more precisely, on ad-hoc MAC for moving networks.

It should be noted that access to ultra-reliable services can be regulated by the Ultra Reliable Communication (URC) framework. In part, this is done by introducing an availability indicator that informs the application whether the current communication link is reliable or not [51]–[53]. More details on the topic of URC can be found in Chapter 4.

Further, it is worth mentioning that other important technology components will help to achieve reliability for V2X. For instance, channel prediction and channel estimation in V2V scenarios provide significant performance improvement compared to state of the art solutions [54][55].

7.5.1   Medium access control for nodes on the move

V2X communication can be over D2D links or over a traditional uplink-downlink arrangement. Certain services, such as traffic safety and traffic efficiency, rely on V2V communication between vehicles that are close to each other. For these services and similar V2X services, it is therefore attractive to use D2D links. Moreover, the safety-critical nature of the services requires the links to be operational also when there is limited or no connectivity to the fixed infrastructure (i.e. fixed base stations). It is therefore desirable to have a flexible MAC scheme that can cope with varying degrees of network connectivity. At one extreme, there is no network connectivity and an ad-hoc MAC scheme is the only alternative, i.e. a scheme with no pre-assigned central entity that controls the medium access. Since the network topologies are quite dynamic in vehicular networks, it is attractive to use schemes that require a minimum of coordination. The current state-of-the-art systems for ad-hoc V2V communication are based on the IEEE 802.11 MAC, i.e. Carrier-Sense Multiple Access (CSMA). Unfortunately, CSMA does not scale well, i.e. the performance degrades quite drastically as the channel load increases. A scheme that scales better is Coded Slotted Aloha (CSA). In the following, it is shown how CSA can be adapted to provide latency guarantees (which are not possible with standard CSA). Furthermore, the broadcast-to-broadcast scenario is studied, i.e. when all vehicles in the vicinity of each other would like to broadcast packets to each other. Since a slotted scheme is used, it is assumed that the nodes (vehicles) are able to establish slot and frame synchronization with sufficient accuracy. The frame duration is adapted to the latency requirement. That is, the frame duration is equal to the MAC layer latency budget. The slot duration is long enough to carry a complete packet. The scheme is similar to the one described in Section 7.6.3, with the difference that this scheme is specifically designed to support broadcast-to-broadcast communication.

Just as in regular CSA, the nodes transmit several packets with identical payloads during a frame. The number of packets a certain node transmits in a frame is randomly selected at the beginning of each frame according to the node degree distribution $\Lambda(x) = \sum_{k=1}^{N} p_k x^k$, where p_k is the probability of transmitting in k slots and N is the number of slots in a frame. The transmission slots are selected randomly and uniformly over the frame. Hence, a slot can contain zero, one, or more than one packet. Further, it is assumed that the receiver can decode a packet if and only if there is no collision in the slot. Such a slot is called a singleton slot. An example of the process is found in Figure 7.23. Here, the five users transmit either 2 or 3 times in the 8-slot frame. As seen, there is a singleton slot starting at time $3T$ and there are two empty slots.

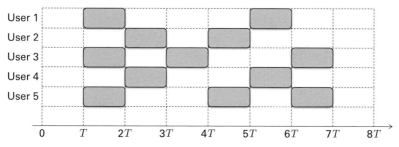

Figure 7.23   Example of CSA randomized slot assignment.

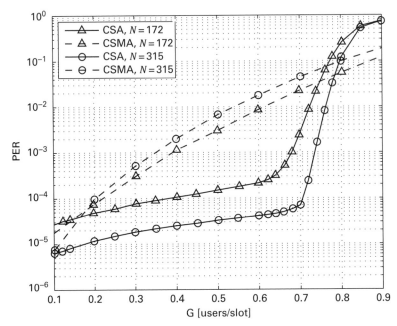

Figure 7.24   Packet error rate versus channel load for CSA with N slots compared to CSMA.

In spite of the significant number of collisions, a receiver can recover all transmitted packets in this example by following the procedure below.

A decoded packet contains pointers to the slots that contain the packet copies. The receiver cancels the packets from these slots, which might reveal new singleton (i.e. decodable) slots. This decode-and-cancel process continues until no more new singleton slots are revealed. The overall packet error rate is, therefore, $P_e = 1 - K_{dec}/K$, where K_{dec} is the number of decoded users and K is the total number of users [56].

The simulation results in Figure 7.24 are for a system with $NT = 100$ ms long frames, where the slot duration T is chosen such that a complete packet fits into one slot. Two packet lengths are considered: 200 bytes and 400 bytes, implying $N = 315$ and $N = 172$, respectively, using the 802.11p protocol overhead and a data rate of 6 Mbps, which is the default rate for 802.11p for traffic safety applications. The node distribution is given by $\Lambda(x) = 0.86x^3 + 0.14x^8$ for $N = 172$ and $\Lambda(x) = 0.867 + 0.143$ for $N = 315$. The channel load is defined as $G = K/N$. As seen in Figure 7.24, CSA can support 125% more users than CSMA at a packet error rate of 10^{-3} for $N = 315$, while the gain for $N = 172$ is about 75% [57].

7.6   Radio access for massive machine-type communication

To address the challenges and characteristics of massive MTC (mMTC) discussed in Chapter 4, novel radio-access technologies are required. The non-coherent multiple-access schemes discussed in Section 7.3 provide a basis for mMTC, but especially the *massive* access problem requires novel MAC protocols and PHY layer algorithms beyond the mere access technology to handle a huge number of users. Therefore, this section focuses on MAC and PHY schemes for contention or reservation based access of a large number of devices. As an extension to the LTE RACH, *coded access reservation* is discussed for large packet mMTC. To reduce signaling overhead for small packet mMTC, scheduled access is avoided and replaced by the alternative access scheme *coded random access*. This alternative access scheme makes use of repeated transmissions following a code pattern to enable resolving collisions. A crucial component of this novel MAC scheme is a suitable physical layer processing at the receiver to resolve potential collisions already on PHY layer. To this end, *Compressed Sensing based Multi-User Detection* (CS-MUD) is discussed as it provides natural synergies with the MAC protocol to achieve a powerful technique for efficient access of a large number of mMTC devices. Evaluation shows an increase in the number of supported devices by a factor of up to 10 compared to LTE RACH.

7.6.1 The massive access problem

The core problem for 5G addressed in this section is medium access control for a massive number of devices. In the following, the limitations of the LTE Random Access Channel (RACH) are analyzed to show the need for a new MAC solution. Thereafter, the MAC signaling overhead in comparison to mMTC payload sizes is considered. Afterwards, the requirements for a 5G MAC solution are shortly outlined. Note that knowledge about the LTE (P)RACH definition and procedures is assumed here, for detailed information please refer to [57]–[60].

7.6.1.1 LTE / LTE-A RACH limitations

To gain access to an LTE base station, every user has to first access the system through the random-access procedure to make himself known the network and obtain resources in which to carry out his communication. In contrast to UMTS, the LTE RACH does not allow for direct data transmission such that even the smallest messages need to be scheduled first. The basic outline of the LTE RACH procedure is well known and shortly summarized in Figure 7.25. The initial message *msg1* is a randomly drawn Zadoff-Chu

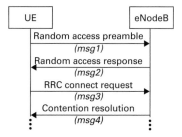

Figure 7.25 Outline of the LTE RACH procedure.

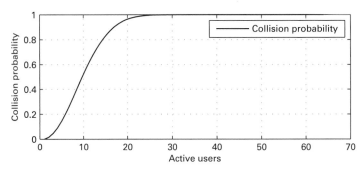

Figure 7.26   Probability of RRC connect request (*msg3*) collisions vs. the number of concurrently active users in one LTE PRACH slot using 64 Zadoff-Chu sequences.

sequence which acts as an ID and robust physical layer scheme at the same time. If the base station (also known as eNodeB) successfully identifies this sequence, it acknowledges its ID and provides some resources for further communication with the user. The second user message *msg3* is then sent using Physical Resource Blocks (PRBs) provided for the ID (not the user). Only after the RRC connect request the user is known and henceforth uses the shared channels.

The main advantage of this approach is the high robustness of the Zadoff-Chu sequences against noise and the accommodation of asynchronous reception through propagation delays and timing offsets in the PRACH design. Once a connection is set up, the base station can fully control access through scheduling. However, if multiple users try to gain access at the same time, each drawing a random sequence, collisions may occur. If two users use the exact same sequence and the base station cannot identify this already after *msg1* (resulting in no acknowledgement), then both users will send *msg3* in the same resources, hence colliding again. Therefore, the LTE RACH can be analyzed in a simplified manner in terms of the number of *msg3* collisions, which provide a worst-case performance estimation.

Figure 7.26 shows the probability of such collisions occurring for a single PRACH slot using 64 Zadoff-Chu sequences over an increasing number of contending users. LTE offers diverse PRACH configurations to accommodate different cell sizes and cell loads, but this simple example already shows that the probability of successful access quickly diminishes if many users are vying for access. Increasing the number of PRACH slots may decrease the number of users per slot and may be a temporary solution for medium-level machine-type device activity. However, even this has its limits considering the massive access problem and the increasing level of collisions which lead to retries and further congestion. Hence, novel strategies are required to solve the access reservation problem, which are discussed in Section 7.6.2.

7.6.1.2   Signaling/control overhead for mMTC

Beside the pure access strategy as discussed, a second aspect has to be considered in the access design: the payload size (see also Section 4.2.1 in Chapter 4). Depending on the amount of data that a user needs to send, the signaling overhead introduced by an access reservation scheme like the LTE-PRACH might become prohibitive.

The exact signaling/control overhead that is required by LTE/LTE-A to set up a connection and send a data packet is hard to quantify due to the large amount of different processes that need to be completed before a user is (securely) connected to the network, has requested resources and finally transmits its data. Furthermore, it depends on the previous state of the user (first connection, wake-up, etc.). Therefore, only *msg3* is considered to model the total relevant overhead of the RACH procedure for mMTC connections on the MAC layer. Note that any comparison in this sense is surely favorable for LTE and practical gains might be even higher.

Coming back to payload size, the 5G MAC for mMTC actually has to solve two problems: (A) an efficient access strategy for large packets and (B) an efficient access strategy for small packets. Problem (A) occurs for high data rate MTC applications, e.g. image or video upload, large sets of measurement data, and is similar to current use cases driven by human-based communication. A potential solution is extended access reservation, which is discussed in Section 7.6.2. However, Problem (B) describes a major part of MTC where short status updates, single measurements, control messages and so on lead to a flood of small packets. Here, a leaner approach than access reservation and scheduling is required to avoid congestion. To this end, a contention-based random-access strategy with direct data transmission will be discussed. This strategy, often termed "direct random access" or "one shot transmission", exploits novel medium-access control schemes as well as novel physical layer algorithms to efficiently solve the massive access problem.

7.6.1.3   KPIs and methodology for 5G performance

The first step to outline new MAC solutions for 5G is the consideration of a performance metric that allows a fair comparison to the current state of the art, i.e. LTE. Due to the two potential access solutions presented here, two approaches are required.

For the extended access reservation strategy presented in Section 7.6.2, the efficiency of the MAC protocol is used and is given by

$$\rho = \frac{pU}{N} \tag{7.12}$$

for a general comparison. Here, U is the number of all users accessing the system in one contention period, p is the success probability of a user, and N is the number of resources used for contention.

However, for the direct random-access scheme presented in Section 7.6.3, this "per slot" measure is insufficient. The scheme facilitates direct data transmission, whereas the LTE access reservation protocol facilitates only the access and uses separate resources for data transmission. Therefore, two aspects have to be considered: (i) actual resource requirements in terms of time and frequency should be considered to give a "per resource" comparison of both schemes, and (ii) the actual physical layer performance is now relevant for the overall MAC performance. The first aspect is solved by fixing the time-frequency product that is available for direct random access to be equal to the LTE PRACH and data transmission, i.e. the required PRBs and PRACH resources as defined in the standard are counted. The second aspect is solved by considering the physical layer performance of the direct random-access scheme through numerical simulations, but by assuming LTE data transmission to be error-free. Note that this once again favors LTE, and presented results can therefore only show "worst-case gains" over LTE.

7.6.2   Extending access reservation

The main limitation of the LTE access reservation scheme considering problem A as illustrated in Figures 7.25 and 7.26 is the limited contention space, i.e. the limited number of orthogonal Zadoff-Chu sequences used for *msg1* per PRACH slot. If the number of contending users U is close to the number of sequences L, the collision probability is close to one and practically no user will be served. On the one hand, simply increasing the number of slots does not solve this problem, as each PRACH slot is independently processed and users will collide again. On the other hand, increasing the number of Zadoff-Chu sequences requires more resources for access reservation and might require a change of the LTE frame structure definition.

To extend the contention space in an LTE frame-compatible way and without requiring additional resources for PRACH, *coded access reservation* has been proposed [61]–[63]. Coded access reservation collects a number of PRACH slots and processes the detected Zadoff-Chu sequences jointly. Thus, each contending user may send multiple random sequences in multiple slots to construct a larger contention codeword extending the contention space. A drawback of this approach is the introduction of so-called "phantom codewords", which are combinations of Zadoff-Chu sequences that are not used by any user, but can still be deducted from the PRACH slot observations.

Figure 7.27 depicts the MAC efficiency as defined in (7.12) for the coded access reservation scheme using 4 PRACH slots with 64 orthogonal Zadoff-Chu sequences each to achieve a larger contention space. Furthermore, the base LTE scheme using the PRACH slots independently is included for comparison. It can be seen that up to a user

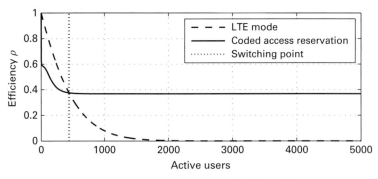

Figure 7.27   MAC efficiency of coded access reservation and LTE PRACH procedure for different user loads using 4 PRACH slots and 64 orthogonal Zadoff-Chu sequences.

load of 440 active users, the standard LTE PRACH provides the best performance. However, beyond 440 active users, coded access reservation is always superior. Most importantly, the coded access reservation shows a stable performance over a large range of active users.

The main advantage of coded access reservation is the potential compatibility with the current LTE standard, allowing it to be adapted in a more evolutionary approach. Performance is, however, limited and only extends the capability of access reservation to roughly four times the number of users at a low efficiency. Furthermore, the signaling overhead introduced by access reservation schemes may become prohibitive for small packets.

7.6.3   Direct random access

To solve the massive access problem for *small packets* (problem B), an alternative to the LTE-like access reservation approach is required. As previously discussed, a random-access scheme, that allows for direct transmission of data while contending with other users, seems to be a good alternative to avoid the reservation overhead. However, most contention-based protocols like, e.g., ALOHA are also quickly congested and are thus questionable candidates for the envisioned massive access. Recent developments in MAC protocols have extended ALOHA to exploit the so-called *capture effect* to at least partly resolve collisions [64]. The capture effect describes that colliding users can still be separated if they can be sufficiently distinguished, e.g. by different receive power levels or through other signal properties. On the physical layer, such techniques are usually described by multi-user detection algorithms. This already hints at the necessity of a joint MAC and PHY scheme.

Based on the capture effect and the observation that graph decoding methods can be applied to slotted ALOHA, multiple enhanced versions have been proposed in literature [65][66]. The most promising approach presented so far is frameless ALOHA, also called *coded random access* [67][68]. The main ideas of coded random access are: (i) all contending users send multiple replicas of their data in random slots of a contention period; (ii) the base station decodes collisions on a graph similar to successive interference cancellation; (iii) the base station controls the average number of replicas per user; and (iv) the contention phase is flexibly stopped by the base station based on a live throughput estimation. The idea here is similar to the one presented in Section 7.5.1, with the difference that this is specifically designed for the reception of multiple data flows in a single receiver (the BS), which can control the number of replicas and the contention phase using a signaling channel.

Figure 7.28 illustrates the protocol schematically. First, the base station sends a beacon that starts the contention period and tells users the average number of slots to be used for replicas. Then, users contend and send multiple replicas in random slots, including pointers to all replica slots in their packets. The graph structure on the right side of the figure illustrates the mapping of users to occupied slots, which is then used to

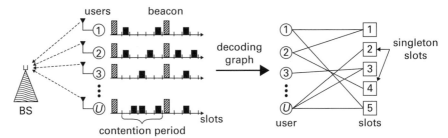

Figure 7.28 Illustration of coded random access and the decoding strategy. Left: random access in slots started by a beacon from the BS. Right: graph representation of the users accessing different slots with replica packages.

decode the contention period as a whole. A critical requirement for this scheme to work is the existence of collision free slots, i.e. slots where only a single user has been active. Without these, the starting point for the graph decoding is missing and decoding of all users fails.

This restriction can be eased through the capture effect or, more generally, through multi-user detection algorithms. A suitable physical layer scheme that is either able to separate users based on power differences (NOMA, cf. Section 7.3.1) or employs non-orthogonal medium access like CDMA, IDMA (cf. Section 7.3.3) and SCMA (cf. Section 7.3.2), can resolve collisions of two or more users in a slot. Thus, it can provide the required starting point for the MAC graph decoding even in congested situations. In contrast to standard medium access assumptions, the set of active users and therefore resources (spreading codes in CDMA) is not known in advance, but has to be estimated by the multi-user detection algorithms. Classical methods like subspace based estimation [69]–[71] are principally able to solve MUD problems with unknown user activity. However, these methods only work for systems where the number of resources (e.g. spreading codes and interleavers, etc.) is larger than or equal to the number of potential users (in contrast to the active ones). This case is often termed "fully loaded" or "under-loaded", and it implies that the number of resources available for random access in a slot needs to be larger than the potential number of concurrently active users. Hence, the system design would be wasteful and require genie knowledge about expected user loads.

Two novel approaches to the MUD problem with unknown user activity are the so-called *Compressed Sensing based Multi-User Detection (CS-MUD)* [72] and *compressive random access* [73][74]. Both incorporate ideas from sparse signal processing and compressed sensing to efficiently solve the posed random access MUD problem for mMTC. With these approaches, highly over-loaded systems can still be solved. Furthermore, CS-MUD has already been investigated in the system context (channel coding, channel estimation, false alarms/missed detections, etc.) and provides many tools to tune the physical layer performance according to the needs of the MAC scheme [75]–[78]. The combination of coded random access and CS-MUD enables robust collision resolution and high MAC performance in terms of the MAC

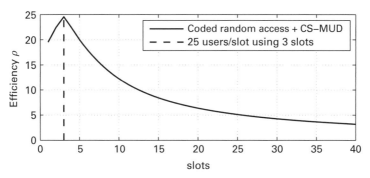

Figure 7.29 MAC efficiency over the number of used slots for the combined MAC coded random access and PHY CD-MUD for a four times overloaded CDMA example with $U = 128$ users, and SNR = 10 dB, assuming optimally controlled user activity levels [16][79].

efficiency [79]. Figure 7.29 illustrates this performance for an exemplary setup with $U = 128$ users in one contention period at a working point of SNR = 10 dB. The number of slots is a design variable in coded random access and should be chosen to maximize the efficiency. Here, the maximum results in 24 users/slot using overall 3 slots, which will provide the basis for further comparison with the LTE RACH. For more results and further details, refer to [16][79].

As discussed in Section 7.6.1.3, a pure MAC layer efficiency comparison to LTE does not make sense here due to the physical layer performance and completely different assumption on required resources. As a comparison, the LTE PRACH performance can be calculated by the number of average users served without collision at *msg3* for 3 consecutive PRACH slots (cf. Section 7.6.1). The average number of collision-free users is given by $U(1 - 1/B)^{U-1}$, where B is the number of orthogonal resources. Assuming 3 slots and 64 Zadoff-Chu sequences per slot, we would for instance have $B = 192$. Therefore, LTE can serve roughly 66 users out of $U = 128$ overall users given 192 orthogonal resources. However, the access is organized in two phases, thus requiring overall 3 PRACH slots and 3 PRBs for the whole PRACH procedure. Furthermore, LTE allocates 800 μs and 1.08 MHz to PRACH slots with Zadoff-Chu sequences of 839 symbols. In this example, coded random access with CS-MUD requires 3328 symbols, i.e. roughly 4 PRACH slots. Finally, data is transmitted during random access, which requires at least 2 PRBs (minimum size in scheduling) per user in LTE after access reservation. Considering the wasted resources in a PRB for very small messages as denoted above, coded random access with CS-MUD serves up to 10x more users per time-frequency resource, including data transmission. Note that this gain is still a "worst-case gain" over LTE due to the previously outlined assumptions, and practical gains can be expected to be even higher. Thus, the novel MAC and PHY schemes outlined in this section may provide an important building block toward fulfilling the mMTC requirements posed in Chapter 4 (see Section 4.3.1).

7.7   Conclusions

The diverse requirements of the various services foreseen for 5G mobile communication systems demand more flexibility and scalability in the system design. Novel technologies for the radio access presented in this chapter can be considered as key enablers for a flexible air interface, allowing an appropriate response to those future challenges. Promising building blocks for the 5G air interface are new waveforms based on filtering, which enable partitioning the transmission band into sub-bands that can be individually configured according to the needs of a service, and multiple-access schemes allowing overloading to take spectral efficiency to its ultimate limit. A novel air interface design tailored for UDN provides scalability with respect to the range of supported carrier frequencies and yields highly efficient data transmission at significantly reduced latency. To address requirements related to V2X and mMTC scenarios, the air interface is complemented with novel MAC schemes for ad-hoc access between vehicular devices, yielding a much higher user capacity compared to CSMA, and a combined MAC/PHY scheme for the massive access of machine-type devices, supporting up to 10 times more devices compared to an equivalent LTE setup.

References

[1] G. Wunder, T. Wild, I. Gaspar, N. Cassiau, M. Dryjanski, B. Eged, et al., "5GNOW: Non-orthogonal asynchronous waveforms for future mobile applications," *IEEE Communications Magazine*, vol. 52, no. 2, pp. 97–105, February 2014.

[2] ICT-318362 Emphatic project, "Flexible and spectrally localized waveform processing for next generation wireless communications," White Paper, 2015, www.ict-emphatic.eu/submissions.html

[3] G. Proakis, *Digital Communications*, New York: McGraw-Hill, 2001.

[4] H. Menouar, F. Filali, and M. Lenardi, "A survey and qualitative analysis of mac protocols for vehicular ad hoc networks," *IEEE Wireless Communications*, vol. 13, no. 5, pp. 30–35, October 2006.

[5] P. W. Baier, "CDMA or TDMA? CDMA for GSM?," in IEEE International Symposium on Personal, Indoor and Mobile Radio Communications, The Hague, September 1994.

[6] E. H. Dinan and B. Jabbari, "Spreading codes for direct sequence CDMA and wideband CDMA cellular networks," *IEEE Communications Magazine*, vol. 36, no. 9, pp. 48–54, September 1998.

[7] M. Costa, "Writing on dirty paper," *IEEE Trans. Information Theory*, vol. 29, no. 3, pp. 439–441, May 1983.

[8] M. Tomlinson, "New automatic equalizer employing modulo arithmetic," *Electron. Lett.*, vol. 7, no. 5, pp. 138–139, March 1971.

[9] ICT-317669 METIS project, "Report on simulation results and evaluations," Deliverable D6.5, February 2015, www.metis2020.com/documents/deliverables/

[10] P. Siohan, C. Siclet, and N. Lacaille, "Analysis and design of OFDM/OQAM systems based on filterbank theory," *IEEE Trans. on Signal Processing*, vol. 50, pp. 1170–1183, May 2002.

[11] J. Nadal, C. A. Nour, A. Baghdadi, and H. Lin, "Hardware prototyping of FBMC/OQAM baseband for 5G mobile communication," in IEEE International Symposium on Rapid System Prototyping, Uttar Pradesh, October 2014.

[12] ICT-211887 PHYDYAS project, "FBMC physical layer: A primer," June 2010, www.ict-phydyas.org/

[13] M. Fuhrwerk, J. Peissig, and M. Schellmann, "Channel adaptive pulse shaping for OQAM-OFDM systems," in European Signal Processing Conference, Lisbon, September 2014.

[14] M. Fuhrwerk, J. Peissig, and M. Schellmann, "On the design of an FBMC based air interface enabling channel adaptive pulse shaping per sub-band," in European Signal Processing Conference, Nice, September 2015.

[15] H. Lin, M. Gharba, and P. Siohan, "Impact of time and frequency offsets on the FBMC/OQAM modulation scheme," IEEE Signal Proc., vol. 102, pp. 151–162, September 2014.

[16] ICT-317669 METIS project, "Proposed solutions for new radio access," Deliverable D2.4, February 2015, www.metis2020.com/documents/deliverables/

[17] D. Pinchon and P. Siohan, "Derivation of analytical expression for low complexity FBMC systems," in European Signal Processing Conference, Marrakech, September 2013.

[18] H. Lin and P. Siohan, "Multi-Carrier Modulation Analysis and WCP-COQAM proposal," EURASIP Journal on Advances in Sig. Proc., vol. 2014, no. 79, May 2014.

[19] M.J. Abdoli, M. Jia. and J. Ma, "Weighted circularly convolved filtering in OFDM/OQAM," in IEEE International Symposium on Personal, Indoor and Mobile Radio Communications, London, September 2013, pp. 657–661.

[20] J. P. Javaudin, D. Lacroix, and A. Rouxel, "Pilot-aided channel estimation for OFDM/OQAM," in IEEE Vehicular Technology Conference, Keju, April 2003, pp. 1581–1585.

[21] Z. Zhao, N. Vucic, and M. Schellmann, "A simplified scattered pilot for FBMC/OQAM in highly frequency selective channels," in IEEE International Symposium on Wireless Comm. Systems, Barcelona, August 2014.

[22] M. Caus and A. Perez-Neira, "Multi-stream transmission for highly frequency selective channels in MIMO-FBMC/OQAM systems," IEEE Transactions on Signal Processing, vol. 62, no. 4, pp. 786–796, February 2014.

[23] Z. Zhao, X. Gong and M. Schellmann, "A novel FBMC/OQAM scheme facilitating MIMO FDMA without the need for guard bands," in International ITG Workshop on Smart Antennas, Munich, March 2016.

[24] T. Wild and F. Schaich, "A reduced complexity transmitter for UF-OFDM," in IEEE Vehicular Technology Conference, Glasgow, May 2015.

[25] ICT-318555 5GNOW project, "5G waveform candidate selection," Deliverable D3.1, 2013, www.5gnow.eu/?page_id=418

[26] V. Vakilian, T. Wild, F. Schaich, S.t. Brink, and J.-F. Frigon, "Universal-filtered multi-carrier technique for wireless systems beyond LTE," in IEEE Global Communications Conference Workshops, Atlanta, December 2013.

[27] F. Schaich, T. Wild, and Y. Chen, "Waveform contenders for 5G: Suitability for short packet and low latency transmissions," in IEEE Vehicular Technology Conference, Seoul, May 2014.

[28] X. Wang, T. Wild, F. Schaich, and S. ten Brink, "Pilot-aided channel estimation for universal filtered multi-carrier," submitted to IEEE Vehicular Technology Conference VTC–Fall '15, September 2015.

[29] T. Wild, F. Schaich, and Y. Chen, "5G air interface design based on universal filtered (UF-)OFDM," in Intl. Conference on Digital Signal Processing, Hong Kong, August 2014.

[30] K. Higuchi and A. Benjebbour, "Non-orthogonal multiple access (NOMA) with successive interference cancellation for future radio access," *IEICE Transactions on Communications*, vol. E98-B, no. 3, pp. 403–414, March 2015.

[31] D. Tse and P. Viswanath, *Fundamentals of Wireless Communication*, New York: Cambridge University Press, 2005.

[32] A. Benjebbour, A. Li, Y. Saito, Y. Kishiyama, A. Harada, and T. Nakamura, "System-level performance of downlink NOMA for future LTE enhancements," in IEEE Global Communications Conference, Atlanta, December 2013.

[33] A. Benjebbour, A. Li, Y. Kishiyama, H. Jiang, and T. Nakamura, "System-level performance of downlink NOMA combined with SU-MIMO for future LTE enhancements," in IEEE Global Communications Conference, Austin, December 2014.

[34] K. Saito, A. Benjebbour, Y. Kishiyama, Y. Okumura, and T. Nakamura, "Performance and design of SIC receiver for downlink NOMA with open-loop SU-MIMO," in IEEE Intl. Conference on Communications (ICC), London, UK, June 2015.

[35] M. Taherzadeh, H. Nikopour, A. Bayesteh, and H. Baligh, "SCMA codebook design," in IEEE Vehicular Technology Conference, Vancouver, September 2014.

[36] H. Nikopour and H. Baligh, "Sparse code multiple access," in IEEE Intl. Symposium on Personal, Indoor and Mobile Radio Communications, London, September 2013.

[37] H. Nikopour, E. Yi, A. Bayesteh, K. Au, M. Hawryluck, H. Baligh, and J. Ma, "SCMA for downlink multiple access of 5G wireless networks," in IEEE Global Communications Conference, Austin, December 2014.

[38] A. Bayesteh, E. Yi, H. Nikopour, and H. Baligh, "Blind detection of SCMA for uplink grant-free multiple-access," in IEEE International Symposium on Wireless Comm. Systems, Barcelona, August 2014.

[39] K. Au, L. Zhang, H. Nikopour, E. Yi, A. Bayesteh, U. Vilaipornsawai, J. Ma. and P. Zhu, "Uplink contention based sparse code multiple access for next generation wireless network," in IEEE Global Communications Conference Workshops, Austin, December 2014.

[40] L. Ping, L. Liu, K. Wu, and W.K. Leung, "Interleave-division multiple–access," *IEEE Trans. Wireless Commun.*, vol. 5, no. 4, pp. 938–947, April 2006.

[41] K. Kusume, G. Bauch, and W. Utschick, "IDMA vs. CDMA: Analysis and comparison of two multiple access schemes," *IEEE Trans. Wireless Commun.*, vol. 11, no. 1, pp. 78–87, January 2012.

[42] Y. Chen, F. Schaich, and T. Wild, "Multiple access and waveforms for 5G: IDMA and universal filtered multi-carrier," in IEEE Vehicular Technology Conference, Seoul, May 2014.

[43] ICT-318555 5GNOW Project, "5G waveform candidate selection," Deliverable D3.2, 2014, www.5gnow.eu/?page_id=418

[44] E. Lähetkangas, K. Pajukoski, J. Vihriälä et al., "Achieving low latency and energy consumption by 5G TDD mode optimization," in IEEE International Conference on Communications, Sydney, June 2014.

[45] K. Haneda, F. Tufvesson, S. Wyne, M. Arlelid, and A.F. Molisch, "Feasibility study of a mm-wave impulse radio using measured radio channels," in IEEE Vehicular Technology Conference, Budapest, May 2011.

[46] C. Gustafson, K. Haneda, S. Wyne, and F. Tufvesson, "On mm-wave multi-path clustering and channel modeling," *IEEE Trans. Antennas Propag.*, vol. 62, no. 3, pp. 1445–1455, March 2014.

[47] C. Gustafson, F. Tufvesson, S. Wyne, K. Haneda, and A. F. Molisch, "Directional analysis of measured 60 GHz indoor radio channels using SAGE," in IEEE Vehicular Technology Conference, Budapest, May 2011.

[48] International Telecommunications Union Radio (ITU-R), "Guidelines for evaluation of radio interface technologies for IMT-Advanced," Report ITU-R M.2135, December 2008, www.itu.int/pub/R-REP-M.2135-2008

[49] 3GPP TR 36.912, "Feasibility Study for Further Advancements for E-UTRA," Technical Report TR 36.912 V10.0.0, Technical Specification Group Radio Access Network, March 2011.

[50] M. Lauridsen, "Studies on Mobile Terminal Energy Consumption for LTE and Future 5G," PhD thesis, Aalborg University, 2015.

[51] H. Schotten, R. Sattiraju, D. Gozalvez, Z. Ren, and P. Fertl, "Availability indication as key enabler for ultra-reliable communication in 5G," in European Conference on Networks and Communications, Bologna, June 2014.

[52] R. Sattiraju and H.D. Schotten, "Reliability modeling, analysis and prediction of wireless mobile communications," in IEEE Vehicular Technology Conference Workshops, Seoul, May 2014.

[53] R. Sattiraju, P. Chakraborty, and H.D. Schotten, "Reliability analysis of a wireless transmission as a repairable system," in Intl. Workshop on Ultra-Low Latency and Ultra-High Reliability in Wireless Communications, IEEE Global Communication Conference, Austin, December 2014.

[54] S. Beygi, U. Mitra, and E. G. Ström, "Nested sparse approximation: Structured estimation of V2V channels using geometry-based stochastic channel model," *IEEE Trans. on Signal Proc.*, vol. 63, no. 18, pp. 4940–4955, September 2015.

[55] R. Apelfröjd and M. Sternad, "Design and measurement based evaluation of coherent JT CoMP – A study of precoding, user grouping and resource allocation using predicted CSI," *EURASIP Journal on Wireless Comm. and Netw.*, vol. 2014, no. 100, 2014, http://jwcn.eurasipjournals.com/content/2014/1/100

[56] M. Ivanov, F. Brännström, A. Graell i Amat, and P. Popovski, "Error floor analysis of coded slotted ALOHA over packet erasure channels," *IEEE Commun. Lett.*, vol. 19, no. 3, pp. 419–422, March 2015.

[57] M. Ivanov, F. Brännström, A. Graell i Amat, and P. Popovski, "All-to-all broadcast for vehicular networks based on coded slotted ALOHA," in IEEE International Conference on Communications Workshop, London, June 2015.

[58] H. Holma and A. Toskala, *LTE for UMTS: Evolution to LTE-advanced*, Chichester: John Wiley & Sons, 2011.

[59] 3GPP TS 36.211, "Evolved Universal Terrestrial Radio Access (E-UTRA); Physical channels and modulation," Technical Specification TS 36.211 V11.6.0, Technical Specification Group Radio Access Network, September 2014.

[60] 3GPP TS 36.213, "Evolved Universal Terrestrial Radio Access (E-UTRA); Physical layer procedures," Technical Specification TS 36.213 V11.9.0, Technical Specification Group Radio Access Network, January 2015.

[61] H. Thomsen, N.K. Pratas, and C. Stefanovic, "Analysis of the LTE access reservation protocol for real-time traffic," *IEEE Communication Letters*, 2013.

[62] H. Thomsen, N.K. Pratas, and C. Stefanovic, "Code-expanded radio access protocol for machine-to-machine communications," *Trans. on Emerging Telecomm. Technologies*, 2013.

[63] N. K. Pratas, H. Thomsen, C. Stefanovic, and P. Popovski, "Code expanded random access for machine-type communications," in IEEE Global Conference on Communications Workshops, Anaheim, December 2012.

[64] A. Zanella and M. Zorzi, "Theoretical analysis of the capture probability in wireless systems with multiple packet reception capabilities," *IEEE Trans. Commun.*, vol. 60, no. 4, pp. 1058–1071, 2012.

[65] E. Cassini, R. D. Gaudenzi, and O. del Rio Herrero, "Contention resolution diversity slotted ALOHA (CRDSA): An enhanced random access scheme for satellite access packet networks," *IEEE Trans. Wireless Commun.*, vol. 6, no. 4, pp. 1408–1419, April 2007.

[66] G. Liva, "Graph-based analysis and optimization of contention resolution diversity slotted ALOHA," *IEEE Trans. Commun.*, vol. 59, no. 2, pp. 477–487, February 2011.

[67] C. Stefanovic and P. Popovski, "ALOHA random access that operates as a rateless code," *IEEE Trans. Commun.*, vol. 61, no. 11, pp. 4653–4662, November 2013.

[68] C. Stefanovic and P. Popovski, "Coded slotted ALOHA with varying packet loss rate across users." in IEEE Global Conference on Signal and Information Processing, Austin, December 2013.

[69] S. Buzzi, A. De Maio, and M. Lops, "Code-aided blind adaptive new user detection in DS/CDMA systems with fading time-dispersive channels," *IEEE Trans. Signal Proc.*, vol. 51, no. 10, pp. 2637–2649, 2003.

[70] M. Honig, U. Madhow, and S. Verdu, "Blind adaptive multiuser detection," *IEEE Trans. Inf. Theory*, vol. 41, no. 4, pp. 944–960, 1995.

[71] D.D. Lin and T.J. Lim, "Subspace-based active user identification for a collision-free slotted ad hoc network," *IEEE Trans. Commun.*, vol. 52, no. 4, pp. 612–621, 2004.

[72] C. Bockelmann, H. F. Schepker, and A. Dekorsy, "Compressive sensing based multi-user detection for machine-to-machine communication," *Transactions on Emerging Telecommunications Technologies*, vol. 24, no. 4, pp. 389–400, April 2013.

[73] M. Kasparick, G. Wunder, P. Jung, and D. Maryopi, "Bi-orthogonal waveforms for 5G random access with short message support," in European Wireless 2014, Barcelona, May 2014.

[74] G. Wunder, P. Jung, and C. Wang, "Compressive random access for post-LTE systems," in IEEE International Conference on Communications, Sydney, June 2014.

[75] H. Schepker, C. Bockelmann, and A. Dekorsy, "Improving group orthogonal matching pursuit performance with iterative feedback," in IEEE Vehicular Technology Conference, Las Vegas, September 2013.

[76] H. Schepker, C. Bockelmann, and A. Dekorsy, "Exploiting sparsity in channel and data estimation for sporadic multi-user communication," in International Symposium on Wireless Communication Systems, Ilmenau, August 2013.

[77] F. Monsees, C. Bockelmann, and A. Dekorsy, "Compressed sensing soft activity processing for sparse multi-user systems," in IEEE Global Communications Conference Workshops, Atlanta, December 2013.

[78] F. Monsees, C. Bockelmann, and A. Dekorsy, "Compressed sensing neyman-pearson based activity detection for sparse multiuser communications," in International ITG Conference on Systems Communications and Coding, Hamburg, February 2015.

[79] Y. Ji, C. Stefanovic, C. Bockelmann, A. Dekorsy, and P. Popovski, "Characterization of coded random access with compressive sensing based multi-user detection," in IEEE Global Communications Conference, Austin, December 2014.

8 Massive multiple-input multiple-output (MIMO) systems

Antti Tölli, Lars Thiele, Satoshi Suyama, Gabor Fodor, Nandana Rajatheva, Elisabeth De Carvalho, Wolfgang Zirwas, and Jesper Hemming Sorensen

8.1 Introduction

As stated in Chapter 2, one of the main 5G requirements [1] is to support 1000 times larger capacity per area compared with current Long Term Evolution (LTE) technology, but with a similar cost and energy dissipation per area as in today's cellular systems. In addition, an increase in capacity will be possible if all three factors that jointly contribute to system capacity are increased: More spectrum, a larger number of base stations per area, and an increased spectral efficiency per cell.

Massive or large Multiple-Input Multiple-Output (MIMO) systems are considered essential in contributing to the last stated factor, as they promise to provide a substantially increased spectral efficiency per cell. A massive MIMO system is typically defined as a system that utilizes a large number, i.e. 100 or more, of individually controllable antenna elements at least at one side of a wireless communications link, typically at the Base Station (BS) side [2][3]. An example of such usage of massive MIMO at the BS side is shown in Figure 8.1. A massive MIMO network exploits the many spatial Degrees of Freedom (DoF) provided by the many antennas to multiplex messages for several users on the same time-frequency resource (referred to as *spatial multiplexing*), and/or to focus the radiated signal toward the intended receivers and inherently minimize intra-cell and inter-cell interference [4]–[7]. Such focusing of radiated signals in a particular direction is possible by transmitting the same signal from multiple antenna points, but with a different phase shift applied to each of the antennas (and possibly a different phase shift for different parts of the system bandwidth), such that the signals overlap coherently at the intended target location. Note that in the remainder of the chapter, the term *beamforming* is used when applying the same phase shift at individual transmit antennas over the entire system bandwidth, while the term *precoding* is used when applying different phase shifts for different parts of the system bandwidth to tackle small-scale fading effects, for instance by applying phase shifts in frequency domain. With this definition, beamforming can be seen as a subclass of precoding algorithms. Regardless of whether precoding or beamforming is applied,

5G Mobile and Wireless Communications Technology, ed. A. Osseiran, J. F. Monserrat, and P. Marsch. Published by Cambridge University Press. © Cambridge University Press 2016.

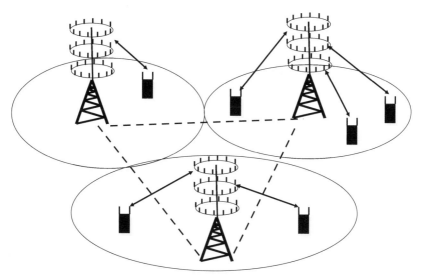

Figure 8.1   Multiuser and single user with massive MIMO base stations.

the gain of obtaining a coherent overlap of signals at the receive point is commonly referred to as *array gain*.

Besides being used for access links, massive MIMO can also play a key role in creating multi-Gbps backhaul links between infrastructure nodes that are deployed in Frequency Division as well as Time Division Duplexing (FDD/TDD) systems.

Although large antenna arrays in radar systems have been in wide use since the late 1960s, the commercial deployment of massive MIMO systems, i.e. for the access and backhaul of mobile communications systems, has been considered only recently. In particular, recent research and experiments with practical implementations have identified the key challenges that must be overcome in order to realize the potential benefits of massive MIMO in cellular communications [7], and which are listed in the sequel.

One of the most severe challenges is given by the need of accurate Channel State Information (CSI) at the transmitter side. In principle, the CSI may be obtained through the transmission of orthogonal pilot signals (also called reference signals) from each transmit antenna element, and the subsequent feedback of the observed spatial channel from the receiver to the transmitter side. This approach has the drawback that the pilot signal overhead in terms of required CSI grows linearly with the number of transmit antennas. Another option for obtaining the CSI at the transmitter side is to utilize channel reciprocity, which is for instance possible in TDD systems. The cost in utilizing reciprocity is that it requires array calibration in order to take the differences in the transmit/receive Radio Frequency (RF) chains of the different antenna elements into account. In time varying channels, the delay between pilot transmission, channel estimation, channel feedback, beamformer calculation and the actual beamformed data transmission will degrade the performance of a massive MIMO. Fortunately, channel prediction techniques could be used to mitigate this delay, as for instance discussed in Section 9.2.1.

Another challenge is to understand the impact of massive MIMO on the design of multi-cell multi-tier networks [8]. One of the problems is the impact of pilot contamination [4][9]. The inherent trade-off between the time, frequency, code, spatial and power resources allocated to pilot signals and data transmission in multi-antenna systems is well known, see for example the classical results in [10]–[13] as well as the more recent contributions in [14]–[16]. In multi-cell multi-user massive MIMO systems, the pilot-data resource allocation trade-off is intertwined with the management of inter-cell interference (also known as contamination) both on the pilot and data signals and calls for rethinking the pilot signal design of classical systems such as the 3rd Generation Partnership Project (3GPP) LTE system. Recent works provide valuable insights into the joint design of pilot and data channels in multi-cell massive MU-MIMO systems [17].

Furthermore, the exploitation of massive MIMO in a network with an ultra-dense deployment of small cells may be difficult in practice due to the size of the antenna arrays. At higher frequencies, such as millimeter Wave (mmW), this is not an issue, as discussed in Section 6.3.2 of Chapter 6. One approach is to deploy massive MIMO at the macro side, and to exploit the spatial DoF to lower the interference between the macro and the small cells, in the case of a co-channel deployment between macro and small cell layer. Another approach is to consider possible massive MIMO deployments at the small cell side, in particular in high-frequency systems, where the small antenna dimensions allow the deployment of large-scale antenna arrays with a realistic form factor.

Based on the general observations above, different technology solution components are currently investigated, which aim to unleash the promising potential and enable an efficient usage of massive MIMO in 5G.

The remainder of this chapter is structured as follows. First, Section 8.2 provides the theoretical background on massive MIMO, investigating the principle capacity scaling behavior that can be expected from massive MIMO systems. The design of pilot signals, covering all associated challenges such as the mentioned pilot contamination, is presented in Section 8.3. A discussion of methods of resource allocation as given in Section 8.4 is essential to understand the utilization of spatial DoFs in a massive MIMO environment. Then, Section 8.5 introduces digital, analog and hybrid beamforming as the principle forms of precoding and beamforming implementation in hardware, providing different trade-offs between the extent of massive MIMO gains captured and the cost-efficiency of the implementation. Finally, Section 8.6 gives some ideas on suitable channel models for massive MIMO investigation.

8.1.1   MIMO in LTE

Historically, the 3GPP standard for LTE was designed with MIMO as a goal to increase capacity. LTE has adopted various MIMO technologies. In LTE Release 8, the downlink transmission supports up to four antennas at the BS. For the uplink transmissions, only a single antenna is supported for transmission from the user. There is an option for performing antenna switching with up to two transmit antennas. Multi-User MIMO

(MU-MIMO) is also supported in the uplink. Further, LTE Release 10 (also known as LTE-Advanced) provided enhanced MIMO technologies. A new codebook and feedback design are implemented to support spatial multiplexing with up to eight independent spatial streams and enhanced MU-MIMO transmissions. In the uplink, single user MIMO is utilized with up to four transmit antennas at the user side. In LTE Release 12, downlink MIMO performance was enhanced thanks to the definition of new CSI reporting schemes, which permitted the BS to transmit with more accurate CSI. The LTE Release 13 looked into higher-order MIMO systems with up to sixty four antenna ports at the BS, to become more appropriate to the use of higher frequencies.

8.2   Theoretical background

This section provides insight on the fundamental bounds and behavior of wireless networks where the BSs are equipped with a large number of antenna elements. In particular, the massive MIMO foundations are reviewed including a brief introduction to the fundamentals of multi-user/cell MIMO communications and its analysis via random matrix theory.

After Marzetta's pioneering work [4], massive MIMO gained significant attention in both the academic and the industrial communities. Assuming the dimensions of a MIMO system grow large, the results of Random Matrix Theory (RMT) can be applied to provide simple approximations, for example, for the user-specific Signal to Interference plus Noise Ratio (SINR) expressions [18][19].

The linear MIMO transmitter-receiver design in a multi-user (multi-cell) setting, which involves resource allocation, has received considerable attention in the literature; see for example [9][20][21]. In general, (near) optimal design to maximize a certain optimization objective often leads to an iterative solution where each sub-problem is presented as an optimization problem, and at each iteration some information needs to be exchanged between adjacent cells. In a massive MIMO setting, the coordinated transceiver design methods can be potentially simplified, as compared to state-of-the-art iterative optimization-based schemes.

In a special case, where the imbalance between the number of independently fading transmit antennas n_t or receiver antennas n_r at the serving BS and the number of users K in the cell becomes large, i.e. $n_t \gg K$, the processing can be simplified in a way that even Matched Filter (MF) and Zero-Forcing (ZF) can be used in an ideal independent and identically distributed (i.i.d.) channel for near optimal detection and precoding [4][5]. However, in a general setting and given practical non-ideal conditions such as non-zero correlation between antennas and physical limitation of antenna array sizes, more complicated precoder designs with some limited coordination between adjacent cells can be still highly beneficial [22][23].

In order to understand the basics, more details of single user and multiple user MIMO cases are considered next. Both scenarios are incorporated in the general single cell MIMO system given in Figure 8.2, where n_t is the number of transmit antennas, n_r the

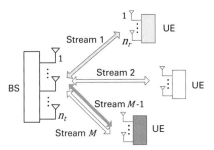

Figure 8.2   Generic single and multiple user MIMO system.

number of receive antennas and M the number of spatial data streams transmitted simultaneously to all users.

8.2.1   Single user MIMO

Assuming a narrowband and time-invariant MIMO channel with n_t transmit (Tx) antennas and n_r receive (Rx) antennas, the received signal vector at symbol time m can be described by

$$y[m] = Hx[m] + n[m], \tag{8.1}$$

where $x \in \mathbb{C}^{n_t}$ is the Tx signal, subject to $Tr(E[xx^H]) \leq P$, $y \in \mathbb{C}^{n_r}$ is the Rx signal, $n \sim CN(0, N_0 I_{n_r})$ denotes the complex white Gaussian noise with noise variance N_0, and $H \in \mathbb{C}^{n_r \times n_t}$ the channel matrix. The achievable rate of the channel H measured in bits per channel use is upper bounded by the mutual information between the input x and output y as

$$C = \max_{Tr(E[xx^H]) \leq P} I(x; y)$$

$$= \max_{K_x: Tr[K_x] \leq P} \log \left| I_{n_r} + \frac{1}{N_0} H K_x H^H \right|, \tag{8.2}$$

where $K_x = Q P Q^H$ is the covariance matrix of $x \sim CN(0, K_x)$, Q is a unitary steering matrix and $P = \text{diag}(p_1, \dots, p_{n_t})$. When the channel is known at the transmitter side, the optimal strategy is to assign Q to the right singular vectors V of H while the power allocation P is found via waterfilling.

On the other hand, when the elements of H are i.i.d. $CN(0, 1)$ and the channel is not known at the transmitter, the optimal K_x is

$$K_x = \frac{P}{n_t} I_{n_t}. \tag{8.3}$$

In such a case, the capacity of the MIMO channel (8.2) is simplified to [24],

$$C = \log\left| I_{n_r} + \frac{P}{n_t N_0} HH^H \right| = \sum_{i=1}^{n_{min}} \log\left(1 + \frac{P}{n_t N_0}\lambda_i^2\right), \quad (8.4)$$

where $n_{min} = \min(n_t, n_r)$, λ_i are the singular values of H and SNR $= \dfrac{P}{N_0}$.
Let us now focus on the square channel $n = n_t = n_r$ and define

$$C_{nn}(\text{SNR}) = \sum_{i=1}^{n} \log\left(1 + \text{SNR}\frac{\lambda_i^2}{n}\right). \quad (8.5)$$

Assume now that the number of antennas becomes large, i.e. $n \to \infty$, then the distribution of $\lambda_i/\sqrt{n}$ becomes a deterministic function

$$f^*(x) = \begin{cases} \dfrac{1}{\pi}\sqrt{4 - x^2} & 0 \le x \le 2, \\ 0 & else. \end{cases} \quad (8.6)$$

For increasing n, the normalized capacity $C(\text{SNR}) = C_{nn}(\text{SNR})/n$ per spatial dimension becomes [24]

$$C(\text{SNR}) = \frac{1}{n}\sum_{i=1}^{n} \log\left(1 + \text{SNR}\frac{\lambda_i^2}{n}\right) \xrightarrow{n \to \infty} \int_0^4 \log(1 + \text{SNR} \cdot x)f^*(x)dx \quad (8.7)$$

and the closed form solution to the integral is [24][25]

$$C(\text{SNR}) = 2\log\left(\frac{1 + \sqrt{4\text{SNR} + 1}}{2}\right) - \frac{\log e}{4\text{SNR}}(\sqrt{4\text{SNR} + 1} - 1)^2. \quad (8.8)$$

Finally, when $n \to \infty$, the capacity of the $n \times n$ point-to-point MIMO link can be approximated as

$$\lim_{n \to \infty} \frac{C_{nn}(\text{SNR})}{n} = C(\text{SNR}) \to C_{nn}(\text{SNR}) \approx nC(\text{SNR}) \quad (8.9)$$

The actual ergodic capacity $E_H[C(H)]$ is compared with the large-n approximation in Figure 8.3 for $n = 2, 4, 8, 16$ and 32. It can be seen that the approximation is very close even for relatively small values of n. The result above can be extended to any fixed ratio between transmit and receive antennas assuming the number of antenna elements grows large at both ends of the transmission link [25]. Even though the capacity expression becomes deterministic for large n, the rate optimal transmission strategy still requires in general right singular vectors of H to be used as optimal transmit directions.

Assume now that $n_r \gg n_t$, and the elements of H are i.i.d. CN$(0, 1)$. When n_r becomes very large, the columns of $H = [h_1, \dots, h_{n_t}]$ become close to orthogonal, i.e.

$$\frac{H^H H}{n_r} \approx I_{n_t}. \quad (8.10)$$

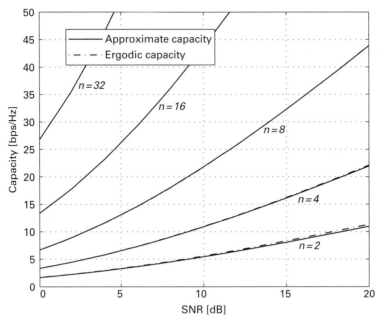

Figure 8.3 Capacity versus SNR.

By plugging in this approximation, the capacity of the MIMO link with or without CSIT can be approximated as

$$\log\left|\boldsymbol{I}_{n_r} + \frac{P}{n_t N_0}\boldsymbol{H}\boldsymbol{H}^H\right| = \log\left|\boldsymbol{I}_{n_t} + \frac{P}{n_t N_0}\boldsymbol{H}^H\boldsymbol{H}\right|$$

$$\approx \sum_{i=1}^{n_t}\log\left(1 + \frac{P\|\boldsymbol{h}_i\|^2}{n_t N_0}\right) \approx n_t\log\left(1 + \frac{Pn_r}{n_t N_0}\right). \tag{8.11}$$

Therefore, the MF receiver is the asymptotically optimal solution for $n_r \gg n_t$.

Similarly, when $n_t \gg n_r$ and the elements of $\boldsymbol{H}$ are i.i.d. $CN(0, 1)$, the columns $\boldsymbol{h}_i$ of $\boldsymbol{H}^T = [\boldsymbol{h}_1, \ldots, \boldsymbol{h}_{n_r}]$ become close to orthogonal, i.e.

$$\frac{\boldsymbol{H}\boldsymbol{H}^H}{n_t} \approx \boldsymbol{I}_{n_r}. \tag{8.12}$$

Then, the rate expression *without CSIT* is simplified to

$$\log\left|\boldsymbol{I}_{n_r} + \frac{P}{n_t N_0}\boldsymbol{H}\boldsymbol{H}^H\right| \approx n_r\log\left(1 + \frac{P}{N_0}\right). \tag{8.13}$$

Thus, no array gain from having n_t transmit antennas is attained. This is due to the lack of channel knowledge at the transmitter, and hence, the power is evenly dissipated from all n_t antennas.

As the rows of H are asymptotically orthogonal, the n_r dominant singular vectors are asymptotically equivalent to the normalized rows of H, and hence the transmit covariance matrix K_x can be approximated (at high SNR, assuming equal power loading) as

$$K_x = VPV^H \approx \frac{P}{n_r n_t} H^H H, \tag{8.14}$$

where the matrix V corresponds to the right singular vectors of H. Thus, using the MF precoders at the transmitter is the asymptotically optimal solution. In such a case, the rate expression *with full CSIT* can be simplified to

$$C = \log\left|I_{n_r} + \frac{1}{N_0} HK_x H^H\right| \approx n_r \log\left(1 + \frac{n_t P}{n_r N_0}\right), \tag{8.15}$$

providing n_t/n_r-fold array gain.

8.2.2   Multi-user MIMO

8.2.2.1   Uplink channel

Assume a time-invariant uplink channel with K single-antenna users and a single BS with n_r receive antennas such that $n_r \gg K$. The received signal vector at symbol time m is described by

$$y[m] = \sum_{k=1}^{K} h_k x_k[m] + n[m] = Hx[m] + n[m], \tag{8.16}$$

where x_k is the Tx symbol of user k, per user power constraints $E[|x_k|^2] \le P_k$, $y \in \mathbb{C}^{n_r}$ is the Rx signal, $n \sim CN(0, N_0 I_{n_r})$ denotes the complex Gaussian noise and $h_k = \sqrt{a_k}\bar{h}_k \in \mathbb{C}^{n_r}$ is the channel vector of user k, where a_k is the large-scale fading factor and $\bar{h}_k$ is the normalized channel.

The sum capacity expression for the multiuser MIMO is equal to the Single User (SU) MIMO without CSIT, i.e.

$$C_{sum} = \log\left|I_{n_r} + \sum_{k=1}^{K} \frac{P_k}{N_0} h_k h_k^H\right| = \log\left|I_{n_r} + \frac{1}{N_0} H K_x H^H\right|, \tag{8.17}$$

where $H = [h_1, \ldots, h_K]$ and $K_x = \text{diag}(P_1, \ldots, P_K)$.

Assume $n_r \gg K$, and that the elements of the normalized channel vectors $\bar{h}_k$ are i.i.d. $CN(0, 1)$, then

$$\frac{H^H H}{n_r} \approx A_K, \tag{8.18}$$

where $A_K = \text{diag}(a_1, \ldots, a_K)$. Then, the sum rate can be approximated as

$$C_{sum} = \log\left|I_{n_r} + \frac{1}{N_0} H K_x H^H\right| = \log\left|I_K + \frac{1}{N_0} K_x H^H H\right|$$

$$\approx \sum_{k=1}^{K} \log\left(1 + \frac{P_k a_k \|\bar{h}_k\|^2}{N_0}\right) \approx \sum_{k=1}^{K} \log\left(1 + \frac{n_r P_k a_k}{N_0}\right), \quad (8.19)$$

and again the matched filter receiver is the asymptotically optimal solution.

8.2.2.2   Downlink channel

The downlink problem is somewhat different due to the sum power constraint across users. Assume a time-invariant downlink channel with K single-antenna users and a single BS with n_t transmit antennas such that $n_t \gg K$. The received signal vector at user k at symbol time m is

$$y_k[m] = h_k^H x[m] + n_k[m]$$

$$= h_k^H u_k \sqrt{p_k} d_k[m] + \sum_{i=1, i \neq k}^{K} h_i^H u_i \sqrt{p_i} d_i[m] w_k[m], \quad (8.20)$$

where $x \in \mathbb{C}^{n_t}$ is the Tx signal vector, subject to power constraint $E(Tr[xx^H]) = \sum_{k=1}^{K} p_k \leq P$, $u_k \in \mathbb{C}^{n_t}$ is the normalized precoder with $\|u_k\| = 1$, $d_k \in \mathbb{C}$ is the normalized data symbol with $E[|d_k|^2] = 1$, $y_k \in \mathbb{C}$ is the Rx signal, $n_k \sim CN(0, N_0)$ is the complex white Gaussian noise and $h_k = \sqrt{a_k} \bar{h}_k \in \mathbb{C}^{n_t}$ is the channel vector of user k, assumed to be ideally known at the transmitter. Maximization of the downlink sum rate can be expressed via the *dual uplink* reformulation, where the role of the transmitter and the receivers is reversed [24]. The sum rate optimal solution is attained from the constrained optimization problem

$$\max_{q_k} \ \log_2\left|I_{n_t} + \frac{1}{N_0} \sum_{k=1}^{K} q_k h_k h_k^H\right| \quad (8.21)$$

subject to

$$\sum_{k=1}^{K} q_k \leq P, \quad q_k \geq 0, \quad k = 1, \dots, K,$$

where q_k is the dual uplink power defined as the uplink power of the dual uplink reformulation mentioned above such that the sum power between downlink and dual uplink powers holds $\sum_{k=1}^{K} q_k = \sum_{k=1}^{K} p_k = P$. When $n_t \gg K$, the objective of (8.21) is simplified to

$$\max_{q_k} \ \log\left|I_K + \frac{1}{N_0} K_x H^H H\right| \approx \max_{q_k} \ \sum_{k=1}^{K} \log\left(1 + \frac{q_k n_t a_k}{N_0}\right), \quad (8.22)$$

where $K_x = \mathrm{diag}\,(q_1, \dots, q_K)$ and $\dfrac{H^H H}{n_t} \approx \mathrm{diag}(a_1, \dots, a_K)$. Since the inter-user interference vanishes as $n_t \gg K$, the dual uplink power allocation is equal to the downlink

Table 8.1 Asymptotic capacity scaling behavior for massive MIMO assuming large imbalance between transmit and receive antennas.

Mode	Transmission	CSIT Feedback	Asymptotic capacity scaling behavior
SU-MIMO, $n_t \gg n_r$	DL & UL	No CSIT	$n_r \log\left(1 + \dfrac{P}{N_0}\right)$
SU-MIMO, $n_t \gg n_r$	DL & UL	CSIT	$n_r \log\left(1 + \dfrac{n_t P}{n_r N_0}\right)$
SU-MIMO, $n_r \gg n_t$	DL & UL	No CSIT	$n_t \log\left(1 + \dfrac{n_r P}{n_t N_0}\right)$
MU-MIMO, $n_t \gg K$	DL	CSIT	$\displaystyle\max_{q_k} \sum_{k=1}^{K} \log\left(1 + \dfrac{q_k n_t a_k}{N_0}\right)$
MU-MIMO, $n_r \gg K$	UL	No CSIT	$\displaystyle\sum_{k=1}^{K} \log\left(1 + \dfrac{n_r P_k a_k}{N_0}\right)$

power allocation, i.e. $p_k = q_k \ \forall k$. Combining above relations, the optimal power allocation can be found via a simple waterfilling principle

$$p_k^* = \max\left(0, \ \mu - \frac{N_0}{n_t a_k}\right), \tag{8.23}$$

where the optimal water level μ is found to satisfy the power constraint $\sum_{k=1}^{K} p_k \leq P$.

8.2.3   Capacity of massive MIMO: a summary

The asymptotic capacity scaling behaviors of massive MIMO systems for various MIMO modes and directions are summarized in Table 8.1, assuming a large imbalance between transmit and receive antennas, and perfect CSI at the receiver. Note that for finite imbalance the accuracy of the approximation depends on the correlation between antennas.

8.3   Pilot design for massive MIMO

Spectral and energy-efficient operation relies heavily on the acquisition of accurate CSI at the transmitters and receivers in wireless systems in general, and in Orthogonal Frequency Division Multiplexing (OFDM) and massive multiple antenna systems in particular. Consequently, channel estimation methods have been studied extensively and a large number of schemes, including blind, data-aided, and decision-directed non-blind techniques, have been evaluated and proposed in the literature. One reason for this is that, for conventional coherent receivers, the effect of the channel on the transmitted signal must be estimated in order to recover the transmitted information. As long as the receiver accurately estimates how the channel modifies the transmitted signal, it can recover the transmitted information. In practice, pilot signal-based data-aided techni-ques are used due to their superior performance in fast fading environments and cost efficiency and interoperability in commercial systems [26]–[28].

As the number of antennas at the BS grows large, it is desirable to have pilot-based schemes that are scalable in terms of the required pilot symbols and provide high-quality CSI for uplink data detection and downlink precoding. To this end, massive MIMO systems rely on channel reciprocity and employ uplink pilots to acquire CSI at BSs. Although solutions for non-reciprocal systems (such as systems operating in FDD mode) are available [29], it is generally assumed that massive MIMO systems can only feasibly operate in TDD mode [4][5].

Pilot reuse in general causes contamination of the channel estimates, which is known as Pilot Contamination (PiC) or pilot pollution. As there are a large number of channels to be estimated in massive MIMO systems, this becomes a significant challenge due to the potentially limited number of pilots available. Consequently, PiC limits the performance gains of non-cooperative MU-MIMO systems [4][30]. Moreover, PiC could cause a saturation effect in the SINR as the number of BS antennas increases to a very large value. This is in contrast to the scenario where the SINR increases almost linearly with the number of antennas without the presence of PiC [30].

In the following subsections the impact of CSI error on the performance of pilots in a massive MIMO setup is shown. Further, two methods for combating pilot contam-ination in massive MIMO systems are presented. The first method, Pilot Power Control (PPC) based on Open Loop Path loss Compensating (OLPC), provides efficient measures against the detrimental effects of PiC. The second method, coded random-access protocol for massive MIMO systems, brings a fundamentally new approach to mitigate pilot contamination, that is, instead of avoiding intra-cell inter-ference it embraces it.

8.3.1   The pilot-data trade-off and impact of CSI

Although pilot-based CSI acquisition is advantageous in fast fading environments, its inherent trade-offs must be taken into account when designing channel estimation techniques for various purposes. These purposes include demodulation, precoding or beamforming, spatial multiplexing and other channel-dependent algorithms such as frequency-selective scheduling or adaptive Modulation and Coding Scheme (MCS) selection [11]–[13]. The inherent trade-offs between allocating resources to pilot and data symbols include the following:

- Allocating more power, time or frequency resources to pilots improves the quality of the channel estimates, but also leaves fewer resources for uplink or downlink data transmission [11]–[13][16][31].
- Constructing longer pilot sequences (for example, employing orthogonal symbol sequences such as those based on the well-known Zadoff-Chu sequences in LTE-Advanced systems) helps to avoid tight pilot reuse in multi-cell systems, which helps

to reduce or avoid inter-cell pilot interference. On the other hand, spending a greater number of symbols on pilots increases the pilot overhead and might violate the coherence bandwidth [13][32][33].

• Specifically in multiuser MIMO systems, increasing the number of orthogonal pilot sequences may increase the number of spatially multiplexed users at the expense of spending more symbols when creating the orthogonal sequences [12].

In addition to these inherent trade-offs, the arrangement of the pilot symbols in the time, frequency and spatial domains have been shown in practice to have a significant impact on the performance of MIMO and massive MIMO systems in particular; see for example [11][12][33].

8.1.1.1   Impact of channel state information errors on the throughput of massive MIMO systems

When CSI is affected by channel estimation errors, the performance of channel adaptive algorithms, such as symbol detection or MIMO transmission mode selection, degrades. This leads to an overall degradation of the system performance in terms of spectral or energy efficiency, as well as provided end-user data rates or Quality of Service (QoS) [34].

For example, a per-cell sum rate degradation due to erroneous CSIT in a system employing 50 antennas at the BS from 17 bps/Hz down to 5 bps/Hz was reported in [34] while [35] noted that the per-cell sum rate degradation depends heavily on whether the BS uses Maximum Ratio Transmission (MRT) or ZF for the downlink transmission; see Figure 8.4. The figure compares the per-cell sum rate performance of a 7-site TDD

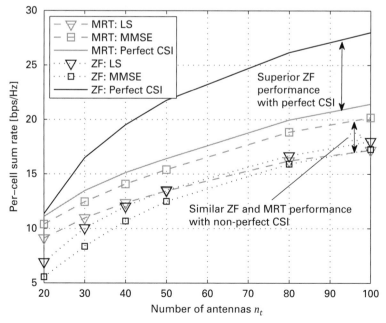

Figure 8.4   Comparison of the impact of channel estimation error on the downlink sum rate with MRT and ZF precoding.

system with an inter-site distance of 500 m, in which each site accommodates 3 cells and serves 12 users per cell. Least Square (LS) and Minimum Mean Square Error (MMSE) estimation methods are compared. It can be noticed that with imperfect CSI, the performance degradation of ZF precoding is more severe than that of MRT precoding, even with a moderate number of BS antennas.

8.3.2   Techniques to mitigate pilot contamination

In order to reduce the pilot overhead, pilot sequences in nearby cells can be reused at the cost of intercell pilot interference; leading to PiC as introduced before. PiC has been shown to limit the achievable performance of non-cooperative MU MIMO systems [36].

Specifically, it has been found that PiC may cause the saturation of the SINR as the number of BS antennas increases to infinity, while the SINR increases approximately linearly with the number of BS antennas n_t in the absence of PiC. More precisely, as pointed out by [36], when the number of users is comparable to the number of antennas, the performance of a simple matched filter with contaminated estimate is limited by the pilot interference. These insights led the research community to find effective measures to mitigate the impact of PiC, both in non-asymptotic and asymptotic regimes [37][38].

A large number of PiC mitigation schemes have been evaluated in the literature due to the importance of accurate CSI and the strong negative impact on the quality of the acquired CSI. These schemes differ in complexity and in the assumptions regarding multi-cell cooperation, as well as in whether they operate in the time, frequency or power domains.

A precoding scheme was proposed in [39], according to which each BS linearly combines messages aimed to users of different cells that reuse the same pilot sequence. This limited collaboration between BSs can resolve the PiC problem and allow for tight pilot reuse.

Another approach to improve the channel estimation involves using a Bayesian estimator that mitigates PiC for spatially well-separated users [34]. However, implementation of the Bayesian estimator relies on the knowledge of the second-order statistics of the useful and crosstalk channels. Acquiring this knowledge entails some overhead of estimating the covariance matrices and computation complexity. An iterative filter may be employed that avoids explicit estimation of covariance matrices, but its convergence remains an open problem [36].

A low-complexity Bayesian channel estimator known as the polynomial expansion channel was proposed in [40]. This method is efficient in the presence of PiC.

Limited cooperation, which is based on the exchange of second-order channel statistics and makes use of Bayesian channel estimation, is proposed in [34]. In this method, PiC can be almost completely eliminated by allocating the same pilots to spatially well-separated users. A different approach based on a less aggressive pilot reuse is proposed in [37]. This approach can be effectively combined by spatially separating users in different cells [38].

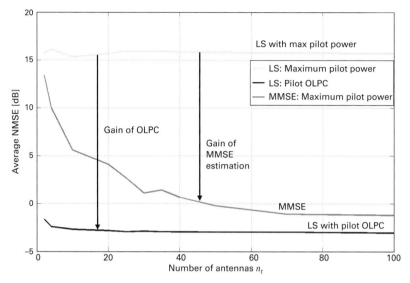

Figure 8.5 The average normalized mean square error (NMSE) performance of LS and MMSE channel estimation.

8.3.2.1 Pilot power control based on open loop path loss compensation

Within the framework of current LTE measurements and employing an LS estimator, pilot power control based on the OLPC scheme of 3GPP LTE systems and pilot reuse schemes can also provide feasible and efficient measures against the detrimental effects of PiC [35]. The positive impact of employing the OLPC scheme on the transmitted pilot symbols is illustrated in Figure 8.5. The figure shows the average normalized mean square error of the estimated channel when LS or MMSE estimation is used by the BS to acquire CSI. A multi-cell system is assumed where either full power or OLPC for pilot transmissions is employed. As the figure shows, LS with OLPC yields lower average NMSE than the much more complex MMSE estimation using maximum pilot power.

8.3.2.2 Coded random access in massive MIMO systems

The rationale of using pilot sequences across the cellular network, in a way that decreases the amount of inter-cell interference, can be challenged [41]. In fact, for a very dense population of users with varying traffic patterns, a centralized pilot allocation may become infeasible. Instead, a random access protocol may be a more suitable choice. Consider a scheme where inside a cell a pool of pilot signals is available for users to select randomly. The pools of different cells are orthogonal. Following this random selection, users from the same cell will choose the same pilot sequence causing a collision in the random access process. Hence, PiC is seen as a collision. However, PiC is limited to within cells, where collisions can be processed and mitigated more easily.

In order to avoid intra-cell interference, conventional massive MIMO systems operate with $\tau \geq K$, where τ is the length of the pilot sequences, and thereby the maximum number of mutually orthogonal sequences, and K is the number of active users in a cell. In [41] a fundamentally different approach to the mitigation of pilot contamination is proposed where $\tau < K$. The philosophy is to move the inevitable interference from outside the cell to within the cell, such that it can be handled through appropriate medium access procedures. The decrease of τ makes it possible to apply larger pilot reuse factors, such that inter-cell interference is virtually non-existent. Hence, instead of considering pilot contamination as an inter-cell interference problem that calls for pilot planning, it is considered as an intra-cell interference problem that calls for Medium Access Control (MAC) protocols. Especially in crowd scenarios (e.g. the large-outdoor-event or stadium use cases presented in Chapter 2) with tens of thousands of users served by a single cell, such an approach is interesting, since insisting on $\tau \geq K$ becomes prohibitively expensive. Furthermore, random MAC protocols, like the different variants of ALOHA, are particularly suited for crowd scenarios with unpredictable traffic patterns. An important motivation for coded random access is that the asymptotic normalized throughput, i.e. decoded messages per resource block for increasing K, approaches 1 [42], which is optimal.

Recently, random access protocols (see Chapter 7) have been further developed with inspiration from the area of erasure coding [43][44]. Instead of considering collisions, i.e. multiple users randomly selecting the same resource, as wasted resources, they are processed using SIC. This potentially results in resolved collisions, which improves the overall throughput. The combination of a random access protocol and SIC resembles an erasure code, which makes the theory developed in this area applicable. In [41] the framework of coded random access is adopted in the proposal of a joint pilot training and data transmission scheme. Hence, the solution to the pilot contamination problem becomes an integral part of the MAC protocol. An example of the coded random access in the uplink transmissions for three time slots is depicted in Figure 8.6.

The joint channel and data acquisition is driven by low-complexity and low memory requirements, which excludes collision resolution before spatial processing, demanding storage of many received vector signals. This means that pilot contamination is not solved: contaminated channels are used as matched filters transferring collisions to the post-processing data domain. Two fundamental properties of massive MIMO are

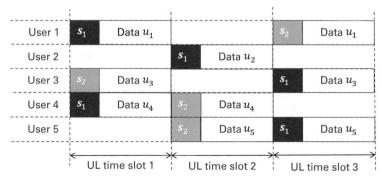

Figure 8.6   Example of uplink transmission using coded random access with 5 users u_i and a set of 2 pilot sequences s_1 and s_2.

exploited to decode the data: (1) asymptotic orthogonality between user channels; and (2) asymptotic invariance of the power received from a user over a short time interval. In the following, the coded random access is analyzed for the uplink and downlink transmissions.

Uplink

An uplink transmission is illustrated in Figure 8.6. Each user is active in a time slot with probability p_a. Active users transmit a pilot sequence $s_k = [s_k(1) \ s_k(2) \ldots s_k(\tau)]$, chosen randomly among a set of size τ, and send a data packet. One given user retransmits the same data at each active time slot, which ensures each message is represented in several encoded messages, similar to erasure codes. $A[v]$ denotes all active users in time slot v and $A^j[v]$ denotes the subset applying s_j. Following the example in Figure 8.6 with $A[2] = \{u_2, u_4, u_5\}$, $A^1[2] = \{u_2\}$, $A^2[2] = \{u_4, u_5\}$, the received uplink pilot signal in time slot v can then be expressed as

$$S[v] = \sum_{j=1}^{\tau} \sum_{k \in A^j[v]} h_k[v] s_j + N_S[v], \qquad (8.24)$$

where $h_k[v]$ is the channel vector from user k in time slot v to the massive array with n_t antennas. Users are assumed to be equipped with a single antenna. $S[v]$ is a $n_t \times \tau$ matrix grouping all the received vectorial signals in the pilot phase. $N_S[v]$ is a matrix of i.i.d. Gaussian noise components containing the noise vectors at the massive array.

All active users transmit a message of length T in the uplink data phase. The message from the kth user is denoted x_k. The received uplink data signal in time slot v is then expressed as

$$Y[v] = \sum_{k \in A[v]} h_k[v] \ x_k + N[v]. \qquad (8.25)$$

If channel estimation is performed based on the received pilot signals, the pilot contamination problem appears. The least squares estimate, $\hat{h}^j[v]$, based on the pilot signal in time slot v from users applying s_j, is found as

$$\hat{h}^j[v] = (s_j s_j^H)^{-1} \ Y[v] \ s_j^H = \sum_{k \in A^j[v]} h_k[v] + N_S^j[v], \qquad (8.26)$$

where $N_S^j[v]$ is the post-processed noise term originating from $N_S[v]$. Instead of achieving the estimate of individual channel vectors, a sum of channel vectors is achieved, i.e. intra-cell interference is experienced. The contaminated estimates are not discarded but instead used as matched filters on the received signals in order to produce linear combinations as follows:

$$f^j[v] = \hat{h}^j[v]^H \ Y[v] = \sum_{k \in A^j[v]} \|h_k[v]\|^2 \ x_k + \overline{N}^j[v], \quad \text{and} \qquad (8.27)$$

$$g^j[v] = \hat{h}^j[v]^H \, S[v] = \sum_{k \in A^j[v]} \|h_k[v]\|^2 \, s_j + \overline{N}_S^j[v]. \tag{8.28}$$

$\overline{N}^j[v]$ and $\overline{N}_S^j[v]$ contain post-processed noise terms as well as cross-user channel scalar products which are null only asymptotically. Both space-time equation systems are jointly exploited to recover the data. The coefficients of the linear combinations in (8.27) and (8.28) are the two-norms, $\|h_k[v]\|^2$, of the involved channels. In a massive MIMO system, these can be assumed slowly fading, contrary to the fast fading of the many individual channel coefficients. This enables a simplified application of SIC on the filtered signals in order to solve the equation systems. Initially, signals with no contamination are identified, which provide the corresponding channel powers and data directly. These signals are then cancelled out, through subtraction from any other signals they may appear in. The iterative process continues until all data has been recovered.

As a simple example, consider time slots 1 and 3 in Figure 8.6 and noise-free reception. The channel energies $\|h_k[v]\|^2$ are stable in time, so that the time index is not written. User 1 and user 4 collide in time slot 1: the channel estimate corresponding to the transmission of pilot s_1 is $h_1 + h_4$. Applying this contaminated estimate as a matched filter in the training and data domain, two signals are obtained $(\|h_1^2\| + \|h_4\|^2)s_1$ and $\|h_1\|^2 \, x_1 + \|h_4\|^2 \, x_4$. In time slot 3, user 1 has an uncontaminated transmission: the channel estimate corresponding to the transmission of pilot s_2 is h_1. Applying it as a matched filter, the following signals are obtained $\|h_1\|^2 \, s_2$ and $\|h_1\|^2 \, x_1$. Knowing h_1 and hence $\|h_1\|^2$, x_1 can be estimated in time slot 3. Removing the contribution of user 1 in time slot 1 decontaminates user 4, enabling an estimation of $\|h_4\|^2$ and thereby x_4. This resembles belief propagation decoding in erasure codes. With appropriate selection of p_a, the degrees, i.e. the plurality of the collisions, can be shaped to follow a distribution, which favors belief propagation decoding.

Existing literature does not consider random MAC protocols specifically designed for massive MIMO systems. Hence, the only reference for performance comparisons is the conventional slotted ALOHA protocol. In [41], it is shown that coded random access approximately doubles the throughput (for $n_t = 500$) compared to slotted ALOHA. See also Section 7.6.3 for more details on coded random access.

Downlink

Conventionally, downlink operation in a massive MIMO system relies on channel estimates achieved in the uplink phase and the assumption of reciprocity. However, such estimates are not guaranteed to be available when applying SIC in the uplink, since only the channel norms are obtained after successful decoding. A solution is to perform data assisted channel estimation. Provided the x_k's are sufficiently long, it is a valid assumption that they are mutually orthogonal for all k. Hence, after successfully decoding x_k, an estimate of $h_k[v]$ can be found as

$$\hat{h}_k[v] = (x_k \, x_k^H)^{-1} \, Y[v] \, x_k^H = h_k[v] + N_k[v]. \tag{8.29}$$

Downlink transmission for user k is thus possible whenever an uplink message has been successfully decoded.

8.4   Resource allocation and transceiver algorithms for massive MIMO

In order to exploit the advantages of massive MIMO, in particular, the spatial multiplexing and array gain offered, the allocation of resources among the users needs to be designed carefully. In particular, a number of issues shall be taken into account: which users and how many shall be scheduled simultaneously, how the antenna weights (at the transmitter and receiver) shall be chosen, shall the transceiver (i.e. antenna weight vector) and scheduling be designed jointly, etc. Moreover, this section investigates specifically two resource allocation schemes designed for a massive MIMO system.

The first scheme suggests how to design the transceiver of a massive MIMO system in order to avoid the full exchange of CSI centrally between BSs. In particular, the optimal minimum power beamformers are obtained locally at each BS relying on limited backhaul information exchange between BSs.

The second scheme is baptized interference clustering and user grouping. It allows serving a large number of users in a massive MIMO system by properly combining user clustering, grouping and precoding.

8.4.1   Decentralized coordinated transceiver design for massive MIMO

In general, coordinated resource allocation problems can be formulated as optimization problems, maximizing a desired utility in the network subject to some constraints, which can be solved iteratively along with exchange of some information between nodes at each iteration [45]–[48]. In a Massive MIMO setting, however, the coordinated transceiver design methods can be potentially simplified, as compared to state-of-the-art iterative optimization based schemes. The coordinated multi-cell minimum power beamforming approach, which has been studied widely in the past, is such a problem that satisfies a given SINR for all users while minimizing the total transmitted power [46][48][49]. This scheme is the basis for more general problems, and various solutions have been suggested for solving this optimization problem, based on uplink-downlink duality and optimization decomposition approaches. Sharing instantaneous CSI between nodes under a delay constraint and limited backhaul capacity becomes an important problem when the dimensions of the problem, i.e. the number of antennas n_t and the number of users K, grow large or when dealing with a fast fading scenario.

A scheme to avoid the full exchange of CSI is based on the decoupling method developed in [46] where the optimal minimum power beamformers can be obtained locally at each BS relying on limited backhaul information exchange between BSs. The original centralized problem is reformulated hereafter and named as decentralized coordinated transceiver design, such that the BSs are coupled by real-valued inter-cell interference terms. The centralized problem can be decoupled by a primal or dual decomposition approach leading to a distributed algorithm. The optimal beamformers are obtained based on convergence of an iterative solution that solves the local problems and exchanges resulted Inter-Cell Interference (ICI) values at each iteration.

The decentralized coordinated transceiver scheme consists of a set of rules how to design the transceiver of a massive MIMO system in order to avoid the full exchange of CSI centrally between BSs. In particular, the optimal minimum power beamformers are obtained locally at each BS relying on limited backhaul information exchange between BSs. This scheme exploits the results of Random Matrix Theory (RMT) while making use of the Stieltjes transform [22][23].

8.4.1.1 System model

The cellular system consists of N_B BSs and K single antenna users; each BS has n_t transmitting antennas. Users allocated to BS b are in set U_b. The signal for user k consists of the desired signal, intra-cell and inter-cell interference. Let $\boldsymbol{h}_{b,k}$ be the channel from the BS b to the user k. Assuming $\boldsymbol{w}_k$ as the transmit beamformer for user k, the following coordinated minimum power precoding problem subject to user-specific SINR constraints $\gamma_k \ \forall k \in U_b, \ b \in B$ is solved [46]:

$$\min \sum_{b \in B} \sum_{k \in U_b} \|\boldsymbol{w}_k\|^2$$

$$\text{subject to} \ \frac{|\boldsymbol{w}_k^H \boldsymbol{h}_{b_k,k}|^2}{\sigma^2 + \sum_{l \in U_{b_k}, \ k} |\boldsymbol{w}_l^H \boldsymbol{h}_{b_k,k}|^2 + \sum_{b \neq b_k} \varepsilon_{b,k}^2} \geq \gamma_k \ \forall k \in U_b, b \in B \ \text{ and}$$

$$\sum_{l \in U_b} |\boldsymbol{w}_l^H \boldsymbol{h}_{b_l,k}|^2 \leq \varepsilon_{b,k}^2 \forall k \neq U_b, b \in B, \tag{8.30}$$

where the optimization variables are $\boldsymbol{w}_k$ and inter-cell interference variable $\varepsilon_{b,k}^2$.

A generic per-user channel correlation model is introduced such that

$$\boldsymbol{h}_{b,k} = \boldsymbol{\theta}_{b,k}^{\frac{1}{2}} \boldsymbol{g}_{b,k}, \tag{8.31}$$

where $\boldsymbol{\theta}_{b,k}$ is the correlation matrix of user k and $\boldsymbol{g}_{b,k}$ is a vector with i.i.d. complex entries with variance $1/n_t$. This per-user channel correlation model can be applied to various propagation environments.

The optimization problem above can be solved iteratively using uplink-downlink duality by first computing the power allocation and beamformers in the dual uplink [46] [48]. It was shown that, by using large system analysis, the approximately optimal uplink powers λ_k for the generic model are given by [22][23]

$$\lambda_k = \frac{\gamma_k}{\nu_{b_k, \theta_{b_k,k}}(-1)}, \tag{8.32}$$

where $\nu_{b_k, \theta_{b_k,k}}(z)$ is the Stieltjes transform of a measure defined for $z \in C \backslash R^+$ by [23], and

$$v_{b_k, \theta_{b_k,k}}(z) = \frac{1}{n_t} tr \boldsymbol{\theta}_{b_k,k} \left(\frac{1}{n_t} \sum_l \frac{\lambda_l \theta_{b_k,l}}{1 + \lambda_l v_{b_k, \theta_{b_k,l}}(-1)} - z \boldsymbol{I}_{n_t} \right)^{-1}. \tag{8.33}$$

The approximation for downlink powers can be derived similarly, see [22][23] for details. The above approximations result in an algorithm that gives the approximately optimal uplink and downlink powers based on local CSI and statistics of other BS channels. However, the error in approximations causes variations in the resulting SINRs and rates. Thus, the SINR constraints cannot be guaranteed and those achieved SINRs might be higher or lower than the target SINRs. In this case, the SINR constraints can be met only asymptotically as the number of users and antennas grow large.

Next, an alternative approach for decoupling the sub-problems at the BSs is described. Following the same logic as in [46], ICI is considered as the principal coupling parameter among BSs, and the large dimension approximation for ICI terms based on statistics of the channels is derived as [22]:

$$\varepsilon_{b,k}^2 \approx \sum_{l \in U_b} \frac{1}{n_t} \frac{v'_{b_k, \theta_{b_k,l}}(-1)}{\left(1 + \lambda_l v_{b_k, \theta_{b_k,l}}(-1) \right)^2}, \tag{8.34}$$

where $v'_{b_k, \theta_{b_k,l}}(-1)$ is the derivative of the Stieltjes transform $v_{b_k, \theta_{b_k,l}}(z = -1)$ (see [23]), δ_l is the downlink power weighting scalar that relates the optimal downlink and uplink precoding/detection vectors of user k i.e. $\mathbf{w}_k = \sqrt{\delta_k} \tilde{\mathbf{w}}_k$, where $\tilde{\mathbf{w}}_k$ is the MMSE uplink detection vector of user k. This approximation allows the derivation of approximately optimal ICI terms based on statistics of the user channels. Each BS needs the knowledge about user-specific average statistics, i.e. user-specific correlation properties and path loss values from other BSs (these statistics can be exchanged over the backhaul between coordinating network nodes). In addition, each BS needs to know the local CSIT to each user within the coordination cluster. Based on the statistics, each BS can locally and independently calculate the approximately optimal ICI values. Plugging the approximate ICI into the primal problem decouples the sub-problems at the BSs and the resulting SINRs satisfy the target constraints with slightly higher transmit power compared to the optimal method. The fact that the coupling parameters depend only on channel statistics results in a reduced backhaul exchange rate and processing load. Moreover, the algorithm can be applied to fast fading scenarios as the channel statistics (path loss, correlation properties) change slower than the instantaneous channel realizations.

8.4.1.2   Performance results

The two algorithms developed in the previous section for a multi-cell system with large dimensions provide good approximations even when the dimensions of the problem (i.e. the number of users and antennas) are practically limited. In order to show the performance of the approximate algorithms, some numerical examples are presented in this section. The results for the algorithm based on ICI approximation are presented. This

algorithm satisfies the target SINRs for all users; however, the error in the approximations results in a somewhat higher transmit power at the BSs. A wrap-around network
with 7 cells is considered and users are equally distributed within cells. An exponential
path loss model is used for assigning the path loss to each user,

$$a_{b,k} = \left(\frac{d_0}{d_{b,k}}\right)^{2.5}, \tag{8.35}$$

where $d_{b,k}$ is the distance between BS b and user k. The path loss exponent is 2.5 and the
reference distance (d_0) is 1m. The path loss from a BS to the boundary of the reference
distance of the neighboring BS is fixed to 60 dB. The correlation among channel entries
is introduced using a simple exponential model

$$[\boldsymbol{\theta}_{b,k}]_{i,j} = \rho^{|i-j|}, \tag{8.36}$$

where ρ represents the correlation coefficient which is 0.8 for the following simulations.
The users are dropped randomly for each trial and in total 1000 user drops are used for
calculating the average transmit power. The number of antennas at each BS varies
from 14 to 84 and the total number of users is equal to half the number of antennas at
each BS. Thus, the spatial loading is fixed as the number of antennas is increased.
Figure 8.7(a) illustrates the transmit SNR versus the number of antennas for a 0 dB
SINR target. It is clear that the gap between the approximated and optimal algorithm
(denoted as centralized) diminishes as the number of antennas and users increase.
The small gap for small dimensions indicates that the approximate algorithm can be
applied to the practical scenarios with a limited number of antennas and users. From
the results, it is clear that the centralized algorithm and the approximated ICI
algorithm outperform the ZF method. Note that the gap between ZF and optimal
and approximated method is fixed, which is due to the fixed ratio of the number of
antennas to the total number of users. The gap in performance is mainly because the
ZF algorithm wastes a degree of freedom for nulling the interference toward the
distant users, while the centralized algorithm finds the optimal balance between
interference suppression and maximizing the desired signal level. MF beamforming
must be dealt with more care since it completely ignores the interference (both intra-
and inter-cell), and hence the SINR target is below the target SINR shown in
Figure 8.7(b). Note that MF beamforming can satisfy the target SINR only asymptotically in a very special case, i.e. when the ratio of the number of antennas to the
number of users approaches infinity.

8.4.2   Interference clustering and user grouping

Although massive MIMO transmission reduces the inter-user interference significantly,
multi-cell interference is still an issue that limits the overall system performance.
A particular interference clustering and user-grouping scheme described in this section
allows tackling the interference by separating the multi-cell interference mitigation from

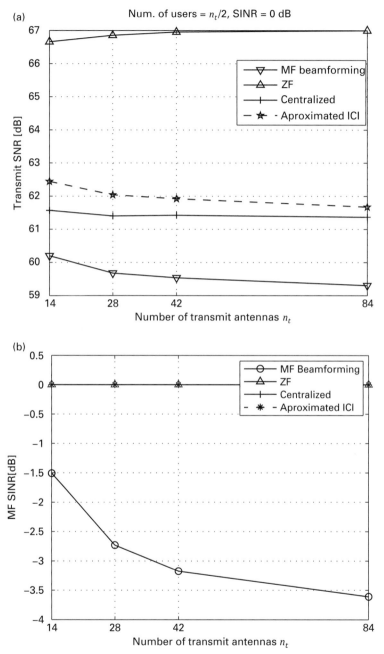

Figure 8.7   Comparison of required transmit SNR for 0 dB SINR target.

the multi-user multiplexing at each BS. In particular, a joint approach combining user clustering, grouping and regularized precoding [50] is considered.

The scheme is depicted in Figure 8.8. It can be divided into three steps applying a two-stage beamforming precoding. It should be noted that the two-stage

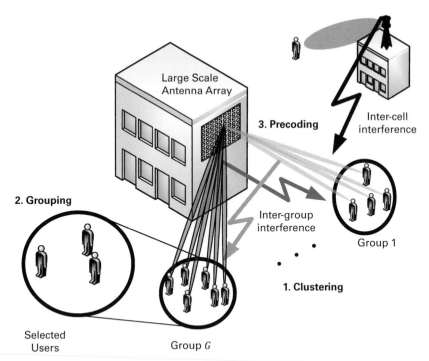

Figure 8.8   Inter-group interference clustering, user grouping and inter-cell interference aware precoding with massive MIMO antenna arrays.

beamforming-precoding concept can be either implemented in digital domain only or in a hybrid form as shown in Section 8.5.3:

- First (beamforming) stage: Inter-group interference aware clustering of co-located users into groups resulting in wide-band multi-user beamforming. The beamforming is extracted from the channel's second order statistics, more precisely channel covariance information.
- Second (precoding) stage (per user group): Optimizes the multi-user SINR conditions, e.g. based on regularized ZF precoding constraints. User selection for downlink transmission on the same time-frequency resource, and inter-cell interference aware precoder design $\mathbf{P}(k)$ are performed for selected users based on small-scale fading CSI feedback.

In the first step, the user clustering, the set of users K is divided into G groups of users with similar second-order channel statistics. The Joint Spatial Division Multiplexing (JSDM) [51] concept of first-stage beamforming is used to spatially separate the channels of each user group. A density-based clustering algorithm called Density-based Spatial Clustering of Applications with Noise (DBSCAN) is employed. This algorithm clusters an adaptive number of groups with respect to a certain user density that is adapted to the level of inter-group interference.

In the second step, which is done independently for each group, Semi-orthogonal User Selection (SUS) [51] based grouping is used to find a subset of users for simultaneous downlink transmission on the same time-frequency resource but on different spatial layers. In combination with the selected groups for downlink transmission, this step covers the resource allocation. Note that the groups can be selected in time and frequency, whereas the SUS corresponds to spatial domain resource allocation. Here, the SUS algorithm is adapted with the maximum sum-rate objective using projection-based rate approximation according to [52]. Hence, it is ensured that the sum-throughput is increased while the limited transmit power budget is divided among all active spatial data streams.

Finally, in the third step, the second stage precoders are designed at the BS. In order to balance the desired signal and the suppression of multiuser interference with remaining non-treated inter-cell interference, regularized ZF precoding is considered. This is an important step as the BS can obtain the channel via reciprocity, but does not have any knowledge about the interference or SINR situation at the user. Since the modulation and coding scheme cannot be selected matching the SINR conditions at the user, regularized zero-forcing precoding is used such that the inter-cell interference is considered for the regularization weights. To obtain knowledge of the SINR condition at the user, a scalar broadband power-value, obtained as the average of the diagonal elements from the interference-covariance matrix measured at the user, is introduced [52]. This power-value is fed back from the users to the BS independently of the time or frequency division duplex system.

Combining all these three steps, large performance gains are shown in the order of 10 times for the sum-throughput compared to a baseline scenario with 8 transmit antennas.

8.4.2.1   Performance results

The following performance evaluation focuses on an FDD downlink. To evaluate the first step of this approach, a single sector with a 1x256 antenna array and 6 physically closely located clusters of users, each consisting of 6 users, is utilized. This clustering into space-orthogonal groups has to be done with respect to inter-group interference. The clustering with the well-known K-means and K-means++ algorithm [53] with a joint spatial-division multiplexing (JSDM) scheme is not practical for this task since the number of groups G is required as an input parameter, which is not known a priori. An exhaustive search for the optimal G of each user constellation is hardly feasible. Therefore, a density based clustering algorithm is introduced by adapting the DBSCAN algorithm.

The K-means++ and DBSCAN algorithms are compared with the baseline scenario in Figure 8.9. For the baseline, 8 transmit antennas and multi-user regularized zero-forcing beamforming were considered. The gain for ZF precoding over JSDM with per-group processing is relatively small and attributes to 11% for DBSCAN user clustering. Note that in JSDM only multi-user interference within in the same user group is mitigated by ZF beamforming (second stage) while inter-group interference is considered in the first precoding stage only.

Assuming perfect channel knowledge of the downlink user channels at the BS, Figure 8.10 shows the multiplexing gain by increasing the number of user candidates $K = |K|$ from 40 to 120. Further, the following two user-grouping modes are analyzed:

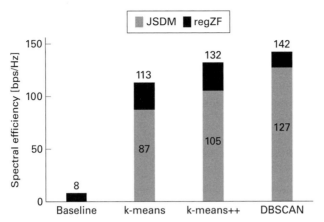

Figure 8.9   JSDM with per group precoding compared to joint regularized ZF for 8 (baseline) and 256 (massive MIMO) transmit antennas.

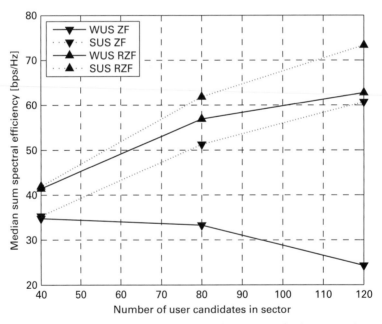

Figure 8.10   Multi-cell performance over users with and without user selection comparing zero-forcing and regularized zero-forcing with interference power feedback.

- **WUS:** Without User Selection (WUS), where the set of scheduled users is $\mathcal{K}_{\text{WUS}} = \mathcal{K}$
- **SUS:** With SUS, where the set of scheduled users is $\mathcal{K}_{\text{SUS}} \subset \mathcal{K}$

The solid and dashed lines with upper triangles in Figure 8.10 consider interference aware Regularized ZF (RZF) precoding. It can be observed that RZF precoding (compared to ZF) yields 150% and 20% gain at $K = 120$ for SUS and WUS, respectively. As a main result, multi-user interference is reduced to the level of noise plus out-of-cell interference resulting in higher signal power per user and smaller loss due to precoder normalization. The gain from SUS over WUS for RZF vanishes for a small number of users, i.e. $K = 40$, and increases by up to 16% for $K = 120$ users. Note that SUS requires a larger set of users to provide feedback and thus significantly increases feedback overhead. For the simple ZF precoding advanced user grouping (such as SUS) is mandatory, as otherwise the system performance degrades significantly, refer to the solid line with downwards pointing triangle.

8.5   Fundamentals of baseband and RF implementations in massive MIMO

8.5.1   Basic forms of massive MIMO implementation

Beside the theoretical analysis for SU and MU massive MIMO, one major limiting aspect of massive MIMO is related to hardware constraints [54]. It is clear that RF hardware complexity scales with the number of active antenna elements n_t in the system. This has to be taken for granted and cannot be avoided in the context of massive MIMO. However, the complexity of massive MIMO implementations can differ strongly depending on the extent to which precoding and beamforming are conducted in the digital frequency/time or analog time domain. As noted at the beginning of this chapter, the term *precoding* here relates to the usage of individual phase shifts for each antenna and each sub-part of the system bandwidth, while the term *beamforming* relates to the usage of common phase shifts over the entire system bandwidth. In this section, the possible basic forms of massive MIMO hardware implementation are shown.

    In order to utilize all degrees of freedom and extract all gains from the massive MIMO channel, it is required to introduce a channel-dependent individual phase shift to each antenna element and each sub-carrier. Consequently, frequency-selective precoding should be conducted in the digital baseband, as illustrated in Figure 8.11. The amount of baseband chains is denoted as $L \geq M$, where M represents the number of MIMO streams for all the users. The precoding is employed to cope with inter-beam interference for the L beams. Note that the beamforming and the precoding can be performed simultaneously by one matrix multiplication combining the precoding matrix with the digital BF matrix. In this setup, the number of baseband chains is equivalent to the number of RF chains, i.e. $L = n_t$. In the figure, the different signal processing blocks refer to the application of an Inverse Discrete Fourier Transform (IDFT) and the introduction of

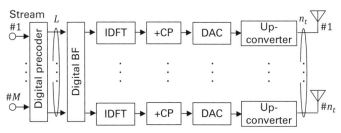

Figure 8.11   Massive MIMO OFDM transmitter employing digital precoding.

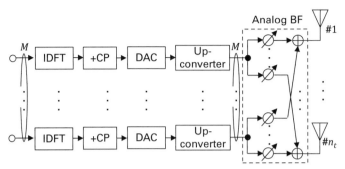

Figure 8.12   Massive MIMO OFDM transmitter employing analog BF.

a Cyclic Prefix (CP) before Digital to Analog Conversion (DAC). This approach is clearly very complex and costly, as the number of baseband signal processing chains have to equal the number of transmit antennas, and one requires CSI for each single channel coefficient on each sub-carrier, which would imply a large pilot and channel estimation and feedback overhead, unless channel reciprocity can be utilized in TDD.

The implementation complexity and cost can be significantly reduced by applying phase shifts over the entire system bandwidth, i.e. by applying beamforming in the analog domain, as depicted in Figure 8.12. Here, only M different baseband signal processing chains are employed, for the purpose of serving up to M different streams, and for each stream configurable phase shifts are introduced among the transmit antennas in the RF circuitry. As the searching time of the phase shifts for the analog beamforming should be shortened, only a very limited set of different phase shift configurations are available. This has the positive effect that the effort for CSI feedback is substantially reduced, as it is sufficient to feed back the index of the preferred phase shift configuration from the receiver to the transmitter. On the other hand, the analog beamforming approach implies that the signals transmitted from the different antennas cannot coherently align (constructively or destructively) as perfectly as in the case of digital, frequency-selective precoding. This means that on one hand the array gains are not as large as they could be in the digital precoding case, and also that a certain extent of residual interference between streams or users is introduced. Further, the fact that all phase shifts are applied equally over the entire system bandwidth is clearly suboptimal in the case the propagation channel is highly frequency-selective.

For these reasons, a more suitable approach to precoding and beamforming in the context of massive MIMO is to use a hybrid beamforming approach [55]–[58], i.e. where some extent of frequency-selective precoding is performed in the digital baseband, and further beamforming is applied in the analog RF circuitry. This approach is depicted in Figure 8.13. Here, the M streams are precoded in digital baseband and on a per-sub-carrier basis to L different signals per sub-carrier to cope with the residual interference between the analog beams and extract additional beamforming gain, and then analog beamforming is used to map these signals to the n_t transmit antennas. In this case, only L baseband signal processing chains are required. The performance is in this case clearly a compromise between the fully digital precoding and fully analog beamforming case.

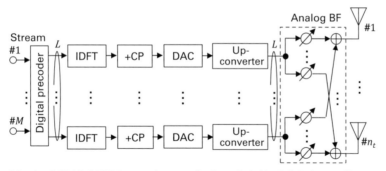

Figure 8.13 Massive MIMO OFDM transmitter employing a hybrid of digital precoding and analog beamforming.

This approach is more suitable to be used for frequency-selective channels, but the limited flexibility of the analog beamforming part still leaves some residual inter-stream or inter-user interference. Also, the effort for CSI feedback is now a compromise between the fully digital precoding and fully analog beamforming approach.

For the latter hybrid approach, there are different mechanisms to determine the precoding weights in the digital baseband domain and phase shifts in the analog domain. In the sequel, a particular scheme based on hybrid Fixed BF and CSI-based Precoding (FBCP) [59], which performs a successive optimization of both beamforming domains, is presented.

8.5.2   Hybrid fixed BF with CSI-based precoding (FBCP)

FBCP is referred sometimes as analog FBCP but was re-baptized herein as hybrid FBCP in order to emphasize that it is implemented using a hybrid of precoding in the digital baseband and analog beamforming. The FBCP approach is composed of two successive steps. First, it selects the analog fixed BF weights whose number is larger than the number of the streams and much less than the number of transmitter antennas. More precisely, L fixed BF weights denoted by $\mathbf{W}$ are initially selected from some analog fixed BF weight candidates based on the steering vector for arbitrary 2D angles and according to a maximum total received power criterion [59].

Next, an Eigenmode (EM)-based precoding matrix $\mathbf{P}(k)$ is computed at the kth subcarrier based on the SVD of an equivalent frequency-domain channel matrix $\mathbf{H}(k)\mathbf{W}$ that multiplies the frequency-domain channel matrix $\mathbf{H}(k)$ by the selected fixed BF weights $\mathbf{W}$. Note that $\mathbf{H}(k)\mathbf{W}$ is estimated by exploiting a pilot signal with the selected fixed BF.

8.5.2.1   Performance of FBCP

Link-level simulation results of single-user massive MIMO operating in TDD mode at a carrier frequency of 20 GHz and employing the FBCP approach are shown in terms of downlink throughput performances. The parameters used for the simulations are given in Table 8.2. The number of transmitter antennas, n_t, is set to 16 or 256, and the number of receiver antennas, n_r, is fixed to 16. The number of the streams, M, is fixed to 16. The UE

Table 8.2 Simulation parameters of Massive MIMO.

Concept	Value
Transmission scheme	Downlink Massive MIMO OFDM
Signal bandwidth	400 MHz
Active subcarriers	Pilot: 32; data: 2000
No. of antennas	n_t: 16, 256; n_r: 16
No. of data streams, M	16
Modulation scheme	QPSK: 16QAM, 64QAM, 256QAM
Channel coding	Turbo code: $R = 1/2, 2/3, 3/4$
Maximum bit rate	31.4 Gbps (256QAM, $R = 3/4$)
Antenna array structure	UPA
Angular power spectrum	Laplacian distribution in θ
	Wrapped Gaussian distribution in ϕ
Average angle (θ, ϕ)	Departure: (90 deg, 90 deg)
	Arrival: (90 deg, 90 deg)
Angular spread (θ, ϕ)	Departure: (5 deg, 5 deg)
	Arrival: (20 deg, 20 deg)
Channel model	Kronecker model
Fading channel	Nakagami-Rice ($K = 10$ dB); 16-path

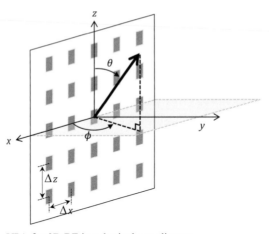

Figure 8.14   UPA for 3D BF in spherical coordinates.

receiver detects the spatially multiplexed streams by using a postcoding (receive weight) matrix that is calculated from the SVD of the equivalent channel matrix $\mathbf{H}(k)\mathbf{W}$. Both the transmitter and the receiver employ a 2D Uniform Planar Array (UPA) as the antenna array structure. As shown in Figure 8.14, θ and ϕ are zenith and azimuth angles. The maximum bit rate reaches 31.4 Gbps according to the parameters set, where the modulation and coding scheme is 256QAM with coding rate, R, of 3/4, using a turbo code. Ideal adaptive modulation and coding (AMC) is assumed. The channel model was based on the Kronecker model including Line of Sight (LOS) and Non-Line of Sight (NLOS) components.

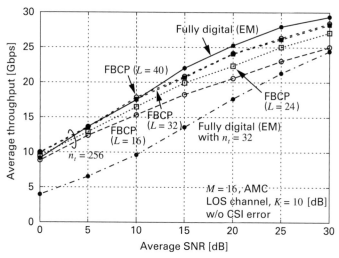

Figure 8.15   Throughput of FBCP for different choices of L.

Figure 8.15 shows the throughput performance of the FBCP for different choices of L. Ideal channel estimation is assumed and the pilot and feedback overhead is not taken into account in the throughput calculation. Angular intervals for the possible analog BF steps in zenith and azimuth are fixed to 5 degrees. For comparison, the throughput performances of the EM precoding that employs the fully-digital Massive MIMO, i.e. with $L = n_t$, are also plotted in Figure 8.15. It is shown (for the $n_t = 256$ case) that as L increases, the throughput of the FBCP approaches the fully digital EM-based precoding and converges for L equal to 32. Moreover, in comparison with the conventional fully digital MIMO with $n_t = 32$, the FBCP with $n_t = 256$ and $L = 32$ can reduce the required SNR for 20 Gbps throughput by more than 9 dB by exploiting higher BF and diversity gains, while requiring the same number of baseband signal processing chains.

Figure 8.16 shows the throughput performance of the FBCP when this is subject to CSI errors. Again, the throughput performance of the fully digital EM-based precoding is added to the figure for comparison. For all schemes, the CSI error is generated according to a complex Gaussian distribution with zero mean and a variance of σ_e^2. σ_n^2 denotes the noise power per antenna, and σ_e^2 is set to $\sigma_n^2 - 10$ dB or $\sigma_n^2 - 20$ dB. Figure 8.16 demonstrates that as the CSI error increases, the throughput of the fully-digital EM-based precoding approach drastically degrades, while the FBCP is robust to the CSI error. This can be explained by the fact that the fully-digital approach requires accurate CSI to exploit the full array gain. It is also found that when $\sigma_e^2 = \sigma_n^2 - 20$ dB, the FBCP can achieve the same throughput as the fully-digital approach, despite the significantly reduced complexity.

Finally, the throughput performance of the FBCP subject to phase errors is shown in Figure 8.17. The phase errors reflect the hardware impairment in the RF phase shifters of the FBCP and are generated by Gaussian distribution with zero mean and variance of σ_p^2. The standard deviation σ_p is set to 3 degrees or 5 degrees. From Figure 8.17, it can be seen that the FBCP achieves the same throughput irrespective of the phase error and is

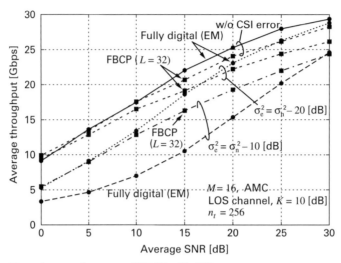

Figure 8.16   Throughput performance of FBCP with CSI error.

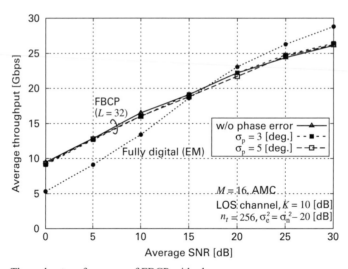

Figure 8.17   Throughput performance of FBCP with phase error.

fairly robust to this hardware impairment. Since the analog BF weights are chosen based on the maximum received power criterion and inherently take the phase errors into account, the FBCP is robust against the phase error.

8.5.3   Hybrid beamforming for interference clustering and user grouping

In this section, the advantage of using hybrid beamforming is shown for the interference clustering and user-grouping scheme that was presented in Section 8.4.2. Figure 8.18 illustrates the interference clustering and user-grouping scheme implemented using

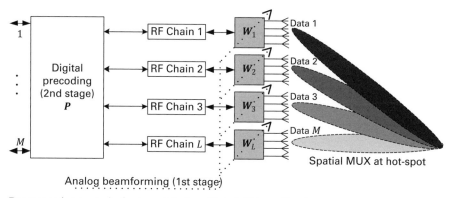

Figure 8.18   Beam steering control of antenna blocks with hybrid beamforming.

hybrid BF, where the number of radio frequency chains is L and the number of data streams M. Furthermore, it is assumed a system with $L = M$ streams for K users and n_t antennas at the BS. In contrast to the hybrid BF shown in Figure 8.13, here it is assumed that each RF chain feeds a single set of subarray antennas. Further, the antenna array is partitioned into groups [17] of $\vartheta = n_t / L$ elements.

Mathematically, this means that the $L \times n_t$ matrix $\mathbf{W}$ is constrained to take on a block-diagonal form with L diagonal blocks of dimension $\vartheta \times 1$, each (column) block being the steering vector for the corresponding beam-steering antenna. Consider for example the case where the target is to serve M users belonging to the same group 1 and at the same time create low interference to users in group 2. For an unconstrained first-stage beamforming matrix $\mathbf{W}$, this is easily achieved by block diagonalisation (see above and details in [40]) in the case of the low-complexity hybrid BF implementation with beam-steering antennas. However, when imposing a block-diagonal constraint on $\mathbf{W}$, the optimization of the beamforming matrix for the multi-user multi-cell downlink is far from obvious. Driven by the LOS intuition this should be possible if the matrix $\mathbf{W}$ defines multiple beams that point in the direction of the desired user-group 1 and put a minimum in the direction of the undesired user-group 2. The multi-user multiplexing in user-group 1 is enabled through the baseband precoding matrix $\mathbf{P}(k)$ and requires CSIT incorporating the analog beamforming. However, whether this is possible and to what extent one can actually achieve a multi-user multiplexing gain for each group of users, has to be verified through an accurate statistical channel model and extensive system simulation.

8.5.3.1   Performance of hybrid BF for interference mitigation

To evaluate the two-stage beamforming-precoding approach with hybrid BF, the following deployment is considered, as depicted in Figure 8.19. Two transmitting BSs facing toward each other and 2 receiver groups at the cell border, each with 8 users, are considered. The focus of this section is to demonstrate the potential of the hybrid BF, so the deployment is kept intentionally simple. User-groups 1 and 2 are forced to be served by BS 1 and 2, respectively.

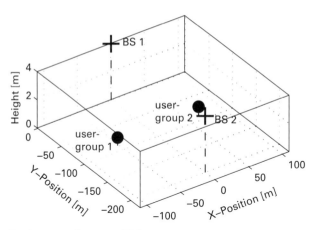

Figure 8.19   Deployment of users and BSs.

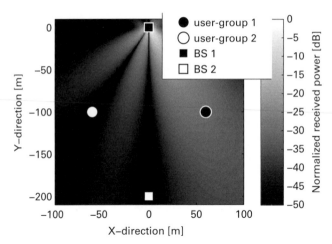

Figure 8.20   Received power of the analog beamformed data stream 1 transmitted from the first antenna block in Figure 8.18.

The BS transmitter antenna is a uniform linear array consisting of $\vartheta = 4$ blocks separated by 2λ, where each block is formed by 4 antennas with $\lambda/2$ spacing as shown in Figure 8.18. For more details on the assumptions, see [17].

In order to utilize the multiplexing gain for a user group, each of the antenna blocks generates the same "broad beam" with the same beamforming vector $\mathbf{w}_L$ selected from a DFT matrix. In a more general form, the beamforming matrix $\mathbf{W} = \mathbf{w}_1 \dots \mathbf{w}_{L-1}\mathbf{w}_L$ is a concatenation of multiple row vectors chosen from any codebook matrix, where each beamforming vector $\mathbf{w}_L$ could have different weight entries. The effective channel $\mathbf{H}(k)\mathbf{W}g(k)$ of such a beam, with $g(k)$ being the channel gain on frequency block k, for a single antenna-block is illustrated in Figure 8.20, where multi-path components are neglected to simplify the visualization. It is seen that the power is steered to user-group 1 and at the same time interference toward group 2 is mitigated.

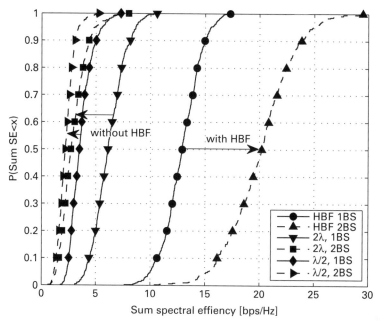

Figure 8.21   Performance comparison of one and two BSs with and without hybrid beamforming.

Performance results from system simulations are shown in Figure 8.21. For comparison, the performance results of a ULA first with $\lambda/2$ and then with 2λ antenna spacing are shown. The optimal performance in this scenario is considered as the interference free single BS case with one user group labeled as 1BS. The case of two BSs and two user groups as in Figure 8.19 is labeled with 2BS. Looking at the bold lines with one transmitter and user group, a beamforming gain from $\lambda/2$ to 2λ spacing can be observed.

Without beamforming and interference mitigation, the sum spectral efficiency (SE) of two transmitters is less than the single transmitter. In contrast, the sum spectral efficiency with the beamforming in Figure 8.21 increases from 13 to 20 bps/Hz. Dividing the sum performance by the number of transmitters results in 10 bps/Hz per BS, which is only a loss of 3 bps/Hz compared to the optimal scenario of two transmitters serving simultaneously the two user groups without any mutual interference. This loss is due to the residual interference from multi-path components in the channel, which cannot be fully eliminated by the analog beamforming projection due to its constrained block-diagonal form.

8.6   Channel models

Massive MIMO and joint transmission have enormous potential to improve the spectral efficiency in mobile communication systems. The realistic performance assessment of such techniques requires having channel models that reflect the true behavior of the radio channel [60]. Existing geometry-based stochastic channel models such as the Spatial Channel Model (SCM), WINNER, and QuaDRiGa model allow the separation of antenna and propagation effects and are thus ideal candidates for the evaluation of new transmission schemes. However, many existing models lack features such as time-evolution and full-3D propagation as well as parameter tables for many important scenarios. In addition, the validation of such models against measured data is still an open issue.

The physical characteristics of the radio channel are fundamentally defining the potential and limitations for a new radio system like 5G. For theoretical analysis, the assumptions in this respect are often quite ideal, like a fully scalable number of Tx antennas, easily tractable Rayleigh fading channels or perfectly calibrated Tx-arrays generating perfect beam patterns with eventually infinitely small half power beam width, negligible multi-user or inter-cell interference, etc.

The definition of new channel models as such, ranging from RF frequencies below 6 GHz to cmW as well as mmW, can be found in Chapter 13. Furthermore, the 3GPP extension of the spatial channel model enhanced to the so-called full dimension MIMO channel model might be of interest [61]. Time-evolutions and full-3D propagation modeling were also included in the QuaDRiGa channel model [62].

For mmW, extensive measurement campaigns have been and are being conducted. The interested reader is referred to [63][64]. For the cmW, interesting results can be found in [65] demonstrating the feasibility of strong spatial multiplexing even in the 28 GHz frequency band over few hundreds of meters.

Massive MIMO arrays in local area scenarios inside of buildings are another very specific application, which is affected by outdoor-to-indoor as well as wall penetration losses (see some results in [66] or [67]).

8.7   Conclusions

This chapter has covered massive MIMO as one of the clear key technology components for 5G. While the analyzed capacity scaling behavior for large numbers of antennas and users appears highly promising, it has become clear that massive MIMO inherits various challenges that have to be overcome.

One key challenge is the aspect of so-called pilot contamination, i.e. the fact that for large massive MIMO constellations with many channel components to be estimated it is unavoidable that channel estimation becomes subject to interference, if pilot overhead is to be kept reasonable. Multiple options to overcome pilot contamination have been discussed, for instance based on pilot power control, coded pilots exploiting sparse channel properties or the random usage of pilot sequences within cells.

Further challenges are related to resource allocation, i.e. the right grouping of users to be served by massive MIMO, and the actual transceiver design, for instance in the context of hybrid beamforming, where part of the precoding is performed in the digital baseband, and part in the analog domain.

References

[1] ICT-317669 METIS project, "Scenarios, requirements and KPIs for 5G mobile and wireless system," Deliverable D1.1, May 2013, www.metis2020.com/docu ments/deliverables/

[2] E. G. Larsson, O. Edfors, Fr. Tufvesson, and T. L. Marzetta, "Massive MIMO for next generation wireless systems," *IEEE Communications Magazine*, vol. 52, no. 2, pp. 186–195, 2014.

[3] H. Q. Ngo, E. G. Larsson, and T. L. Marzetta, "Energy and spectral efficiency of very large multiuser MIMO systems," *IEEE Transactions on Communications*, vol. 61, no. 4, pp. 1436–1449, 2013.

[4] T. Marzetta, "Noncooperative cellular wireless with unlimited numbers of base station antennas," *IEEE Trans. Wireless Commun.*, vol. 9, no. 11, pp. 3590–3600, November 2010.

[5] F. Rusek, D. Persson, B. K. Lau, E. G. Larsson, T. L. Marzetta, O. Edfors, and F. Tufvesson, "Scaling up MIMO: Opportunities and challenges with very large arrays," *IEEE Signal Processing Mag.*, vol. 30, no. 1, pp. 40–60, January 2013.

[6] J. Hoydis, S. T. Brink, and M. Debbah, "Massive MIMO: How many antennas do we need?," in Annual Allerton Conference on Communication, Control and Computing, Monticello, IL, September 2011, pp. 545–550.

[7] L. Lu, G. Li, A. Swindlehurst, A. Ashikhmin, and R. Zhang, "An overview of massive MIMO: Benefits and challenges," *IEEE Journal of Selected Topics in Signal Processing*, vol. 8, no. 5, pp. 742–758, October 2014.

[8] M. Kountouris and N. Pappas, "HetNets and Massive MIMO: Modeling, Potential Gains, and Performance Analysis," in IEEE-APS Topical Conference on Antennas and Propagation in Wireless Communications, Torino, September, 2013.

[9] J. Jose, A. Ashikhmin, T. Marzetta, and S. Vishwanath, "Pilot contamination and precoding in multi-cell TDD systems," *IEEE Transactions on Wireless Communications*, vol. 10, no. 8, pp. 2640–2651, August 2011.

[10] B. Hassibi and B. M. Hochwald, "How much training is needed in multiple antenna wireless links?," *IEEE Transactions on Information Theory*, vol. 49, no. 4, pp. 951–963, 2003.

[11] T. Kim and J. G. Andrews, "Optimal pilot-to-data power ratio for MIMO-OFDM," in IEEE Global Telecommunications Conference, St. Louis, December 2005, pp. 1481–1485.

[12] T. Kim and J. G. Andrews, "Balancing pilot and data power for adaptive MIMO-OFDM Systems," in IEEE Global Telecommunications Conference, San Francisco, December 2006.

[13] T. Marzetta, "How much training is needed for multiuser MIMO?," in Asilomar Conference, Pacific Grove, October 2006, pp. 359–363.

[14] N. Jindal and A. Lozano, "A unified treatment of optimum pilot overhear in multipath fading channels," *IEEE Transactions on Communications*, vol. 58, no. 10, pp. 2939–2948, October 2010.

[15] K. T. Truong, A. Lozano, and R. W. Heath Jr., "Optimal training in continuous block-fading massive mimo systems," in European Wireless Conference, Barcelona, May 2014.

[16] G. Fodor and M. Telek, "On the pilot-data power trade off in single input multiple output systems," in European Wireless Conference, Barcelona, May 2014, pp. 485–492.

[17] M. Kurras, L. Thiele, and G. Caire, "Interference mitigation and multiuser multiplexing with beam-Steering antennas," in International ITG Workshop on Smart Antennas, Ilmenau, March 2015, pp. 1–5.

[18] R. Couillet and M. Debbah, *Random Matrix Methods for Wireless Communications*. Cambridge, UK: Cambridge University Press, 2011.

[19] S. Wagner, R. Couillet, M. Debbah, and D. T. Slock, "Large system analysis of linear precoding in correlated MISO broadcast channels under limited feedback," *IEEE Trans. Inform. Theory*, vol. 58, no. 7, pp. 4509–4537, 2012.

[20] D. Gesbert, S. Hanly, H. Huang, S. Shamai Shitz, O. Simeone, and W. Yu, "Multi-cell MIMO cooperative networks: A new look at interference," *IEEE Journal on Selected Areas in Communications*, vol. 28, no. 9, pp. 1380–1408, December 2010.

[21] A. Tölli, M. Codreanu, and M. Juntti, "Cooperative MIMO-OFDM cellular system with soft handover between distributed base station antennas," *IEEE Trans. Wireless Commun.*, vol. 7, no. 4, pp. 1428–1440, April 2008.

[22] H. Asgharimoghaddam, A. Tölli, and N. Rajatheva, "Decentralizing the optimal multi-cell beamforming via large system analysis," in IEEE International Conference on Communications, Sydney, June 2014, pp. 5125–5130.

[23] H. Asgharimoghaddam, A. Tölli, and N. Rajatheva, "Decentralized multi-cell beamforming via large system analysis in correlated channels," in European Sign. Proc. Conf., Lisbon, September 2014, pp. 341–345.

[24] D. Tse and P. Viswanath, *Fundamentals of Wireless Communication*, Cambridge, UK: Cambridge University Press, 2005.

[25] A. M. Tulino and Sergio Verdú, "Random matrix theory and wireless communications," *Foundations and Trends in Communications and Information Theory*, vol. 1, no. 1, pp. 1–182, 2004.

[26] M. K. Ozdemir and H. Arslan, "Channel estimation for wireless OFDM systems," *IEEE Communications Surveys and Tutorials*, vol. 9, no. 2, pp. 18–48, 2007.

[27] S. Coleri, M. Ergen, A. Puri, and A. Bahai, "Channel estimation techniques based on pilot arrangement in OFDM systems," *IEEE Trans. Broadcasting*, vol. 48, no. 3, pp. 223–229, September 2002.

[28] Y-H. Nam, Y. Akimoto, Y. Kim, M-i. Lee, K. Bhattad, and A. Ekpenyong, "Evolution of reference signals for LTE-advanced systems," *IEEE Communications Magazine*, vol. 5, no. 2, pp. 132–138, February 2012.

[29] J. Choi, T. Kim, D. J. Love, and J-Y. Seol, "Exploiting the preferred domain of FDD massive MIMO systems with uniform planar arrays," in IEEE International Conference on Communications, pp. 3068–3073, London, June 2015.

[30] B. Gopalakrishnan and N. Jindal, "An analysis of pilot contamination on multi-user MIMO cellular systems with many antennas," in International Workshop on Signal Processing Advances in Wireless Communications, San Francisco, June 2011, pp. 381–385.

[31] K. Guo, Y. Guo, G. Fodor, and G. Ascheid, "Uplink power control with MMSE Receiver in multi-cell MU-massive-MIMO systems," in IEEE International Conference on Communications, Sydney, June 2014, pp. 5184–5190.

[32] N. Jindal and A. Lozano, "A unified treatment of optimum pilot overhead in multipath fading channels," *IEEE Trans. Communications*, vol. 58, no. 10, pp. 2939–2948, October 2010.

[33] G. Fodor, P. Di Marco, and M. Telek, "Performance analysis of block and comb type channel estimation for massive MIMO systems," in International Conference on 5G for Ubiquitous Connectivity, Levi, Finland, November 2014, pp. 1–8.

[34] H. Yin, D. Gesbert, M. Filippou, and Y. Liu, "A coordinated approach to channel estimation in large-scale multiple-antenna systems," *IEEE Journal of Selected Areas in Communications*, vol. 31, no. 2, pp. 264–273, February 2013.

[35] V. Saxena, G. Fodor, and E. Karipidis, "Mitigating pilot contamination by pilot reuse and power control schemes for massive MIMO systems," in IEEE Vehicular Technology Conference, Glasgow, May 2015, pp. 1–6.

[36] N. Krishnan, R. D. Yates, and N. B. Mandayam, "Uplink linear receivers for multi-cell multiuser MIMO with pilot contamination: Large system analysis," *IEEE Trans. Wireless Communications*, vol. 13, no. 8, pp. 4360–4373, August 2014.

[37] Y. Li, Y.-H. Nam, B. L. Ng, and J. C. Zhang, "A non-asymptotic throughput for massive MIMO cellular uplink with pilot reuse," in IEEE Global Telecommunications Conference, Anaheim, December 2012, pp. 4500–4504.

[38] M. Li, S. Jin, and X. Gao, "Spatial orthogonality-based pilot reuse for multi-cell massive MIMO transmissions," in International Conference on Wireless Communications and Signal Processing, Hangzhou, October 2013, pp. 1–6.

[39] A. Ashikhmin and T. Marzetta, "Pilot contamination precoding in multicell large scale antenna systems," in IEEE International Symposium on Information Theory, Cambridge, July 2012, pp. 1137–1141.

[40] N. Shariati, E. Björnsson, M. Bengtsson, and M. Debbah, "Low complexity channel estimation in large scale MIMO using polynomial expansion," in IEEE International Symposium on Personal, Indoor and Mobile Radio Communications, London, September 2013, pp. 1–5.

[41] J.H. Sørensen, E. de Carvalho, and P. Popovski, "Massive MIMO for crowd scenarios: A solution based on random access," in IEEE Global Telecommunications Conference Worshops, Austin, December 2014, pp. 352–357.

[42] E. Paolini, G. Liva, and M. Chiani, "Graph-based random access for the collision channel without feedback: Capacity bound," in IEEE Global Telecommunications Conference, Houston, December 2011, pp. 1–5.

[43] G. Liva, "Graph-based analysis and optimization of contention resolution diversity slotted ALOHA," *IEEE Transactions on Communications*, vol. 59, no. 2, pp. 477–487, February 2011.

[44] C. Stefanovic, P. Popovski, and D. Vukobratovic, "Frameless ALOHA protocol for wireless networks." *IEEE Communications Letters*, vol. 16, no. 12, pp. 2087–2090, December 2012.

[45] Q. Shi, M. Razaviyayan, Z.-Q. Luo, and C. He, "An iteratively weighted MMSE approach to distributed sum utility maximization for a MIMO interfering broadcast channel," *IEEE Trans. Signal Processing*, vol. 59, no. 9, pp. 4331–4340, September 2011.

[46] A. Tölli, H. Pennanen, and P. Komulainen, "Decentralized minimum power multi-cell beamforming with limited backhaul signaling," *IEEE Trans. Wireless Commun.*, vol. 10, no. 2, pp. 570–580, February 2011.

[47] P. Komulainen, A. Tölli, and M. Juntti, "Effective CSI signaling and decentralized beam coordination in TDD multi-cell MIMO systems," *IEEE Trans. Signal Processing*, vol. 61, no. 9, pp. 2204–2218, May 2013.

[48] H. Dahrouj and W. Yu, "Coordinated beamforming for the multicell multiantenna wireless system," *IEEE Transactions on Wireless Communications*, vol. 9, no. 5, pp. 1748–1759, 2010.

[49] S. Lakshminarayana, J. Hoydis, M. Debbah, and M. Assaad, "Asymptotic analysis of distributed multi-cell beamforming," in IEEE International Symposium on Personal, Indoor and Mobile Radio Communications, Istanbul, 2010, pp. 2105–2110.

[50] M. Kurras, L. Raschkowski, M. Talaat, and L. Thiele, "Massive SDMA with large scale antenna systems in a multi-cell environment," in AFRICON, Pointe-Aux-Piments, September 2013, pp. 1–5.

[51] A. Adhikary, J. Nam, J.-Y. Ahn, and G. Caire, "Joint spatial division and multiplexing: The large-scale array regime," *IEEE Transactions on Information Theory*, vol. 59, no. 10, pp. 6441–6463, 2013.

[52] M. Kurras, L. Thiele, and T. Haustein, "Interference aware massive SDMA with a large uniform rectangular antenna array," in European Conference on Networks and Communications (EUCNC) Bologna, 2014, pp. 1–5.

[53] D. Arthur and S. Vassilvitskii, "K-means++: The advantages of careful seeding," in *ACM-SIAM Symposium on Discrete Algorithms*, New Orleans, January 2007, pp. 1027–1035.

[54] U. Gustavsson, C. Sanchez-Perez, T. Eriksson, F. Athley, G. Durisi, P. Landin, K. Hausmair, C. Fager, and L. Svensson, "On the impact of hardware impairments on massive MIMO," in IEEE Global Telecommunications Conference Workshops, San Diego, December 2014, pp. 294–300.

[55] A. Alkhateeb, O. El Ayach, G. Leus, and R. W. Heath, Jr. "Hybrid precoding for millimeter wave cellular systems with partial channel lnowledge," in *Information Theory and Applications*, San Diego, February 2013.

[56] O. El Ayach, R. W. Heath, Jr., S. Abu-Surra, S. Rajagopal, and Z. Pi, "Low complexity precoding for large millimeter wave MIMO systems," in *IEEE International Conference on Communications*, pp. 3724–3729, Ottawa, June 2012.

[57] A. Alkhateeb, J. Mo, N. G. Prelcic, and R. W. Heath, Jr., "MIMO precoding and combining solutions for millimeter-wave systems," *IEEE Communications Magazine*, vol. 52, no. 12, pp. 122–131, December 2014.

[58] A. Alkhateeb, O. El Ayach, G. Leus, and R. W. Heath, Jr., "Channel estimation and hybrid precoding for millimeter wave cellular systems," *IEEE Journal of Selected Topics in Signal Processing*, vol. 8, no. 5, pp. 831, 846, October 2014.

[59] T. Obara, S. Suyama, J. Shen, and Y. Okumura, "Joint fixed beamforming and eigenmode precoding for super high bit rate massive MIMO systems using higher frequency bands," in IEEE International Symposium on Personal Indoor and Mobile Radio Communications, Washington, September 2014, pp. 1–5.

[60] International Telecommunications Union Radio (ITU-R), "Requirements related to technical performance for IMT-Advanced radio interface(s)," Report ITU-R M.2134, November 2008, www.itu.int/pub/R-REP-M.2134-2008.

[61] T. A. Thomas, F. W. Vook, E. Mellios, G. S. Hilton, and A. R. Nix, "3D extension of the 3GPP/ITU channel model," in IEEE Vehicular Technology Conference, Dresden, June 2013, pp. 1–5.

[62] S. Jaeckel, L. Raschkowski, K. Borner, and L. Thiele, "QuaDRiGa: A 3-D multi-cell channel model with time evolution for enabling virtual field trials," *IEEE Transactions on Antennas and Propagation*, vol. 62, pp. 3242–3256, June 2014.

[63] G. R. MacCartney and T. S. Rappaport, "73 GHz millimeter wave propagation measurements for out- door urban mobile and backhaul communications in New York City," in IEEE International Conference on Communications, Sydney, June 2014, pp. 4862–4867.

[64] T. A. Thomas, H. C. Nguyen, G. R. MacCartney, Jr., and T. S. Rappaport, "3D mmWave channel model proposal," in IEEE Vehicular Technology Conference, Vancouver, September 2014, pp. 1–6.

[65] S. Suyama, J. Shen, H. Suzuki, K. Fukawa, and Y. Okumura, "Evaluation of 30 Gbps super high bit rate mobile communications using channel data in 11 GHz band 24x24 MIMO experiment," in IEEE International Conference on Communications, Sydney, June 2014, pp. 5203–5208.

[66] A. O. Martınez, E. De Carvalho, and J. Ø. Nielsen, "Towards very large aperture massive MIMO: A measurement based study," in IEEE Global Telecommunications Conference Workshops, Austin, December 2014, pp. 281–286.

[67] S. Dierks, W. Zirwas, B. Amin, M. Haardt, and B. Panzner, "The benefit of cooperation in the context of massive MIMO," in International OFDM Workshop, Essen, August 2014.

9 Coordinated multi-point transmission in 5G

Roberto Fantini, Wolfgang Zirwas, Lars Thiele, Danish Aziz, and Paolo Baracca

9.1 Introduction

The performance of a wireless network strongly depends on the user positions in a cell. More precisely, the UEs (User Equipments) at the cell border typically experience much lower throughput than those nearer to the transmitting Base Station (BS). This is mainly due to the presence of inter-cell interference, generated by concurrent transmissions in other cells. Inter-cell interference is particularly relevant for modern wireless communication systems like Universal Mobile Telecommunications System (UMTS) or Long-Term Evolution (LTE), and also 5G, where the frequency reuse factor is one or very close to one. In such scenario the system is primarily interference limited, and the performance cannot be improved by simply increasing the transmitted power. Hence, techniques are necessary in order to (1) target inter-cell interference and (2) reduce the gap between the cell edge and average throughput. Consequently, these alternative techniques allow a more even user experience throughout the whole network.

In principle, the following techniques can be pursued to tackle inter-cell interference:

- Interference can simply be treated as white noise. This is clearly suboptimal, as it ignores properties of the interfering signals that could be exploited in order to improve signal reception quality.
- Interference can be avoided through statically leaving some transmit resources in some cells muted (e.g. fractional frequency reuse), or otherwise constraining the usage of resources, or through coordinated scheduling among cells, as investigated in Chapter 11.
- The impact of interference can be alleviated at the receiver side through e.g. Interference Rejection Combining (IRC), where multiple receive antennas and subsequent receive filters are used to attenuate the interference to a certain extent.
- Interference may be decoded and cancelled, a technique that is for instance studied in 3GPP in the context of Network-Assisted Interference Cancelation (NAIC).
- At the transmitter side, interference can also be partially avoided by performing interference-aware precoding, i.e. applying precoding such that the interference caused toward adjacent cells is reduced.

5G Mobile and Wireless Communications Technology, ed. A. Osseiran, J. F. Monserrat, and P. Marsch. Published by Cambridge University Press. © Cambridge University Press 2016.

- Ultimately, signals from other cells can in fact be treated as a useful signal energy instead of interference, if (in the downlink) multiple nodes jointly transmit signals that coherently overlap at the intended receiver, and destructively overlap at interfered receivers. In the uplink (UL), multiple nodes can jointly receive and decode the signals from multiple UEs, and in this form also exploit interference rather than seeing it as a burden.

The latter two techniques are typically grouped under the term Coordinated Multi-Point (CoMP) [1], which in general refers to techniques where multiple nodes in the network coordinate or cooperate to alleviate the impact of interference, or actually exploit interference on physical layer.

From an information-theoretic point of view, a joint transmission from multiple nodes (e.g. BSs) to multiple UEs resembles a broadcast channel, and a joint detection of multiple UEs by multiple nodes resembles a multiple access channel, where the capacity regions are well-known for Gaussian channels [1]. In the context of cellular systems, a form of joint detection was in fact already introduced in Code Division Multiple Access (CDMA) systems through the so-called soft and softer handover. In [2], joint detection between multiple cells was considered via a centralized unit. The centralized unit acted as a receiver that exploited all UEs' signals collected by the BSs, treating the whole system as a network-wide Multiple-Input Multiple-Output (MIMO) scheme. Joint transmission in the downlink (DL) was studied in [3] and showed to provide more than 10-fold increase in spectral efficiency if applied across a large number of cells [4].

Motivated by these promising results, CoMP has been widely studied in 3GPP [5] as one of the features for LTE Advanced (LTE-A), i.e. LTE Releases 11 and 12. In 3GPP, CoMP techniques have been classified into three groups [6]:

- Joint Transmission (JT), where the data related to a UE is available at several transmitting nodes and is transmitted simultaneously by each node over a frequency/time resource. Such transmission can be coherent (and in this case it is sometimes referred to as Network MIMO) or non-coherent. Coherency refers to the ability to precode in a way that exploits the phase and the amplitude relations between channels associated with different transmission points.
- Dynamic Point Selection (DPS), where the data related to a UE is transmitted by a single transmitting node for a frequency/time resource, while the other nodes can be either dedicated to transmit data of other UEs or be muted exploiting Dynamic Point Blanking (DPB). However, the data should still be available at all cooperating transmitters, since the selected point may change dynamically from one transmission time interval to another.
- Coordinated Scheduler/Coordinated Beamforming (CS/CB), in this case, the data related to a single UE is available and transmitted only by one node. Nonetheless, neighboring cells share Channel State Information (CSI) in order to coordinate their scheduling, power control and beamforming decisions, and reduce mutual interference.

A further classification used in literature is to distinguish CoMP techniques that require both the exchange of CSI and user data (sometimes referred to as **cooperative** approaches) and those that involve only CSI exchange (**coordination** approaches). While JT and DPS fall in the first category, CS/CB belongs to the latter. In the last few years, other novel approaches have arisen in this second category. These approaches exploit the fact that it is possible to precode and transmit signals such that interferences are always constrained to confined signal subspaces at each receiver. This allows the receiver to reject efficiently the interference. This idea is known as Interference Alignment (IA) [7] and aims to manage the interference by combining an "align" and "suppression" strategy.

Although CoMP looks promising, there are several practical impairments that could prevent it from reaching the full gain suggested by theoretical bounds. Backhauling limitations in terms of bandwidth and latency, imperfection on the CSI due to realistic estimation procedures, quantization effects, signaling delays or limitations, and imperfect frequency/time synchronization are all aspects that have an impact on potential CoMP gains and should hence be carefully taken into account in the design of a CoMP scheme [1].

In this respect, it is more difficult to take all these aspects into consideration in a matured wireless communications system such as LTE, which was never designed to support CoMP in the first place, and which due to its standards matureness does not allow introducing major changes on e.g. physical layer. In this chapter, hence the emphasis is on how a clean-slate 5G system could natively better support CoMP techniques than any legacy system, by taking all above-stated aspects already into account in its initial design.

The remainder of the chapter focuses on JT CoMP as the most promising, yet also most challenging, CoMP technique, and is structured as follows: Section 9.2 investigates the key JT CoMP enablers, in particular elaborating on how a 5G system could better provide native support for CoMP than legacy systems. Then, Section 9.3 discusses how CoMP can be applied specifically in the context of small cells. Section 9.4 looks into distributed CoMP schemes that relax the need for full data and CSI exchange between cooperating cells. Finally, Section 9.5 deals with JT CoMP in the context of interference-aware receivers, and Section 9.6 concludes the chapter.

9.2   JT CoMP enablers

CoMP, and especially JT CoMP, has been investigated for many years [8]–[10]. This long lasting interest is due to the unique capability of JT CoMP to cancel a relatively high number of moderate to strong interferers. Hitherto results from 3GPP [5] indicated that a straight forward implementation of JT CoMP can achieve only moderate gains compared to a smartly scheduled Multi-User MIMO (MU-MIMO) solution, still leaving a large gap between the achievable performance and the theoretically predicted potential. Therefore, in order to be effective, JT CoMP has to be integrated into an overall interference mitigation framework [11].

Further, the following key enablers are required for JT CoMP [1]:

- **Low-latency backhaul or fronthaul**. JT CoMP requires the exchange of user data and CSI between cooperating nodes. While user data may be buffered to some extent, CSI has to be exchanged at very low latency in order to avoid it to be outdated before being used for e.g. precoder calculation. It is expected that in the 5G era there will be a larger extent of powerful fiber backhaul available at least for macro and micro cells, or multiple nodes will even be served through Centralized RAN (C-RAN) involving fiber fronthaul and Remote Radio Heads (RRHs). Such deployments will inherently provide the low-latency backhaul or fronthaul architecture needed for CoMP. In deployment scenarios where such low-latency backhaul or fronthaul is not available, one may want to use distributed CoMP, as investigated in Section 9.4.
- **Synchronization in time and frequency**. Synchronization of cooperating nodes in time and frequency is essential for JT CoMP [12]. With regards to time synchronization, the main problem are in fact the signal propagation delays within a CoMP cooperation area (inherently limiting the size of cooperation areas) rather than the accuracy of time synchronization of the involved nodes. However, it is required that nodes are synchronized very accurately in frequency domain, or that carrier frequency offset estimation and compensation is performed on the UE side [12]. In C-RAN deployments, synchronization in both time and frequency can easily be obtained over the infrastructure, whereas for distributed architectures over the air synchronization may be employed as for example in [13] and [14].
- **Provision of accurate CSI to the transmitter**. JT CoMP performs a combination of constructive and destructive superposition of signal components, with the goal to maximize the desired received signal and, at the same time, minimize mutual interference. Especially the destructive superposition requires accurate alignment of phases and amplitudes of many channel components, and consequently requires very accurate channel state information at the transmitter side. In a Frequency Division Duplex (FDD) system, this is typically realized through channel estimation in the downlink, and feedback of quantized version of this information through the uplink to the BS side. In the 5G era, it is expected that higher transmission bandwidths will be used, which means that higher-resolution channel estimation may be performed [15]. Furthermore, the higher carrier frequencies that will be used will experience more frequency-flat channels, which again can be estimated more accurately and fed back to the BS side more efficiently. Shortened Transmission Time Interval (TTI) lengths in 5G will allow for faster CSI feedback than in legacy systems. In fact, Time Division Duplex (TDD) transmission (see Chapter 7) will further allow avoiding the need for uplink feedback, as base stations can infer the required channel information from the reciprocity of uplink and downlink channels. However, even with perfect channel estimation, feedback and exchange between cooperating nodes could make the time between channel estimation and application for precoding be so long that JT CoMP performance is clearly impaired. For this reason, channel prediction is seen as one of the important enablers for real world JT CoMP implementations, and it is detailed in Section 9.2.1.

- **Clustering and interference floor shaping**. The most relevant issue is the setup of cooperation clusters dividing large cellular radio networks into manageable sub areas. Theoretical results, promising a linear relation between achievable performance and the number of cooperating nodes, typically assume network wide cooperation [16]. Recently, Massachusetts Institute of Technology (MIT) verified in a real world office WLAN that performance can increase linearly with the number of transmit nodes, when network wide JT CoMP is used, as expected from theory [17]. Nonetheless, providing cooperation on a network scale is not feasible in a cellular network, as this would, e.g. require the distribution of user data and CSI among a huge number of cells. For this reason, cellular networks have to form cooperation clusters within which JT CoMP is enabled. Clearly, limiting cooperation to clusters implies that there is a certain extent of inter-cluster interference that cannot be combatted with CoMP. Hence, smart clustering and interference floor shaping should be used, which is described in detail in Section 9.2.2.
- **User scheduling and precoding**. In particular, when JT CoMP is applied among co-located transmit antennas of a single cell, channel matrices among jointly served UEs may be badly conditioned, which may lead to degradation in terms of UE throughput compared to the non JT CoMP case. Hence, it is important to schedule groups of UEs within the previously-mentioned clusters in order to create suitable compound channel properties. The user scheduling and precoding technique is covered in detail in Section 9.2.3.

9.2.1   Channel prediction

An accurate channel prediction has been identified as one of the key enablers for a robust JT CoMP implementation as it offers several benefits like:

- minimum degradations due to CSI outdating in case precoders are calculated for the predicted instead of the reported CSI;
- relaxed backhauling as well as scheduler latency requirements;
- reduced frequency synchronization requirements by including the phase rotations into the CSI prediction. Alternatively, frequency offsets might be explicitly estimated and de-rotated;
- accurate frequency selective CSI (when using model based channel prediction [18]) for large frequency bands and high number of relevant channel components with low to very low feedback overhead.

Backhaul latencies typically range from 1 or few ms to 20 ms, where the latter value is often used when referring to "non-ideal backhaul". Modern backhaul architectures can in principle achieve 1 to few ms (even under inclusion of several routers or switches). Nonetheless, for JT CoMP solutions involving a central unit, one typically has to consider an overall delay in the range of 10 ms. This delay shall account for CSI estimation, reporting, fiber latencies, scheduling, precoding and finally transmission.

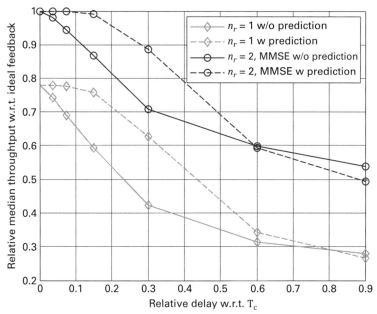

Figure 9.1   RLS-based channel prediction and multi-antenna receivers as a tool to contrast delays prior to CSI feedback and JT precoding [23].

Channel prediction techniques like Wiener or Kalman, and Recursive Least Squares (RLS) filtering have been applied in [19] and have shown to be suitable for real world JT CoMP systems. In particular, the adaptive RLS filtering technique, aiming at low-complexity channel prediction at the UE, can combat impairments due to transmission delays for up to 20% of the coherence time T_c as it is shown in Figure 9.1 (see also [20]). It should be noted that 20% of coherence time corresponds to 20 ms transmission delay at 3 km/h mobility. In case of multiple n_r receive antennas, CSI feedback based on Multi-user Eigenmode Transmission (MET) [21] can be assumed. For larger prediction ranges going beyond a few tenth of the RF wavelength λ, the prediction quality might suffer. In order to stabilize the system further, channel prediction feedback should be accompanied by reliability information. Based on this information, robust precoders may adapt their weights ensuring the best possible performance for the given prediction quality [22]. Alternatively, a higher feedback rate could reduce CSI outdating, but at higher overhead costs. Often, prediction errors vary over different frequency subbands in a different manner. For this reason, a low-rate and low-latency feedback link may be introduced to report subbands where channel evolution deviates significantly from the prediction. This information can be easily obtained at the UE side, and BSs may use this information for re-scheduling of the affected UEs.

9.2.2   Clustering and interference floor shaping

With respect to system-level gains, clustering of UEs into cooperation areas is the most important challenge and one of the main reasons for the large discrepancy between theoretical and practical performance gains. Clusters or cooperation areas are formed as subparts of the full large cellular networks in order to avoid that channel information and user data are exchanged among a large number of transmit nodes. Consequently, a certain level of inter-cluster interference is introduced, which cannot be addressed via CoMP [24] and may hamper system performance.

There is a strong drop in CoMP performance when going from a single network wide cooperation to e.g. two cooperation areas. For simple clustering schemes, one might fall back quite close to the performance without any cooperation. On the other hand, it has been found that performance gains get asymptotically smaller as the size of cooperation area increases [25]. In fact, the additionally cancelled interferers are typically slightly above the interference floor. Consequently, the network full cooperation is not fully beneficial.

Dynamic clustering solutions (partly iterative) have been proposed in [26]. These solutions start with one UE and search for the best companion UE benefiting most from mutual cooperation. This iteration is continued until adding a UE decreases performance. Despite its simplicity compared to an exhaustive search, such dynamic clustering solutions are already a good foundation for reasonable performance gains.

A fixed clustering with threshold-based CSI reporting [20][26] is a simple implementation that is desired in the case of large clusters. The fixed clustering limits cooperation clusters to adjacent sites (see Figure 9.2(a)). Moreover, each UE only reports its strongest channel components out of those included in the enlarged cooperation area, resulting in the definition of overlapping sub clusters inside each cooperation cluster. Afterwards, during the user grouping process, UEs are added to the pool of active UEs in a greedy fashion, i.e. UEs are added as long as the sum-rate is increased. Figure 9.2(b) shows the median throughput per sector as a result from a fixed sub-cluster size of a UE k, $M_{c,k}$ selected out of an enlarged cooperation area with increasing size M_c. Further, $M_{c,k} = 3$ means that only the three strongest channels components are of interest, and $M_c = \{6, 9, 12\}$, given as cross, diamond and square marks, respectively. Overall, the joint clustering and user grouping improves the system performance by 10% compared to the static clustering approach. This gain is achieved since increasing the size of the cooperation areas ensures that a sufficient number of UEs will have the set of strongest channel components belonging to the same cooperation area. It should be noted that spanning cooperation areas over typically three adjacent sites results in a cooperation area of nine cells.

Many UEs typically reside for geometrical reasons at the border of a cooperation cluster, so even for enlarged cooperation clusters these are affected by inter-cluster interference, especially from the adjacent cooperation areas.

To overcome this issue the so-called "cover shift" concept (sometimes denominated as super cells [9]) organizes a different setup of enlarged cooperation areas in different frequency subbands or time slots. Figure 9.3 shows an example of two possible cover shifts, corresponding to different ways to organize the same set of cells surrounding two UEs in different cooperation areas setup. Cooperation areas as defined in cover shift 1 could be used to serve UEs in a certain subband (or time slot), while cooperation areas

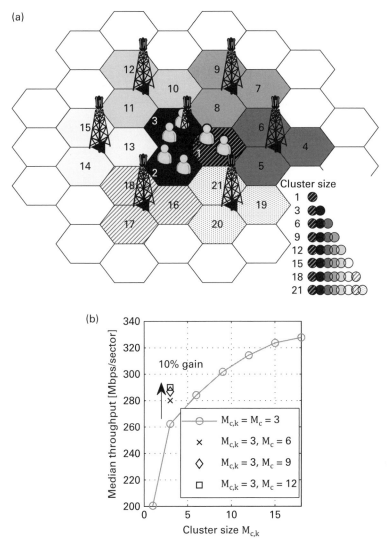

Figure 9.2 (a) Fixed clustering with adaptive threshold-based CSI reporting at UE. (b) Median achievable system rate as a function of sub-cluster size $M_{c,k} = M_c$ (solid line).

corresponding to cover shift 2 could be used in another subband (or time slot). In the upper part of the figure, cells belonging to the cooperation area defined in cover shift 1 are surrounded with a dark thick line, while in the lower part cooperation areas according to cover shift 2 are highlighted with a grey thick line. The two UEs in the figure receive the strongest signals from the cells shown by the arrows. In this scenario, the two UEs would not be served effectively in the frequency subband (or time slot) corresponding to the cooperation areas defined in cover shift 1, since with this configuration there is no cooperation area that can include all their three strongest channel components. Therefore, there will always be at least one strong interfering component coming from

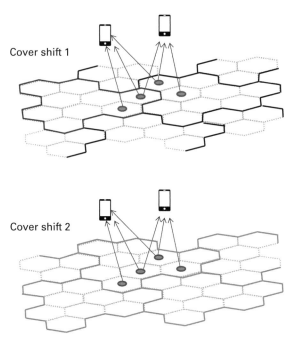

Figure 9.3   The cover shift concept.

outside the cooperation area selected to serve the UE. On the other hand, if the two UEs are served in the frequency subband (or time slot) that uses cooperation areas as defined in cover shift 2 all the strongest channel components can be included in the cooperation. With this approach, UEs will be scheduled into that frequency subband (cover shift) containing the highest number of strongest interferers for the UE [27]. The result is that most UEs are scheduled in the center of their serving cooperation area with high probability that the strongest interferers are part of the selected cover shift.

Cover shifts and enlarged cooperation clusters increase the probability that UEs are served from their three strongest cells by up to 90% [28], meaning that 90% of the UEs will benefit from JT CoMP.

However, there is still a significant portion of interference left from neighboring cooperation areas. In order to improve the performance of UEs at the area borders, the cover shift concept may be extended with a so-called interference floor shaping mechanism [8]. In essence, this mechanism reduces the interference level coming from other cooperation areas by using a 3D beamforming approach. Therefore, an increased antenna tilt in combination with a slightly reduced transmit power in neighboring cells leads to a fast decline of interference power in the desired cooperation area. The beam pattern is shifted according to the given cover shift. In consequence, the number of UEs experiencing bad SINR conditions due to inter-cluster interference can be strongly reduced. Due to the shape of the overlayed received power distribution from multiple cover shifts, this scheme has been termed "tortoise" concept [8]. According to simulations and real world measurements, the interference floor seen by UEs in other cooperation clusters is reduced by 10 dB or more.

9.2.3   User scheduling and precoding

Most theoretical works on JT CoMP assume independent identically distributed Rayleigh channels that achieve optimum MU-MIMO performance without complex scheduling considerations. In reality, for antennas placed at a single site, large correlations in channel coefficients are observed. For the cells of one site, mainly the large-scale parameters like the path loss are correlated. More importantly, the small-scale parameters of typical $\lambda/2$ spaced uniform linear arrays of a single cell will be correlated as well. Consequently, the channel rank is reduced and, quite often, the channel matrices per cell become badly conditioned. It should be noted that cooperation over several sites is helpful as it improves the condition of the overall channel matrix by rank enhancement [29].

For the scheduling point of view, it means that severe correlations are expected mainly within cells, which have to be minimized by sophisticated per cell scheduling, i.e. by searching for best-fitting user groups per cell. It can be shown that one bad user constellation in one cell may already degrade the precoding performance of the overall cooperation cluster significantly. A possible solution is the following two-step scheduling approach, which happens to be nicely compatible with typical LTE schedulers [8].

In the first step or stage, the optimum MU-MIMO user group per cell is found by iterative algorithms or even by exhaustive search to minimize the correlation issues as mentioned above. For the first stage, the corresponding complexity is still manageable due to the limited number of active UEs per cell.

In the second stage, the scheduling decisions of the cells forming the cooperation cluster are taken and combined to calculate an overall cooperation area wide precoder.

In addition to the two-stage scheduling, an important component for a practical system is the fast adaptation of the UE receive beamformers in order to i) partly overcome precoding errors, e.g. due to CSI outdating (Figure 9.1), and ii) to cancel potential far-off interferers not sufficiently suppressed by the interference floor shaping [30]. These far-off interferers are typically the result of wave-guiding effects in long streets. Hence, each UE will typically have only one to few dominant interferers from one direction. IRC processing is an effective solution to combat these dominant interferers.

9.2.4   Interference mitigation framework

The techniques described from Section 9.2.1 to Section 9.2.3 have been combined into a so-called interference mitigation framework [8]. This framework provided gains in the order of up to 100% for 4x2 MIMO at 2 GHz, for a typical 3GPP case 1 urban macro scenario with 500 m inter-site distance. In fact, the spectral efficiency rose from about 3 for pure 3GPP LTE MU-MIMO to 5–6 bps/Hz/cell [8]. Note that these values take into account the overall overhead due to LTE control or reference signals as well as guard intervals that is approximately equal to 47%. Other publications [24] have found similar JT CoMP gains.

9.2.5   JT CoMP in 5G

As mentioned earlier, the following general technology trends in the context of 5G will inherently impact the usage of JT CoMP, for instance:

1. Massive MIMO as described in Chapter 8 leads to a sparse overall channel matrix, so that each UE is served from a limited set of relevant channel components. This reduces the effort for channel estimation and reporting as well as improves JT CoMP precoding robustness. Additionally, it increases the rank and condition of the overall channel matrix of a cooperation area, thereby supporting efficient MU-MIMO modes with a high number of simultaneously served UEs. Furthermore, due to strong beamforming gains, almost all UEs will be interference-limited [31] and accordingly benefit from JT CoMP gains.
2. Ultra-dense networks e.g. of small cells will play a prominent role for 5G, leading to heterogeneous networks with more transmit nodes per macro site with uncorrelated channel realizations to UEs, and hence better-suited channels for CoMP. Small cells generate further inter-cell interference, which can be ideally exploited/overcome by CoMP. However, there are some challenges due to inhomogeneous placement above and below rooftops as well as due to large transmission power differences between macro and small cells, see [26] and [31].
3. For 5G and RF frequencies, there is a trend to more bandwidth going up to several hundred MHz. This can be exploited for improved channel estimation techniques like e.g. bandwidth enlargement, as from theory it is clear that a higher bandwidth allows for more accurate CSI estimation and prediction.
4. Shorter TTI lengths introduced for higher carrier frequencies, as e.g. described in Chapter 7, will allow for faster CSI feedback loops, which will inherently improve JT CoMP performance.
5. The usage of a larger extent of TDD in 5G will allow exploiting uplink-downlink reciprocity for JT CoMP precoder calculation without, or requiring a lesser extent of, explicit CSI feedback. Furthermore, TDD will enable particular distributed CoMP schemes as described in Section 9.4.

Beyond this, a clean-slate approach in 5G may allow learning from the many lessons in the last years on CoMP, and building a system that is natively designed to exploit interference rather than treating it as noise. For instance, a clean-slate approach could provide:

- **Native support for dynamic multi-cell connectivity**. More precisely, a 5G system could to some extent break with the paradigm of "cells" and from the beginning enable a UE to be served by multiple transmit nodes, in a way that is transparent to the UE. This could be possible by, e.g. a dynamic usage of sounding reference (i.e. pilot) signals and CSI reference signals, allowing the system to "probe" which set of transmit points (i.e. antennas across multiple transmit nodes) can best serve a single UE in the form of one multi-antenna virtual cell, without the UE being aware which transmit nodes are involved in forming the virtual cell. Such a concept would inherently allow for a better dynamic setup and usage of CoMP clusters.

- **More resource-efficient channel estimation, prediction and feedback**. In particular, a 5G system could allow dynamically adjusting the density of CSI reference signals to the channel conditions (e.g. coherence bandwidth and time) and the channel estimation needs for the particular CoMP scheme to be used. Similarly, the form, granularity and frequency of CSI feedback from the UE to the network side could be flexibly configured to allow optimizing the trade-off between CoMP performance and overhead. In general, a key requirement in 5G should be that CSI feedback schemes are agnostic of any particular JT CoMP setup (e.g. number of cells involved, antenna configuration at each cell, etc.), so that a larger variety of CoMP setups can be used without requiring implementation changes at the UE side.

- **Better means to handle inter-cluster and intra-cluster interference**. Novel frame structures in 5G may allow for a better rejection or cancelation of inter-cluster interference while dealing with the intra-cooperation area interference. For instance, some 5G radio concepts (see Chapter 7) foresee that cells are synchronized and that Demodulation Reference Signals (DMRS) are always contained in the same OFDM symbol. In this case, interfered entities can rely on the fact that the interference covariance measured on the DMRS OFDM symbol will remain fairly constant throughout the rest of the sub-frame, so that efficient IRC can be applied (see Section 9.5.1). Similarly, a 5G system could provide better means for Network Assisted Interference Cancelation (NAIC, see Section 9.5.2). This would be especially beneficial in the case of one or very few strong inter-cluster interferers, as discussed in Section 9.2.3.

In conclusion, a 5G system that natively supports the aspects listed above may allow unleashing the overall potential of JT CoMP far beyond what is possible in LTE.

9.3   JT CoMP in conjunction with ultra-dense networks

The performance gain of any interference mitigation framework is more or less upper bounded by the loss resulting from inter-cell interference. For a typical urban macro MU-MIMO scenario with an inter-site distance of 500 m the spectral efficiency will be typically a factor of two to three higher for a single standalone cell compared to a cell integrated into a cellular network with corresponding inter-cell interference, which is accordingly defining the potential CoMP gains.

Therefore, for 5G systems targeting spectral efficiency gains by a factor of ten or more, interference mitigation is only one ingredient. Other techniques like massive MIMO and ultra-dense networks (see Chapters 7 and 8) e.g. of small cells integrated into macro networks, have to be added as additional capacity boosters. Moreover, combining these techniques can be motivated by the vision to better concentrate receiver power directly at the UE receivers. UDNs e.g. of small cells reduce path loss due to small distances between the transmitter and the receiver, and higher LOS probability, massive MIMO focuses energy by strong beamforming gains, and JT CoMP by constructively super-positioning of transmit signals.

It is important to stress that the integration of UDNs into an JT CoMP interference mitigation framework of homogeneous macro BSs is not straightforward and needs some further considerations:

- The transmit power of macro BSs is typically around 49 dBm, which is much higher than that of small cells, with typically less than 30 dBm.
- The deployment of macro BSs is typically above rooftop level, which supports a large coverage zone, while small cells are located below rooftop or even at lamppost level. This results in a limited coverage area for small cells and, furthermore, UEs will experience quite often less frequency-selective radio channels.
- The backhaul-connection between macro-sites can be expected to be low-latency and high-capacity fiber systems, while connection of small cells to a central unit might be of varying quality, ranging from fiber over DSL lines to in- or out-band relaying.
- With large number of small cells per cooperation area as intended for UDN, complexity for CSI reporting as well as precoding might easily become prohibitive.

Finally, an example of intergrating JT CoMP in UDN is the opportunistic CoMP (OP CoMP) concept [31], where UDNs of small cells are activated in the macro layer only on a need basis. Furthermore, the power imbalance between small and macro cells is reduced by small-cell power boosting, being limited to certain frequency subbands to fulfill the small-cell power constraints.

9.4   Distributed cooperative transmission

The previous sections of this chapter focused on JT CoMP techniques, assuming that data and CSI are shared between cooperating transmit nodes over a backhaul infrastructure involving decently low latency. However, in some of the 5G scenarios and use cases it is challenging to deploy infrastructure with the high capabilities requested by JT CoMP. For example, consider the use cases of large outdoor and stadium events (see Chapter 2). These use cases are characterized by a very dense population of UEs requesting high capacity for a limited amount of time. Hence, installing a permanent infrastructure to provide a high performance backhaul would not be a very cost effective solution.

Consequently, in order to fulfill the service requirements in such scenarios, new approaches that work with limited backhaul are needed. These approaches shall nevertheless exploit the gains of coordination and cooperation of the transmission nodes.

Most CoMP schemes based on limited information exchange focus solely on interference mitigation through coordination, e.g. CoMP with coordinated scheduling and

CoMP with only CSI sharing (see [32][33]). The interference mitigation is achieved e.g. by the joint design of transmit precoding within the CoMP cluster (see Section 9.2.2). In case the receivers are equipped with multiple antennas, joint optimization can be done for the design of both the transmit precoding and receive processing filters.

Finally, the exchange of limited information in CoMP system limits the achievable gains of cooperation desired in several of the 5G scenarios. Hence, new approaches should be developed. In the sequel, distributed precoding and interference alignment schemes will be described that enable to circumvent the hitherto known limitations of a CoMP system with a constrained backhaul.

9.4.1   Decentralized precoding/filtering design with local CSI

A TDD structure supports the joint optimization of transmit and receive filters. In fact, a 5G frame structure based on dynamic TDD optimization has been widely studied (see e.g. [34]). Based on the dynamic TDD, it has been shown that one can design distributed transmit precoding and receive processing schemes for 5G.

The TDD transmission mode allows the usage of the same frequency for DL and UL by allocating different time slots for respective transmission. In static TDD, the allocation of the slots for DL and UL in a sub-frame/frame is pre-fixed by the system independent of the traffic in DL and UL. In dynamic TDD, the system is able to adapt the allocation of DL/UL slots with respect to the respective traffic demand. This provides higher degrees of freedom and an efficient use of transmission resources. Due to the multi-path delay, and potentially large differences between UL and DL transmit power, dynamic TDD is difficult to realize in macro cells [35]. However, the future networks with small cells and advanced switching technologies drive the application of dynamic TDD in 5G networks as presented in Chapter 7. Moreover, the need of dynamic TDD for efficient resource utilization is even higher in small cells as they typically have much less active UEs per cell and consequently more asymmetric traffic in DL and UL [36].

On the one hand, dynamic TDD offers attractive gains in terms of optimum resource utilization. However, there are also challenges to realize these gains in practical systems. In a multi-cell system, with cell-specific dynamic TDD operation, one of the main challenges is to mitigate the additional UL-to-DL interference UE-to-UE and DL-to-UL interference (BS-to-BS), which are also known as cross-link interferences (see Figure 9.4). In the literature, the main focus is to develop algorithms for efficient time slot allocation with the objective of cross link interference minimization (see for example studies in [37]–[38] and references therein). The solution proposed hereafter relies on the dynamic TDD structure presented in Chapter 7 (see Section 7.4 and [39]). It focuses on the joint optimization of the allocation of resources (space, frequency and time, including the allocation to UL and DL) and coordinated precoder/beamformer (CB CoMP) design across the entire network, to maximize various system performance measures.

Based on dynamic TDD, it is possible to define a bi-directional signaling concept for multi-cell MU-MIMO 5G systems. This signaling concept allows designing

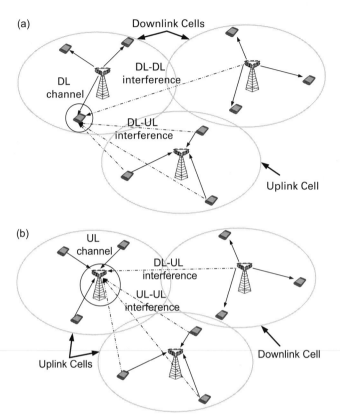

Figure 9.4 Interference types present at UL and DL in dynamic TDD mode.

decentralized coordinated precoder/decoder with minimum amount of iterations. The idea is to use the optimized receivers from the previous backward/forward iterations as pilot precoders for the next forward/backward iteration. With this concept, two-fold gains are achieved. Most importantly, no information exchange is required between coordinated nodes. In addition to that, it can be used to carry out implicit UE selection for each frame by letting the iterative algorithm to decide the optimal set of UEs/streams to be served at any given time instant. Obviously, there is some overhead involved for the bidirectional signaling. Hence, for the achievable gains in throughput, the training period should be large enough to come up with efficient precoder/decoders while short enough to guarantee sufficient resources for the actual data transmission. This multi-objective optimization can be achieved with Weighted Sum Rate (WSR) max-imization [40], where WSR maximization is considered with per transmitter node power constraints.

For further elaboration of this concept, consider a multi-cell multi-user MIMO system operating in dynamic TDD mode with traffic-aware user scheduling. Assume that each TDD frame is allocated either to UL or DL depending on the instantaneous traffic load of the cell, similarly to the frame structure specified in Chapter 7 (see Figure 7.21 in

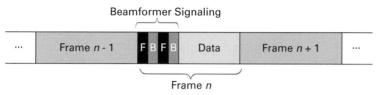

Figure 9.5 Sequence of TDD frames with bi-directional training.

Section 7.4). Within this structure, it is possible to perform a bi-directional signaling/training for each TDD frame [41] as illustrated in Figure 9.5. Let B be the set of BSs serving K UEs in the considered network, so that $B_U \subseteq B$ will be a subset of BSs serving uplink traffic in a given time slot, and $B_D \subseteq B$ will be the subset of the base stations serving downlink traffic, such that $B_U \cup B_D = B$. A BS will be part of B_U or B_D based on the average traffic demand on the UEs within that particular BS. Assume that a linear Minimum Mean Squared Error (MMSE) algorithm is used by the BS and the UEs for the estimation of received signal. Then, optimal transmit precoders can be obtained by solving a WSR maximization problem using the minimizing Weighted Minimum Mean Squared Error (WMMSE) approach as in [40][42]. Although the optimization problem is non-convex NP hard problem, it can be solved iteratively by fixing some of the optimization variables until the weighted sum rate converges. The performance of this approach [36], i.e. dynamic TDD with decentralized beamforming, is summarized hereafter.

9.4.1.1 Performance

The numerical analysis is carried out for three cells with 4 UEs in each cell. The path loss from a BS to in-cell UEs is normalized to be 0 dB. The power constraint for the DL transmission is fixed so that a certain received SNR, e.g. 10 dB or 20 dB, is achieved at each UE. Similarly, the UL transmit power constraint is selected such that the sum power of the UL UEs is equal to the power constraint of the DL BS. All the priority weights are assumed to be 1. The number of antennas at each BS is $n_t = 4$ and the number of antennas at each UE is $n_r = 2$. The simulation environment is defined by three types of separations (that are shown in Figure 9.6): the path loss between the two DL cell edge UEs is α, an UL UE to a DL UE is β, and an UL BS to a DL BS is δ.

It is assumed a fixed arrangement of BSs for DL and UL transmissions. However, in practice, the UL/DL BS allocation may vary in every TDD frame based on the instantaneous traffic demand. Consequently, the DL/UL precoders might need to be recalculated in every TDD frame. Therefore, it is important to observe the impact of over-the-air signaling overhead originated from the decentralized approach presented in [36] on the system performance. Hence, in order to analyze the scheme, the following parameters are introduced

- λ, the number of bi-directional precoder signaling iterations per TDD frame.
- γ, the signaling overhead per one signaling iteration.
- $\rho = \lambda \times \gamma$, the total signaling overhead.
- R_{BDT}, the achieved weighted sum rate after bi-directional signaling.

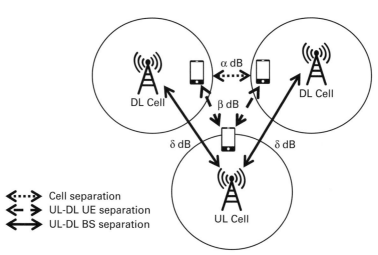

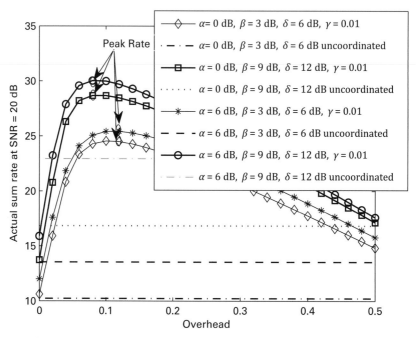

Figure 9.6 Multi-cell multi-user simulation model.

Figure 9.7 Actual sum rate vs. overhead at SNR = 20 dB, with different α, β and δ values.

Thus, the actual achievable weighted sum rate of the system is simply $(1 - \rho)R_{BDT}$.

Figure 9.7 demonstrates the actual sum rate versus the total overhead of the three cells network with two DL BSs and one UL BS. Assuming λ iteration is equal to two OFDM symbols. Thus, $\gamma = 0.01$ and $\gamma = 0.02$ correspond to frame lengths up to 200 and 100 OFDM symbols, respectively. Also, uncoordinated precoder design is considered as the reference case (precoders calculated locally without considering the inter-cell interference). The results in Figure 9.7 are shown for the TDD frame length 200 ($\gamma = 0.01$), and different α, β and δ values. The figure shows considerable gains in sum rate for lower α, β and δ values compared to the uncoordinated system. The peak rate can be reached for 8 times less overhead (i.e. 12% of overhead) compared to a fully synchronized scenario. Hence, the decentralized precoding constitutes a solid alternative to JT CoMP in 5G networks when infrastructure allows only limited backhaul capabilities.

9.4.2   Interference alignment

Interference Alignment (IA) is a powerful technique that can be applied to a distributed cooperative scenario, which has limited backhaul capacity. In particular, the Multi-User Inter-Cell Interference Alignment (MUICIA), that is based on IA, is an alternative to the distributed beamforming, with the advantage of not being limited to a 5G TDD system.

IA manages the interference efficiently by using an "align" and "suppression" strategy [43]. With the help of multiple antennas at the transmitters, transmission precoding can be designed in a way that the spatial dimension can be used for interference alignment. Similarly, with the help of multiple antennas at the receivers, the spatial dimension can be used for interference suppression. IA becomes particularly relevant in the 5G Ultra-Dense Network (UDN) (see [44] and Chapters 2 and 7) scenario, where the UEs (especially those which are located at the cell edges) experience very high interference.

Hitherto, the State of the Art (SoA) provides interference coordination mechanisms for mitigating such high-interference scenarios (for example in LTE [36] enhanced inter-cell interference coordination mechanisms in time domain and frequency domain). However, the SoA based solutions require backhaul and mostly they work on the principle of limiting the resource usage in some parts of the network. Moreover, the SoA solutions mainly aim to mitigate the Inter-Cell Interference (ICI) whereas in UDN scenarios intra-cell/Multi-User Interference (MUI) has also significant impact on the system performance. In addition to that, these solutions are based on the coordination, which is mainly managed by the upper layers (e.g. MAC layer) in contrast to other available solutions that are based on the digital transmission techniques (e.g. at physical layer).

Figure 9.8 represents a typical UDN scenario with multiple active UEs in each cell. One can see that UE1 is co-scheduled with UE2 in cell 1 therefore experiences both MUI and ICI.

9.4.2.1   Multi-user inter-cell interference alignment

As mentioned above, MUICIA is based on IA. It can be applied effectively to a backhaul with limited bandwidth or too high latency. MUICIA requires no coordination between the BSs and relies only on the local CSI within each cell [45]. Assuming a limited number of spatial dimensions at the transmitter and the receiver (e.g. a 2x2 MIMO system), the Degree of Freedom (DoF) is limited. Hence, it is not possible to align all the interferences

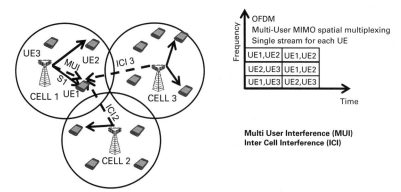

Figure 9.8   A multi user multi cell 2x2-system where multiple independent data streams are sent to different users (UEs) using the same OFDM resource element.

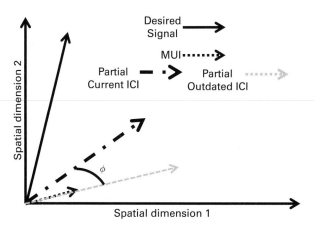

Figure 9.9   Representation of signal space in terms of vector space at the antennas of UE1.

in one subspace and to have the desired signal in an orthogonal space. An alternative approach is to align the MUI with only partial ICI. The partial ICI information is based on the direction of maximum average ICI, which is represented by the eigen-vector. The eigen-vector corresponds to the maximum eigen-value of the ICI covariance matrix estimated by the UE. However, estimating the ICI covariance matrix prior to the transmission in the current TTI is difficult for the UE. One possible solution is that each BS exploits the outdated information based on the estimation of ICI by the UEs from the previous transmission. The alignment with partial and outdated ICI is shown in Figure 9.9 with limited number of DoF in terms of only 2 spatial dimensions at the UE-receiver. Note that the BS requires only the local information, i.e. no inter-BS coordination is needed. Each UE estimates and sends the required information to its serving BS. Based on this design principle, this scheme is suitable for limited bandwidth or no-backhaul scenarios. Finally, the BS designs the transmit precoding such that the partial and outdated ICI subspace is aligned at the intended receiver with the MUI subspace of the current transmission.

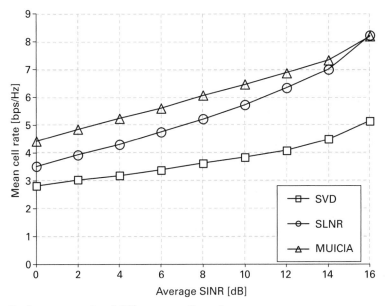

Figure 9.10   Performance results of different precoding schemes.

Figure 9.9 shows an instance of the current partial ICI that differs with the outdated partial ICI with an angle ϕ. This misalignment would cause leakage interference in the desired signal subspace even after the suppression at the receiver. However, this leakage interference can be minimized/suppressed with proper receiver design.

9.4.2.2   Performance

Figure 9.10 presents the simulation results for the system shown in Figure 9.8 with 3 cells where each cell is equipped with 2 antennas and serving two UEs with independent data streams. The UEs are also equipped with two antennas and they are dropped in the coverage area of the cell according to a given input average Signal to Interference plus Noise Ratio (SINR). Full buffer traffic and spatial channel model are assumed [6], and it is also assumed that perfect CSI and ICI information is available at the transmitter without any feedback delay. The UEs are using MMSE algorithm for the estimation and suppression of interference. Further, MUICIA performance is compared with other two baseline approaches: the first approach uses the singular value decomposition (SVD) of the serving channel for the design of precoding; the second approach designs the precoding using the optimization of the desired signal to leakage interference plus noise ratio (SLNR) [46]. The results are plotted against the average SINR of the UEs in a cellular network (e.g. LTE). The average SINR represents the location of the UE in the cell and it is an input parameter. The results show that MUICIA outperforms the baseline precoding approaches in lower and average range of SINR. In fact, in these conditions ICI is very strong and it is possible to find dominant ICI direction for alignment as well as for suppression. At higher SINRs, the system is more MUI-limited and the precoding approach with SLNR seems to be an equally good choice for the system.

9.5   JT CoMP with advanced receivers

As explained in Chapters 1 and 5, a more active role is expected by the UEs and Device-to-Device (D2D) communications will be part of 5G in order to increase the network coverage [47][48]. Moreover, for services like video on demand and video streaming, caching at the cell edge, for instance at the UEs with high memory storage like nowadays smartphones, has been recently recognized as a way to increase system throughput [49]. In the context of coordinated multi-point schemes, most of the existing works in the literature assume that UEs are equipped with only one or two antennas, even if LTE-A already assumes that UEs might be equipped with up to eight antennas [50]. Although this number seems nowadays a bit optimistic, in the near future the technological innovation will allow building smartphones and tablets with a high number of antennas. In particular, this will be easier in the context of higher carrier frequencies and consequently smaller wavelengths and antenna form factors. Then, 5G networks will be designed by considering a huge variety of UEs, for instance cars, which can be easily equipped with more than one antenna [44]. Therefore, it is fundamental to develop CoMP schemes that take into account that UEs are equipped with multiple antennas and can exploit these additional DoF to either implement a proper combiner to tackle the interference (resulting in a power gain) or receive and detect multiple streams of data (resulting in a multiplexing gain) [51].

In the following, Section 9.5.1 presents a dynamic joint BS clustering and UE scheduling algorithm for downlink JT CoMP with multiple antenna UEs. Then, Section 9.5.2 shows how the network can assist and improve the interference cancellation at the UEs.

9.5.1   Dynamic clustering for JT CoMP with multiple antenna UEs

In this section, it is shown that dynamic clustering for JT CoMP (with multiple antenna UEs) improves the performance of a CoMP system. Thanks to the additional DoF at the UE, system throughput can be increased by (1) suppressing the residual interference with an IRC combiner and (2) allowing multi-stream transmission.

The dynamic clustering for downlink JT CoMP algorithm [52] considers the scenario in Figure 9.11, where a Central Unit (CU) coordinates the BSs and uses a resource allocation algorithm to jointly perform dynamic BS clustering and UE scheduling, under the assumption of multi-antenna UEs. Moreover, the UEs perform Successive Interference Cancellation (SIC) with IRC. The algorithm first defines the set of candidate BS clusters that mainly depend on the position of each UE with respect to the interfering BSs. These sets are updated relatively to the large-scale fading (i.e. in the order of hundreds of milliseconds). As shown in Figure 9.11, the cluster size for a given UE is limited to a number of BSs since in practice most of the interference at the UE comes from the closest BSs. In this figure, two clusters are shown. The five BSs depicted will be split between the two clusters: A and B. The top two BSs will be part of the cluster A whereas the three others will belong to the cluster B. Afterward, the CU schedules a subset of these candidate clusters (with associated UEs) for downlink transmission

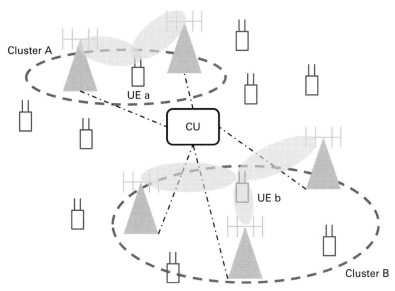

Figure 9.11   Candidate BS cluster selection with the dynamic clustering algorithm for downlink JT CoMP.

based on a two-step procedure that follows the fast fading time scale (i.e. in the order of few milliseconds).

The two-step procedure, in which the CU organizes BSs into clusters and schedules UEs in each cluster, can be summarized as follows:

1. A weighted sum rate is estimated within each candidate cluster by performing UE selection and optimizing precoders, powers and transmission ranks. In particular, the Multiuser Eigenmode Transmission (MET) scheme [21] is used, with equal power allocation among the streams and a greedy iterative eigenmode selection algorithm.
2. The CU selects the set of non-overlapping BS clusters to be used for transmission by maximizing the system weighted sum rate. As the optimization problem is NP-hard, a greedy iterative algorithm can be used to reduce the implementation complexity.

Note that the dynamic clustering for JT allows the flexibility of scheduling a given UE in different BS clusters across successive time slots and with different transmission ranks, depending on the signal to interference plus noise ratio conditions.

9.5.1.1   Performance of dynamic clustering

The dynamic clustering for JT has been analyzed for the following two scenarios:

- A standard homogeneous hexagonal grid scenario with BSs organized in sites [52].
- A Madrid grid scenario with 3 macro sectors and 18 micro sectors, for a total of 21 BSs, deployed in a grid of 3x3 buildings, modeling a dense urban information society (more details in Section 8.2 of [26]).

Table 9.1 Average cell rate and cell border throughput in the homogeneous hexagonal grid scenario.

	$n_r = 1$	$n_r = 2$	$n_r = 4$
Baseline average cell rate [bps/Hz]	6.8	8.5	10.8
Baseline cell border throughput [bps/Hz]	0.23	0.3	0.4
CoMP average cell rate [bps/Hz]	8.3	9.9	12
CoMP cell border throughput [bps/Hz]	0.33	0.41	0.52

Table 9.2 Average cell rate and cell border throughput in the Madrid grid scenario.

	$n_r = 1$	$n_r = 2$	$n_r = 4$
Baseline average cell rate [bps/Hz]	11.1	15.3	21.2
Baseline cell border throughput [bps/Hz]	0.25	0.33	0.45
CoMP average cell rate [bps/Hz]	19.2	22.6	26.9
CoMP cell border throughput [bps/Hz]	0.66	0.8	0.91

The numerical results are shown for these two scenarios in Tables 9.1 and 9.2 respectively, which show the average cell rate and the cell border throughput, i.e. the 5th percentile of the UE rate, for the dynamic clustering (with a maximum cluster size of 3 BSs) and for a baseline setup (without BS coordination). Each UE is equipped with n_r antennas. First, an important performance improvement can be observed by adding antennas at the UE. Two factors contribute to this gain: (a) UEs with lower SINR use IRC to limit the impact of residual ICI not managed at the transmit side and (b) UEs with higher SINR can be served by multiple streams of data. Moreover, the throughput gain, achieved by dynamic clustering over the baseline, decreases by adding more antennas at the UE side. In fact, as the gain of using multiple antenna UEs is mainly due to the IRC, the benefits of increasing n_r are seen more in a non-cooperative scenario, where the residual interference is higher. For more details about the results, the reader can be referred to [52] and [9].

Two important comments should be made when comparing the results achieved in the two different scenarios. First, substantially higher rates are achieved in the Madrid grid: for instance, the average cell rate is almost doubled. Second, the gain achieved by dynamic clustering over the baseline is much higher in the Madrid grid than in the homogeneous hexagonal grid: for instance, in terms of the cell border throughput, when comparing the cases $n_r = 4$ to $n_r = 1$. In fact, the gain (of CoMP over the baseline case) varies from approximately 30% (for $n_r = 4$) to 45% (for $n_r = 1$) in the homogeneous hexagonal grid, whereas the gain varies from approximately 100% (for $n_r = 4$) to 160% (for $n_r = 1$) in the Madrid grid. These two effects can be explained by observing that the Madrid grid is a heterogeneous scenario with a higher density of BSs. Therefore, the UEs are typically closer to their anchor BS, they measure a better SINR and, as a consequence, the achieved rates are also higher. Then, in heterogeneous networks the structure of the interference is typically different with respect to the one in a homogeneous

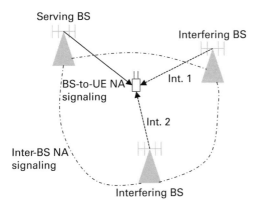

Figure 9.12   Overview picture showing the NA signaling components.

scenario. More precisely, the interference comes mainly from few strong interferers, typically the macro BSs.

9.5.2   Network-assisted interference cancellation

In Section 9.5.1, it was shown how IRC can be beneficial to the system performance by suppressing the residual interference at the UE. A further performance improvement can be obtained by allowing the UE to implement Interference Suppression (IS)/ Cancellation (IC) mechanisms with the aid of Network Assistance (NA) [53]. The key idea is to make the UE receiver aware of additional information characterizing the dominant interference, i.e. information related to the transmission performed by the neighboring cells. An example of a NA-IS/IC implementation is shown in Figure 9.12, where the interfering BSs share with the serving BS some specific information regarding their transmission format. In turn, the serving BS signals these parameters to the scheduled UE (for more details, see Section 8.16 in [9]).

These interference signal parameters may include (but are not limited to) information on the cell ID, reference signal antenna ports, power offset values, precoder selection, transmission rank and modulation order. Such additional knowledge on the dominant interference enables advanced IS/IC with enhanced channel estimation accuracy and improved MIMO detection of interferer. In order to obtain a proper balance between the overhead associated with NA signaling and the processing complexity of a UE receiver, certain interference signal parameters can be blindly detected by the UE receiver.

While IRC is a linear combiner that simply suppresses the interference, this additional information can be used by the UE to detect, reconstruct and cancel the interference signal, thus further improving the achieved spectral efficiency. The UE receiver can perform a Symbol-Level IC (SLIC), (reduced-complexity) Maximum Likelihood (ML) symbol-level joint detection of desired and interferer signals or, when more information about the channel coding is available, a CodeWord-level IC (CWIC) [53].

Network-Assisted Interference Cancelation (NAIC) is likely to be even more relevant in the 5G era, as e.g. TDD with a flexible usage of resources for uplink and downlink and novel communication forms such as direct D2D that will lead to novel interference constellations, possibly involving the notion of very strong interference (i.e. interference signals that are much stronger than the desired signal to be decoded).

NAIC may also be interesting in a JT CoMP context. For instance, joint transmission from multiple transmit nodes could be performed such that the receiver side sees one strong interferer which may be efficiently cancelled, while the remaining inter-user interference is very low. Furthermore, NAIC may be used to cancel inter-cluster interference in the context of CoMP (see also Section 9.2.2).

9.6   Conclusions

CoMP, in particular JT CoMP, has been investigated since many years, as it promises the potential to turn interference from a burden into useful signal energy, hence overcoming the main limitation in today's cellular systems. However, the standardization of JT CoMP in 3GPP LTE Release-10 and beyond has seen difficult times, as the gains from a straightforward application of JT CoMP in a legacy cellular communications system are quite limited.

In the last few years, research has shown that most of the expected large gains for JT CoMP approaches can in fact be unleashed, if the overall solution is constructed carefully and a set of additional enablers is provided:

- improvements of the channel estimation and feedback, namely channel prediction with feedback accompanied by reliability information;
- advanced clustering options, with enlarged cooperation clusters exploiting the cover-shift and tortoise concepts;
- novel scheduling approaches, in particular the two-stage strategy that address scheduling first on a cell level, and then on a cooperation cluster level.

It should be noted that most of these features could actually be part of a further evolution of existing 4G networks. Improved feedback design, for instance, could be integrated in LTE evolution if there is sufficient support for it. The existing CSI reference signals in LTE-A could be enhanced to accommodate advanced channel estimation approaches, designed with CoMP in mind. Moreover, the existing air interface in 4G, OFDM, has proven to work well with CoMP.

Nonetheless, even if an evolution of 4G could in principle support many of the above features, a clean slate design for 5G shall natively support all these features in a more efficient way. Moreover, a change of the paradigm of "cells" could allow the network to serve UEs through virtual CoMP cells, without the UEs realizing which transmit nodes are involved. Further, CSI reference signals and their reporting could be dynamically configurable, such that the best trade-off between CoMP performance, reference signal and reporting overhead can be obtained for different scenarios and channel conditions. Furthermore, a clean-slate 5G could provide better means for inter-cluster interference rejection or cancelation. Some general technology trends in 5G will in fact inherently better facilitate the usage of CoMP. For example, larger bandwidths expected in the 5G era will allow for a better channel estimation, and reduced TTI lengths will allow for faster CSI feedback loops. Other trends, such as a substantial network densification of small cells, will require special attention in the context of CoMP.

When the available network infrastructure does not allow supporting full JT CoMP, one may use distributed cooperative transmission techniques. Moreover, one may also consider that the receivers take an active role in dealing with the interference in the network. The availability of smarter user devices, capable of interference suppression or cancellation, opens the way to new approaches where both the network and UEs cooperate in order to overcome the limitations of interference.

References

[1] P. Marsch and G. Fettweis, *Coordinated Multi-Point in Wireless Communications: From Theory to Practice.* Cambridge, UK: Cambridge University Press, 2011.

[2] S. V. Hanly and P. A. Whiting, "Information-theoretic capacity of multi-receiver networks," *Telecommunication Systems*, vol. 1, no. 1, pp. 1–42, March 1993.

[3] S. Shamai and B. Zaidel, "Enhancing the cellular downlink capacity via co-processing at the transmitting end," in IEEE Vehicular Technology Conference, pp. 1745–1749, Rhodes, Greece, May 2001.

[4] M. K. Karakayali, G. J. Foschini and R. A. Valenzuela. "Network coordination for spectrally efficient communications in cellular systems," *IEEE Wireless Communications*, vol. 13, no. 4, pp.56–61, August 2006.

[5] 3GPP TR 36.819, "Coordination Multi-Point Operation for LTE Physical Layer Aspects (Release 11)," Technical Report TR 36.819 V1.0.0, Technical Specification Group Radio Access Network, June 2011.

[6] 3GPP TR 36.814, "E-UTRA: Further advancements for E-UTRA physical layer aspects (Release 9)," Technical Report TR 36.814 V9.0.0, Technical Specification Group Radio Access Network, March 2010.

[7] V. R. Cadambe and S. A. Jafar, "Interference alignment and degrees of freedom of the K-user interference channel," *IEEE Transactions on Information Theory*, vol. 54, no.8, pp. 3425–3441, August 2008.

[8] ICT-247223 ARTIST4G project, "Interference Avoidance techniques and system design," Deliverable D1.4, July 2012.

[9] ICT-317669 METIS project, "Final performance results and consolidated view on the most promising multi-node/multi-antenna transmission technologies," Deliverable D3.3, February 2015, www.metis2020.com/documents/deliverables/

[10] A. Osseiran, J. Monserrat, and W. Mohr, *Mobile and Wireless Communications for IMT-Advanced and Beyond.* Chichester: Wiley, 2011.

[11] N. Gresset, H. Halbauer, W. Zirwas, and H. Khanfir, "Interference avoidance techniques for improving ubiquitous user experience," *IEEE Vehicular Technology Magazine*, vol. 7, no. 4, pp. 37–45, December 2012.

[12] V. Kotzsch and G. Fettweis, "On synchronization requirements and performance limitations for CoMP systems in large cells," in IEEE International Workshop on Multi-Carrier Systems & Solutions, Herrsching, pp. 1–5, May 2011.

[13] V. T. Wirth, M. Schellmann, T. Haustein, and W. Zirwas, "Synchronization of cooperative base stations," in IEEE International Symposium on Wireless Communication Systems, Reykjavik, October 2008, pp. 329–334,.

[14] K. Manolakis, C. Oberli, and V. Jungnickel, "Synchronization requirements for ofdm-based cellular networks with coordinated base stations: Preliminary results," in International OFDM Workshop, Hamburg, Germany, September 2010.

[15] D. T. Phan-Huy, M. Sternad, and T. Svensson, "Adaptive large MISO downlink with Predictor Antennas Array for very fast moving vehicles," in International Conference on Connected Vehicles & Expo, Las Vegas, USA, December 2013, pp. 331–336.

[16] G. Foschini, K. Karakayali, and R. Valenzuela, "Coordinating multiple antenna cellular networks to achieve enormous spectral efficiency," *IEE Proceedings-Communications*, vol. 153, no. 4, pp. 548–555, August 2006.

[17] H. S. Rahul, S. Kumar, and D. Katabi, "JMB: Scaling wireless capacity with user demands," in ACM conference on Applications, technologies, architectures, and protocols for computer communication, Helsinki, August 2012, pp. 235–246.

[18] W. Zirwas and M. Haardt, "Channel prediction for B4G radio systems," in IEEE Vehicular Technology Conference, Dresden, June 2013, pp. 1–5.

[19] L. Thiele, M. Kurras, M. Olbrich, and B. Matthiesen, "Channel aging effects in CoMP transmission: Gains from linear channel prediction," in IEEE Annual Asilomar Conference on Signals, Systems and Computers, Monterey, November 2011.

[20] L. Thiele, M. Kurras, M. Olbrich, and K. Börner, "On feedback requirements for CoMP joint transmission in the quasi-static user regime," in IEEE Vehicular Technology Conference, Dresden, June 2013.

[21] F. Boccardi and H. Huang, "A near-optimum technique using linear precoding for the MIMO broadcast channel," in IEEE International Conference on Acoustics, Speech and Signal Processin, Honolulu, April 2007.

[22] R. Apelfröjd, M. Sternad, and D. Aronsson, "Measurement-based evaluation of robust linear precoding for downlink CoMP," in IEEE International Conference on Communications, Ottawa, Canada, June 2012.

[23] L. Thiele, "Spatial interference management for OFDM-based cellular networks," PhD thesis, TUM 2013.

[24] L. Thiele, M. Kurras, K. Borner, and T. Haustein, "User-aided sub-clustering for CoMP transmission: Feedback overhead vs. data rate trade-off," in Asilomar Conference on Signals, Systems and Computers, Pacific Grove, November 2012, pp. 1142–1146.

[25] V. Jungnickel, K. Manolakis, W. Zirwas, B. Panzner, V. Braun, M. Lossow, M. Sternad, R. Apelfröjd, and T. Svensson, "The role of small cells, coordinated multi-point and massive MIMO in 5G," *IEEE Communications Magazine*, vol. 52, no. 5, pp. 44–51, May 2014.

[26] ICT-317669 METIS project, "First performance results for multi-node/multi-antenna transmission technologies," Deliverable D3.2, April 2014, www.metis2020.com/docu ments/deliverables/

[27] P. Marsch and G. Fettweis, "Static clustering for cooperative multi-point (CoMP) in mobile communications," in IEEE International Conference on Communications, Kyoto, June 2011, pp. 1–6.

[28] W. Zirwas, W. Mennerich, and A. Khan, "Main enablers for advanced interference mitigation," *European Transactions on Telecomunications*, vol. 24, no. 1, January 2013.

[29] V. Jungnickel, S. Jaeckel, S. Jaeckel, L. Thiele, U. Krueger, and A. C. von Helmolt, "Capacity measurements in a multicell MIMO system," in IEEE Global Communications Conference, San Francisco, November 2006.

[30] L. Thiele, T. Wirth, T. Haustein, V. Jungnickel, E. Schulz, and W. Zirwas, "A unified feedback scheme for distributed interference management in cellular systems: Benefits and challenges for real-time tmplementation," in European Signal Processing Conference, Glasgow, August 2009.

[31] W. Zirwas, "Opportunistic CoMP for 5G massive MIMO multilayer networks," in ITG Workshop on Smart Antennas, Ilmenau, March 2015, pp. 1–7.

[32] D. Gesbert, S. Hanly, H. Huang, S. Shamai Shitz, O. Simeone, and W. Yu, "Multi-cell MIMO cooperative networks: a new look at interference," *IEEE Journal on Selected Areas in Communications*, vol. 28, no. 9, pp. 1380–1408, December 2010.

[33] E. Bjornson, R. Zakhour, D. Gesbert, and B. Ottersten, "Cooperative multicell precoding: Rate region characterization and distributed strategies with instantaneous and statistical CSI," *IEEE Transactions on Signal Processing*, vol. 58, no. 8, pp. 4298–4310, August 2010.

[34] E. Lähetkangas, K. Pajukoski, J. Vihriälä, et al., "Achieving low latency and energy consumption by 5G TDD mode optimization," in IEEE International Conference on Communications, Sydney, June 2014, pp. 1–6.

[35] P. Komulainen, A. Tolli and M. Juntti. "Effective CSI signaling and decentralized beam coordination in TDD multi-cell MIMO systems," *IEEE Trans. Signal Processing*, vol. 61, no. 9, pp.2204–2218, May 2013.

[36] P. Jayasinghe, A. Tölli, J. Kaleva, and M. Latva-aho, "Bi-directional signaling for dynamic TDD with decentralized beamforming," in IEEE International Conference on Communications, London, June 2015.

[37] M. S. El Bamby, M. Bennis, W. Saad, and M. Latva-aho, "Dynamic uplink-downlink optimization in TDD-based small cell networks," in IEEE International Symposium on Wireless Communication Systems, Barcelona, August 2014, pp. 939–944.

[38] I. Sohn, K. B. Lee, and Y. Choi, "Comparison of decentralized timeslot allocation strategies for asymmetric traffic in TDD systems," *IEEE Trans. Wireless Commun.*, vol. 8, no. 6, pp. 2990–3003, June 2009.

[39] ICT-317669 METIS project, "Components of a new air interface: Building blocks and performance," Deliverable D2.3, April 2014, www.metis2020.com/docu ments/deliverables/

[40] P. Komulainen, A. Tolli, and M. Juntti, "Effective CSI signaling and decentralized beam coordination in TDD multi-cell MIMO systems," *IEEE Trans. on Signal Processing*, vol. 61, no. 9, pp. 2204–2218, May 2013.

[41] C. Shi, R. A. Berry, and M. L. Honig, "Bi-directional training for adaptive beamforming and power control in interference networks," *IEEE Trans. on Signal Processing*, vol. 62, no. 3, pp. 607–618, February 2014.

[42] Q. Shi, M. Razaviyayn, Z. Q. Luo, and C. He, "An iteratively weighted MMSE approach to distributed sum-utility maximization for a MIMO interfering broadcast channel," *IEEE Trans. on Signal Processing*, vol. 59, no. 9, pp. 4331–4340, September 2011.

[43] V. Cadambe and S. Jafar, "Interference alignment and spatial degrees of freedom for the k user interference channel," in IEEE International Conference on Communications, Beijing, May 2008, pp. 971–975.

[44] ICT-317669 METIS project, "Initial report on horizontal topics, first results and 5G system concept," Deliverable D6.2, March 2014, www.metis2020.com/docu ments/deliverables/

[45] D. Aziz and A. Weber, "Transmit precoding based on outdated interference alignment for two users multi cell MIMO system," in IEEE International Conference on Computing, Networking and Communications, San Diego, January 2013, pp. 708–713.

[46] M. Sadek, A. Tarighat, and A. Sayed, "A leakage-based precoding scheme for downlink multi-user mimo channels," *IEEE Trans. on Wireless Communications*, vol. 6, no. 5, pp. 1711–1721, May 2007.

[47] F. Boccardi, R. W. Heath Jr., A. Lozano, T. L. Marzetta, and P. Popovski, "Five disruptive technology directions for 5G," *IEEE Commun. Mag.*, vol. 52, no. 2, pp. 74–80, February 2014.

[48] G. Fodor, E. Dahlman, G. Mildh, S. Parkvall, N. Reider, G. Miklós, and Z. Turányi, "Design aspects of network assisted device-to-device communications," *IEEE Commun. Mag.*, vol. 50, no. 3, pp.170–177, March 2012.

[49] M. Ji, A. M. Tulino, J. Llorca, and G. Caire, "On the average performance of caching and coded multicasting with random demands," in IEEE International Symposium on Wireless Communications Systems, Barcelona, August 2014, pp. 922–926.

[50] F. Boccardi, B. Clercks, A. Ghosh, E. Hardouin, K. Kusume, E. Onggosanusi, and Y. Tang, "Multiple-antenna techniques in LTE-advanced," *IEEE Commun. Mag.*, vol. 50, no. 3, pp. 114–121, March 2012.

[51] I. Hwang, C. B. Chae, J. Lee, and R. W. Heath, "Multicell cooperative systems with multiple receive antennas," *IEEE Wireless Commun. Mag.*, vol. 20, no. 1, pp. 50–58, February 2013.

[52] P. Baracca, F. Boccardi, and N. Benvenuto, "A dynamic clustering algorithm for downlink CoMP systems with multiple antenna UEs," *EURASIP Journal on Wireless Commun. and Networking*, 2014, vol. 125, August 2014.

[53] 3GPP TR 36.866, "Network-assisted interference cancellation and suppression for LTE (Release 12)," Technical Report TR 36.866 V1.1.0, Technical Specification Group Radio Access Network, November 2013.

10 Relaying and wireless network coding

Elisabeth de Carvalho, Mats Bengtsson, Florian Lenkeit, Carsten Bockelmann, and Petar Popovski

Relaying and network coding are powerful techniques that improve the performance of a cellular network, for example by extending the network coverage, by increasing the system capacity or by enhancing the wireless link reliability. This chapter focuses on relaying and wireless network coding in 5G. After reviewing the history of relaying, the key envisioned scenarios for relaying in 5G are highlighted, namely the provisioning of wireless backhaul in Ultra-Dense Networks (UDNs), for nomadic cells or for data aggregation in the context of massive machine-type communications. While full-duplex technology is slowly gaining maturity, it is expected that due to complexity reasons most relaying scenarios in 5G will be based on half-duplex devices. Therefore, finding solutions to overcome the half-duplex limitation remains critical. The chapter describes the following three key innovations for efficient half-duplex relaying:

- By applying the principles of wireless network coding to distributed multi-way traffic, in-band relaying becomes a spectrally efficient solution for wireless backhaul in ultra-dense networks of small cells[1], despite conventional views.
- Non-orthogonal multiple access techniques, as required by physical-layer network coding, are essential for increased spectral efficiency when simultaneous multi-flows are exchanged through a same relay. Here, Interleave-Division Multiple-Access (IDMA) is put forward for its ability to support flexible rate requirements.
- Buffer-aided relaying is featured where different ways to exploit buffering are described for improved diversity and increased rates. This technique targets delay tolerant applications having high data rate requirements.

10.1 The role of relaying and network coding in 5G wireless networks

Relaying was a common technique used to convey messages over large distances in ancient empires such as Egypt, Babylon, China, Greece, Persia and Rome [1]. The

---

[1] For conciseness, UDNs of small cells will be simply referred to as small cells.

5G Mobile and Wireless Communications Technology, ed. A. Osseiran, J. F. Monserrat, and P. Marsch. Published by Cambridge University Press. © Cambridge University Press 2016.

messages were transmitted in various forms, such as beacon fires relayed by towers or mountain peaks. A more common method was sending messengers on horseback between Relay Stations (RSs) until the final destination was reached. With the advent of science, communication techniques improved. In 1793, the Chappe brothers of France proposed a telegraph system relying on RSs[2] equipped with telescopes and lighted by lamps.

In modern times, RSs were initially simple devices that amplify a signal and forward it immediately, and were mainly intended to extend the coverage of the wireless system. These were low-cost devices, compared to Base Stations (BSs), that did not include any baseband processing, and hence no network protocol operation was possible. The backhaul connection was usually implemented with microwave links in separate frequency bands to avoid interference with the access link, or in-band with appropriate (receive and transmit) antenna isolation.

10.1.1   The revival of relaying

Relays have long been considered to have a limited applicability in wireless networks. The conventional view is that a relay is deployed for coverage extension, hence, it is placed near the cell edge in a cellular network and forwards messages to or from a cell edge Mobile Station (MS) that is out of reach of the BS. The benefit of relays in terms of data rate is generally seen as a marginal one because of poor spectral efficiency. In fact, an in-band relay, using the same radio spectrum as the BS and operating in half-duplex, causes a loss in spectral efficiency. As two transmission phases are required to send a message, roughly speaking, a spectral loss by a factor of two is created compared to a direct transmission. Note that out-of-band relaying occupying a different radio spectrum from the BS can compensate for this loss at the cost of additional spectrum usage.

Over the past ten years, research in relay technology has progressed significantly, outlining novel domains and modes of applicability. The first game changer came from cooperative diversity [2]. In cooperative diversity, depicted in Figure 10.1(a), both the BS and the relay cooperate to improve the communication quality, meaning capacity or diversity. It targets a scenario where the direct link between the BS and the MS is able to carry some information, excluding scenarios where the MS is out of reach of the BS. The information to be sent from the BS to the MS passes through two

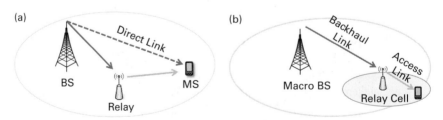

Figure 10.1   (a) Cooperative relaying, (b) self-backhauling in LTE release 10.

---

[2] The term RS and relay are used interchangeably in the chapter.

paths: the direct link and the link through the relay. At the MS, the information from the two paths is combined in a coherent way. Although the communication through the relay still takes two time slots, cooperation yields a better spectral efficiency compared to a single relayed link but also an increased diversity if the two communication paths are independent.

Another game changer came with the introduction of Two-Way Relaying (TWR) [3][4] (see Section 10.2.1). TWR refers to a scenario where two-way communication, i.e. two data flows, is considered: in a cellular scenario, this means one uplink (UL) flow and one downlink (DL) flow. Instead of conventionally considering the optimization of a one-way communication, the two data flows are jointly optimized and simultaneously served through the relay. Combined with the ideas of Wireless Network Coding (WNC), the spectral efficiency loss caused by the half-duplex nature of the relay can be regained. The lesson learned from the TWR scenario is that a joint optimization of uplink and downlink flows expands the design space for increased spectral efficiency.

These research milestones have been among the main reasons to revive the interest in relaying and have set the stage for the introduction of relays as a mainstream architectural component in 5G.

10.1.2   From 4G to 5G

Relays have been introduced in Release 10 of the Long-Term Evolution (LTE) standard [5]. Relaying is described through the concept of self-backhauling depicted in Figure 10.1(b). The relay, seen as a small-cell BS, defines a so-called relay cell. It is linked to the macro BS by a backhaul link and to the MS by an access link. Both in-band and out of-band relays are considered as well as half-duplex and full-duplex relays. For in-band relays, the accent is set on interference avoidance between the access and backhaul links, for which a solution is based on time-multiplexing of the two links.

Half-duplex self-backhauling is the structure underlying the technology components forming the basis of this chapter. A major difference compared with LTE is the treatment of the interference between the access and backhaul links. This chapter considers a multi-way communication scenario, where the principles behind wireless network coding are exploited to embrace interference. Additionally, non-orthogonal multiple-access design allows a further compensation for the half-duplex limitations of the relay and approach the performance of full-duplex relaying. Backed up by buffer-aided relaying, the proposed techniques provide an intermediate technological step toward full-duplex relaying.

10.1.3   New relaying techniques for 5G

From the recent research in relaying and network coding, three new techniques have been identified. These techniques promise to turn relaying into a driving technology for throughput boosting in 5G. Those key techniques constitute the core of this chapter and are described in Sections 10.2, 10.3 and 10.4.

- **Multi-flow relaying**: The principles behind wireless network coding have been extensively applied to the two-way relaying scenario, where two devices exchange messages through a relay. Those principles can be extended to more general scenarios involving multiple BSs, relays and MSs, where multiple communication flows, down-link and uplink, are simultaneously scheduled so that the interference created can be cancelled using side information. The main domain of applicability of multi-flow relaying is foreseen in UDNs of small cells and nomadic nodes. UDNs, nomadic nodes and moving relay nodes are elaborated in Chapter 11.
- **Non-orthogonal multiple access**: When considering multiple TWR communication pairs aided by a common relay, the question of medium access comes into play. In 4G, medium access is based on Orthogonal Frequency Division Multiple Access (OFDMA) and, thus, happens in an orthogonal manner. For TWR that means each communication pair needs to be assigned dedicated resources for transmission. However, for 5G, a massive rise in communicating nodes as well as a much higher demand in flexibility is predicted. Under these assumptions, orthogonal channel access becomes prohibitive as it introduces a large signaling overhead due to the scheduling and only offers limited flexibility in terms of rate requirements and power constraints. To this end, non-orthogonal channel access is a very promising candidate for the 5G air interface as discussed in Chapter 7. The combination of non-orthogonal access with TWR or, more general, multi-flow communication, unleashes new possibilities for system design.
- **Buffer-aided relaying**: If the relays are equipped with buffers, it is possible to do opportunistic scheduling, which provides link selection diversity gains against chan-nel fading. If, in addition, there are several such relays available, instead of exploiting the additional diversity gains, it is possible to bypass the half-duplex limitation, by letting one relay listen to the source, while another relay simultaneously forwards buffered data to the destination, see Section 10.4.

The main thread in this chapter is about regaining the spectral efficiency loss caused by in-band half-duplex relays in one-way communication. This is achieved by optimizing the communication considering a two-way traffic or resorting to buffering at multiple relays. Recent years have seen successful hardware implementations [6][7] of in-band full-duplex radios, able to transmit and receive simultaneously in the same band, hence showing that full-duplex devices are likely to be a part of 5G wireless networks. A natural question is whether full-duplex relays would solve the spectral inefficiency of half-duplex relays in a simpler way. Indeed, a full-duplex relay has the capability of receiving a signal from the previous node while transmitting to the next node. The perspective in this chapter is that full-duplex relays will not pervade in 5G. Half-duplex relays can be expected to dominate as full duplexing comes with a much higher cost and power usage.

The enabling factor for full-duplex radios is the capability for a very large suppression of self-interference between the transmit path and the receive path of the same device. Suppression is achieved by a combination of antenna design, analog and digital cancellation methods [8], with suppression levels exceeding 100 dB. Although this suppression level might not be sufficient for high power BSs [9], it appears adequate to enable the deployment of full-duplex radios in smaller stations, such as RSs or access points in small cells. Cost and power limitations as well as the difficulty in tracking the self-interference channel in a changing environment are currently a serious impediment to full-duplex radios in mobile devices. However, a significant spectral gain can be achieved with full-duplex radios solely at the small-cell base stations, hence appearing as primary candidates for full-duplex in 5G.

The applicability of full-duplex radios in wireless architectures is still not fully understood. A general consensus is that, without a proper communication protocol, the potential of full duplexing cannot be achieved. The main reason is that full-duplex stations double the interference present in the network compared to having only half-duplex stations, and introduce new interference patterns, especially potentially large interference levels between RSs.

10.1.4   Key applications in 5G

The role of relays and network coding can be foreseen in the following technologies:

- **Small cells with wireless backhaul**: Small cells or more precisely UDNs of small cells promise to play a central role in 5G as a powerful answer to network densification and the need for a massive increase in area spectral efficiency or bit rate per unit area (cf. Chapter 7, Chapter 11 and [4][10][11]). One of the key elements in the deployment of small cells is the backhaul that connects the small-cell BS to the infrastructure. Traditionally, the backhaul is wired. However, a wireless backhaul is a preferred solution when it comes to cost efficiency, flexible and rapid deployment, and increased connectivity. The key observation is that using a wireless backhaul turns the small-cell BS into a relay. The advantage of in-band backhauling compared to out-of-band backhauling (e.g. using microwave links) is that extra spectrum is not required only for relaying. Relaying can reuse the same spectrum as a cellular system without relays. Furthermore, the techniques that recover spectral efficiency, such as the techniques based on wireless network coding, help bring forward in-band relaying as a viable solution for wireless backhaul. In other words, the relays are not only an answer to coverage extension problems, but also a key ingredient for network densification.
- **Wireless backhauling for nomadic nodes**: The main feature of a nomadic node is that its BS does not have a fixed location [12]. For example, a parked vehicle may create a cell for the MSs in its proximity, such that these MSs are connected to the Internet through a relaying node mounted on the vehicle. By definition, a nomadic node cannot be connected to the fixed infrastructure through a wired backhaul. Nomadic nodes are seen as important ingredients of the 5G system concept (cf. Chapter 2 and [13]). A distinctive feature of the nomadic relays is that they make the wireless infrastructure highly dynamic: their use should be opportunistic as it cannot be based on network planning, as for fixed relays. A car-mounted device has

a dual role. It can act as a BS for a nomadic node, and can act as a MS for the fixed infrastructure. This puts forward in-band relaying as a natural modus operandi for nomadic nodes.

- **Device-to-Device (D2D) communication**: D2D communication will be a vital part in 5G as these offer many advantages ranging from reduced latency for the end nodes to traffic offloading from the core network (cf. Chapter 5). In fact, D2D communications do not require any infrastructure nodes, and can play a role in supporting D2D links and improving the network performance significantly. Ranging from the support of multiple simultaneous D2D connections by a single relay to cooperation of multiple parallel relays, the possibilities of relays in the D2D context are manifold.

- **Millimeter wave communications**: The large bandwidths available in the mmWave band come at the expense of a significantly reduced coverage area, especially in indoor scenarios. Therefore, relays will be even more essential for these frequency bands, to provide sufficient coverage. See Chapter 6 for more details.

- **Machine-Type Communications**: Traffic aggregation using a relay, Machine-Type Communication (MTC) or Machine-to-Machine (M2M) communication represents an emerging class of devices and services that have vastly different requirements from the usual, human-centered traffic. One feature is that a massive number of MTC devices (e.g. 100000) can be connected to the same BS. Although each MTC device requires only a low data rate, the massive number creates problems for the access protocol. In such a setting, a relaying node is useful to lessen the access burden put on the BS and deal with an access problem of a lower scale, i.e. only the MTC devices in its proximity. The relay aggregates the inputs from its associated MTC devices and relays it to the BS. Another feature of MTC is the ambitious requirement for energy efficiency. A step in that direction is the use of relays, which brings the infrastructure close to the MTC devices and thus allows them to decrease the transmit power. More details on MTC can be found in Chapter 4.

Table 10.1 summarizes which relay-based techniques can be exploited in the 5G use cases and 5G services defined in Chapter 2.

Figure 10.2 encompasses the key 5G applications promoted in the chapter along the techniques that are described in the next sections. Table 10.2 summarizes the three techniques, their application fields and gains. For multi-way wireless backhaul, a gain of up to 100% in spectral efficiency can be achieved as compared with one-way communication based wireless backhaul when practical impairments such as relayed noise amplification or decoding errors are not considered (see [14]–[16] and Section 10.2). Non-orthogonal multiple access provides an increased efficiency compared to orthogonal access, while the use of IDMA provides specific advantages, like the ability to support asymmetric data rates and robustness against timing offsets. Gains brought by buffer-aided techniques are given comparing with techniques that do not use buffers. More details about the gains can be found in Sections 10.2, 10.3 and 10.4.

Table 10.1 Relay-based techniques in 5G use cases and services.

Relay-based technique	5G use case(s)	5G service*
UDNs of small cells with wireless backhaul	Media on demand	xMBB
Relay-supported D2D	Virtual and augmented reality	xMBB
Multi-hop ad-hoc networks	Emergency communication	uMTC
Small cells with wireless backhaul	Smart City	mMTC, xMBB
Multi-hop or mesh networks	Tele-protection in smart grid	uMTC
Mesh networks	Factory automation	uMTC
Traffic aggregation through relaying	Massive number of devices	mMTC
Multiple buffer-aided relays with relay selection	Shopping mall	xMBB, mMTC
Relays distributed around a stadium for wireless backhaul	Stadium	xMBB
Nomadic nodes with wireless backhaul	Traffic jam	xMBB
Relays distributed across an event area for wireless backhaul	Large outdoor event	xMBB, mMTC
Nomadic nodes with wireless backhaul	High-speed train	xMBB

* The 5G services are Extreme Mobile BroadBand (xMBB), massive Machine Type Communication (mMTC), and ultra-reliable MTC (uMTC).

Table 10.2 Application and gains of the techniques described in the chapter.

Technique	Relay type	Application	Gain
Multi-way wireless backhaul	In-band half-duplex	UDNs of small cells Nomadic nodes Shared relay at cell edge	Spectral efficiency gain up to 100%
Non-orthogonal multiple access	In-band half-duplex	Multi-way relaying UDNs of small cells Nomadic nodes Shared relay at cell edge Relay-supported D2D	Increased spectral efficiency Asymmetric data rates Robustness against timing offsets
Buffer-aided relaying	In and out-of band half-duplex	All delay-insensitive applications with relaying	Increased spectral efficiency Diversity gain

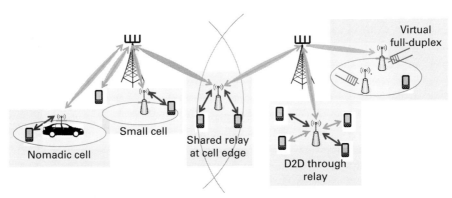

Figure 10.2 Use of relays, wireless network coding and buffers in the 5G landscape.

10.2   Multi-flow wireless backhauling

In Figure 10.3, the main idea behind TWR is shown. In TWR, two devices communicate through a RS and both have traffic to send. The conventional one-way communication requires four transmission phases as the RS is assumed to function in half-duplex mode. When transmission is optimized accounting for the two communication flows, two transmission phases suffice. This transmission scheme is denoted as TWR-2 and comprises the two following transmission phases:

- Phase 1, where Device 1 and Device 2 transmit simultaneously. RS receives over a multiple-access channel, but does not need to decode the signal.
- Phase 2, where the RS transmits a function of the signal received in phase 1, e.g. by Amplify-and-Forward (AF). Device 1 and device 2 decode their desired signal using the signal it transmitted in Phase 1 as a side information.

In a simplified example where the channel coefficients are all equal to 1 and AF is used, the RS receives $x_1 + x_2$ and broadcasts this signal to both devices. Device 1 receives $x_1 + x_2$ from which it removes the contribution of x_1 to decode x_2, likewise for device 2.

The mechanism behind TWR-2 can be generalized through the following two principles: (1) simultaneous service of multiple flows over the wireless medium and (2) cancellation of interference based on previously gathered side information. Using these two principles, transmission building blocks can be devised for traffic patterns that are more general than TWR. Those building blocks can be applied to a cellular architecture with small cells and wireless backhaul, and can be combined to give rise to new optimization problems. In the following, three building blocks are described:

- CDR, Coordinated Direct and Relay transmission: in a network with one small-cell MS and one direct MS, the UL and DL flows are coupled.
- FWR, Four-Way Relaying: in a network with two small cells and two UL and DL flows, both two-way flows are coupled.
- WEW, Wireless-Emulated Wired: in a network with multiple small cells and multiple UL/DL flows, transmission schemes are devised to emulate the performance of a wired backhaul using a wireless backhaul.

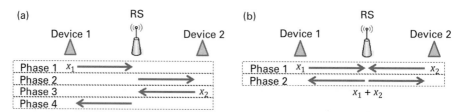

Figure 10.3   Two-way relaying scenario where two devices (BS or MS) communicate through a RS: (a) Conventional i.e. 4 transmission phases, (b) TWR-2 i.e. 2 transmission phases.

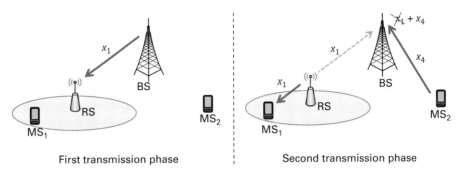

Figure 10.4   CDR1: relayed DL – direct UL.

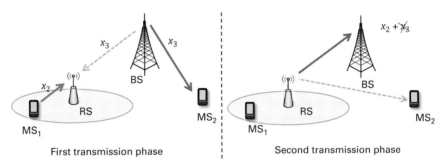

Figure 10.5   CDR2: relayed UL – direct DL.

In all building blocks, RS stands for a small-cell BS. For simplicity, the description of CDR and FWR is limited to single antenna nodes and AF relaying, but the principles can clearly be generalized. In WEW, the emulation of a wired backhaul, relying on the ability of the BS to create multiple beams, assumes multiple antennas at the BS as well as Decode-and-Forward (DF) processing at the RS. Furthermore, a system based on Time-Division Duplexing (TDD) operations is assumed.

10.2.1   Coordinated direct and relay (CDR) transmission

The basic scenario in CDR transmission is depicted in Figures 10.4 and 10.5. It consists of one macro BS, one small cell served by one small-cell BS that acts as a RS and two devices (MS_1 and MS_2). MS_1 has a connection to the BS via the RS only, while MS_2 has a direct connection to the BS. For simplicity, MS_1 and MS_2 are assumed to be distant from each other and not to interfere (e.g. interference cancellation can be applied when the nodes are equipped with multiple antennas). MS_1 and MS_2 have UL and DL traffic. Following the two principles stated above, two schemes can be obtained. These schemes correspond to two different ways of coupling the UL and DL traffic. The first scheme CDR1 in Figure 10.4 couples the relayed DL and direct UL traffic and the second scheme CDR2 couples the relayed UL and direct DL traffic. Decoupling the UL and the DL would require 3 time slots, while CDR1 and CDR2 take only two time slots.

In CDR1, during the first transmission phase, the BS transmits a DL signal x_1 to the RS. During the second transmission phase, two concurrent transmissions are scheduled together: (1) RS transmits x_1 to MS_1, which causes interference at the BS, and (2) MS_2 transmits x_4 to the BS. As the BS knows x_1, its contribution can be removed from the received interfered signal so that decoding of x_4 is performed without the presence of interference.

In CDR2, during the first transmission phase, two concurrent transmissions are scheduled together: (1) MS_1 transmits an UL signal x_2 to the RS and (2) BS transmits x_3 to MS_2, which causes interference to the RS. During the second transmission phase, RS transmits a signal containing x_2 and x_3 to the BS. As the BS knows x_3, the interference caused by x_2 at the BS can be cancelled.

A detailed analysis of these CDR schemes can be found in [14], where large spectral efficiency gains are shown. As a summary, in Figure 10.6, the achievable rate region for a scenario where both MSs have two-way traffic is shown. Note that the ratio between DL and UL rate is assumed fixed to allow for an easy visualization of the rate regions. In order to fulfill the MS rate requirements, in the scheme called S_{CDR} both CDR1 and CDR2 are time-multiplexed with an optimized time multiplexing ratio. Also displayed is S_{ref}, the optimized time-multiplexing of TWR-2 (between BS and MS_1) and UL/DL

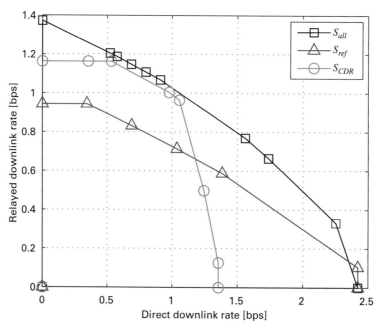

Figure 10.6   Rate region for 3 schemes: (a) S_{CDR} where both CDR1 and CDR2 are time-multiplexed, (b) S_{ref} where both TWR-2 (between BS and MS_1) and UL/DL direct links (between BS and MS_2) are time-multiplexed, (c) S_{all} where S_{CDR} and S_{ref} are time-multiplexed.

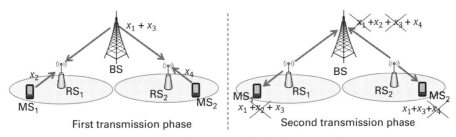

Figure 10.7   Four-way relaying. Two TWR traffic flows are scheduled simultaneously.

direct links (between BS and MS_2), as well as S_{all}, the time-multiplexing of all the schemes cited. It can be observed that the different schemes define different rate regions while the combination of all schemes provide an increased rate region compared to S_{ref} and S_{CDR}. It should be noted that the comparison with conventional transmission (involving one-way communication) is provided in [14]. The CDR schemes outperform conventional schemes in regions where the achievable rate for the direct MS is smaller or comparable to the achievable rate for the relayed MS. On the other hand, if the achievable rate for the direct MS is significantly larger than the achievable rate for the relayed MS, then from the rate viewpoint it is optimal to use only the direct transmission. However, the time multiplexing of the conventional schemes, such as direct transmission, and the CDR schemes always leads to an enlarged region of achievable rates.

10.2.2   Four-way relaying (FWR)

The basic scenario of FWR [15] is depicted in Figure 10.7. It consists of one macro BS, two small cells controlled by the BS, two RSs and one MS in each cell. The two small cells are selected such that they do not interfere with each other (e.g. situated in two different buildings). Each MS is connected to one RS. Both the MSs have two-way traffic. Decoupling the UL/DL in the two small cells would require 8 time slots. In each cell independently, the UL/DL traffic can be coupled and WNC can be applied. Then 4 time slots are required.

In FWR, the UL/DL transmissions of the two small cells are coupled so that only 2 transmission time slots are sufficient, offering superior spectral efficiency. FWR relies on superposition coding to handle simultaneous transmission to both cells while interference within the small cells is handled using WNC. During the first transmission phase, four transmissions are scheduled relying on superposition coding. The BS broadcasts DL signals x_1 and x_3 using superposition coding while MS_1 transmits UL signal x_2 and MS_2 transmits UL signal x_4.

During the second phase, RS_1 transmits a signal containing x_1, x_2 as well as the interfering signal x_3. RS_2 transmits a signal containing x_3 and x_4 as well as the interfering signal x_3. As the BS knows x_1 and x_3, the interference at the BS can be cancelled: the resulting transmission corresponds to a multiple-access process at the BS and decoding

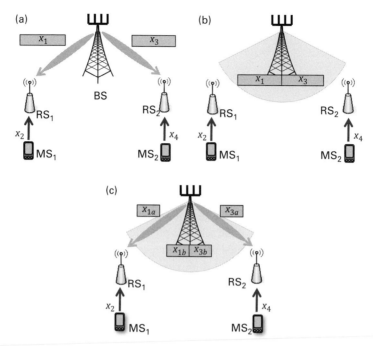

Figure 10.8 Emulating a wired backhaul with a wireless backhaul with two-way traffic, the first transmission phase and the three possible transmission methods. In (a), the BS transmits message x_1 and message x_3 to the RSs using two ZF beams. In (b), the BS broadcasts the concatenated messages to both RS using a common beam. WEW is shown in (c), wherein the BS transmits the messages using both ZF beams and a common beam.

is performed in consequence. As MS_1 knows x_2, it cancels the interference caused by x_2. The remaining signal contains x_1 and x_3, which, transmitted using superposition coding, is decoded accordingly. A similar procedure is performed at MS_2.

10.2.3   Wireless-emulated wire (WEW) for backhaul

The previous subsections focused on regaining the loss caused by half-duplex small-cell BSs with wireless backhaul. This section investigates whether wireless backhauling has the potential to attain the performance of wired backhauling. Again, two-way traffic is a fundamental assumption to achieve this goal, along with the presence of multiple antennas at the macro BS allowing beamforming toward the small cells.

The basic scenario in WEW [16] is depicted in Figure 10.8. It is the same as the FWR scenario with non-interfering small cells and one MS per cell. In the reference system, the backhaul is wired and a fixed rate is assumed for the UL and the DL. The goal is to determine under which conditions the wired backhaul can be replaced by a wireless backhaul while preserving the rate requirements. In the wired backhaul-based system, transmission is organized in TDD mode. In the first transmission phase, the MSs transmit the uplink data while, simultaneously, the downlink data is transmitted through the wire. In the second transmission phase, the RSs transmit their DL data to the MSs while the UL data is transported though the wired backhaul to the macro BS. One essential observation is that the DL data rate is limited by the quality of the wireless access channel between the MSs and the RSs. Although a wired connection has a larger capacity in general than a wireless connection, it is useless to send data at a data rate larger than the wireless channel can support, unless it is beneficial to buffer large amounts of data at the RS (see Section 10.4). This tells us that having a wireless backhaul channel is not a limitation as long as its capacity is larger than the wireless access channel capacity.

It is clear that the traffic pattern in the wired system is a type of two-way traffic where one of the flows passes through wire. To emulate this transmission using wireless backhauling, one natural solution is to form one individual beam per small cell to transmit the DL flow while simultaneously the MSs transmit the UL flow. This procedure, depicted in Figure 10.8(a), is based on a Successive Interference Cancellation (SIC) and DF strategy at the RSs. Below, for ease of presentation, the case of a small cell is first considered before generalizing to multiple cells. Our focus is mostly on the first transmission phase as it is the one determining the equivalence between the wired and the wireless backhaul system.

In the single small-cell case, single UL and single DL flows are transmitted during the first phase. The transmission at the macro BS is optimized in such a way that the created multiple access channel at the RS has the following property with respect to SIC. At first, the DL signal is decoded at the RS considering the UL signal as noise. Hence the transmit power at the macro BS has to be sufficiently large to compensate for the presence of the UL signal. After the DL signal is decoded, it is removed from the received signal at the RS. In theory, this strategy is equivalent to having an UL signal without any interference, which makes it equivalent to the case of wired backhaul. The main objective is reiterated to get equivalent performance as in the case of the wired backhaul, while not changing the power and the modulation/coding for the uplink transmission made by the MS.

The more general case with multiple cells gives rise to new transmission schemes. An intuitive transmission strategy is to use Zero-Forcing (ZF) beamforming at the BS, see Figure 10.8 (a), where the power in each beam is adjusted to perform ordered SIC. As the UL signal acts as noise in the first stage of SIC, the equivalent SNR is low. In such conditions, ZF beamforming is detrimental when the channel is ill-conditioned as it results in noise enhancement. MMSE beamforming is usually seen as a viable alternative to ZF. However, it cannot be used as it leaves a residual interference at the RS that cannot be decoded, hence violating the paradigm that wireless backhauling operations should not affect MS rates.

In the extreme case where all the backhaul channels are identical or linearly dependent, a solution is to send a common message, which is created by concatenation of the data bits to be sent to each MS. The BS then broadcasts the common message to the RSs using a broad beam that can be called common beam, see Figure 10.8(b). Afterwards, the RSs must decode the entire common message, and therefore remove its contribution before decoding the UL signal. The disadvantage is

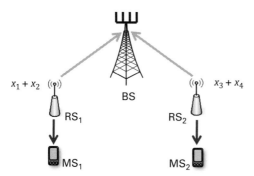

Figure 10.9   Second transmission phase. The messages decoded during the first phase are broadcast using bit-wise XOR operation from each RS toward the macro BS and the MS.

that, by decoding the common message, each RS needs to decode the bits unintended for the MS associated with it.

This brings forward the following idea which is illustrated in Figure 10.8(c). The BS sends part of the bits from a message intended for MS_1 by using ZF beamforming, i.e. spatial separation, and it does the same for part of the bits intended for MS_2. It takes the remaining bits for MS_1 and MS_2, concatenates them into a common message and sends that message through a common beam. This is partially inspired by the private/public messages from the Han-Kobayashi scheme [17] for interference channels. Then a suitable optimization problem can be formulated to determine the right ratio by which the BS mixes the private message (sent by ZF) and the common message (sent by a common beam). The scheme is baptized Wireless-Emulated Wired (WEW). More details can be found in [16].

The second transmission phase is common to all the schemes depicted in Figure 10.8. It involves broadcasting of the downlink and uplink signals using a bit-wise XOR operation from each RS toward the macro BS and the MS. Figure 10.9 shows this broadcast operation. The BS, MS_1 and MS_2 decode the signal sent by their respective RS and apply XOR to recover the desired message from the broadcast message.

Figure 10.10 compares the minimal power that the BS needs to use if it sends the data by ZF beamforming only, a common beam only, and WEW, which is the optimized combination of both. It can be seen that there is a benefit of using a common beam over ZF when the downlink data rate is low, as the same amount of data is sent by using less power. On the other hand, as expected, the WEW scheme always outperforms the individual ZF beamforming and the common beam transmission.

10.3   Highly flexible multi-flow relaying

10.3.1   Basic idea of multi-flow relaying

As explained before, TWR is a powerful concept to overcome the loss in spectral efficiency introduced by half-duplex RSs. TWR reduces the number of required

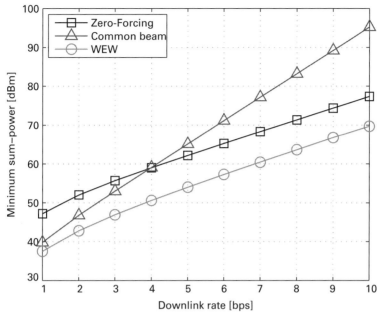

Figure 10.10   The minimal power requirement at the BS is displayed for WEW, the zero-forcing and common based message scheme.

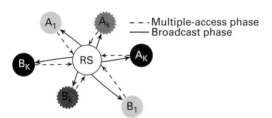

Figure 10.11   Topology of multi-flow relaying: Multiple pairs of nodes (A_k,B_k) exchange messages through a common relay RS.

transmission time slots from four to two, effectively compensating for the half-duplex loss.

However, if multiple pairs of nodes (A_k,B_k) share the same RS, the question of medium access arises. Assuming a half-duplex constraint on the RS, the overall communication can be divided into two communication phases as depicted in Figure 10.11. In the multiple access phase, all nodes transmit their data to the RS. After processing the data, the RS broadcasts the processed data to the corresponding receiving nodes. Thus, there are K communication pairs, corresponding to $2K$ signals in the multiple access phase. Further, adopting the principles of network coding will lead to K compound signals in the broadcast phase. Since, in 5G, manifold devices are expected with diverse requirements in terms of rates and power to coexist in the same network, a high flexibility in the system design is required. The current solutions, e.g. OFDMA in

LTE, do not accommodate for these requirements. To this end, it seems reasonable to approach multi-pair TWR with non-orthogonal medium access as it offers a significantly larger amount of flexibility compared to orthogonal approaches.

Another advantage of non-orthogonal access are the significantly eased requirements on synchronization. While orthogonal schemes require very precise synchronization in time and frequency, non-orthogonal schemes usually do not suffer that severely from imperfect synchronization. In contrast, they might even benefit from asynchrony. Furthermore, non-orthogonal access allows an overloading of the spectrum, effectively leading to a significant system throughput.

A first approach to TWR with non-orthogonal medium access has been presented in [18], in which the two nodes of a communication pair apply the same CDMA spreading sequence while among different pairs different spreading sequences are used. This leads to a reduction in the required spreading sequence length and effectively to an increased spectral efficiency. CDMA, however, is very sensitive to timing offsets between nodes and, hence, not a good choice when aiming at reduced signaling [19].

Another non-orthogonal access scheme is Interleave-Division Multiple-Access (IDMA) [20], see Section 7.3.3 in Chapter 7. IDMA can be seen as a generalized form of the well-known CDMA in which the complete bandwidth expansion is devoted to coding. As CDMA, and in contrast to orthogonal medium access schemes, IDMA is theoretically capacity achieving. Furthermore, IDMA has shown to outperform CDMA in most scenarios in terms of robustness against fading and in coding gain [19]. In addition, timing offsets among nodes do not pose an issue as they do for CDMA and other medium-access schemes. On the contrary, they can even lead to an improved overall performance.

Transmission and detection in IDMA is based on a layering technique. That is, every transmitting node is transmitting one or multiple superimposed layers simultaneously [21]. The layers of multiple nodes again superimpose during simultaneous transmission, composing the received signal. In order to recover the transmitted data, all layers have to be resolved from the received compound signal. This is achieved by invoking an iterative soft Interference Cancellation (IC) process at the receiver [20]. Figure 10.12 depicts the system structure. Here, as an example, two MSs apply IDMA, each transmitting a different number of layers according to their rate requirements. At the receiver, an iterative detection process consisting of soft IC and A Posteriori Probability (APP)-decoding (DEC) is invoked in order to recover the transmitted data.

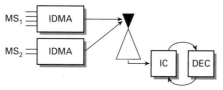

Figure 10.12   Two MSs using IDMA to transmit to a RS, where MS_1 and MS_2 transmit four and two layers, respectively.

Exploiting this layering nature of IDMA delivers a powerful tool for rate adaptation, which was presented in [22] and will be explained in the next subsection. In order to optimize the achievable throughput and identify bottlenecks within the system, both transmission phases, i.e. multiple access phase and broadcast phase, have to be considered jointly.

10.3.2   Achieving high throughput for 5G

Going from orthogonal access schemes as OFDMA in 4G to non-orthogonal access schemes in 5G, the applied channel code becomes even more crucial to the overall system performance. In IDMA systems, e.g. an iterative IC is performed at the receiver in order to separate the MS signals and suppress the inter-symbol interference and, thus, to detect all transmitted data layers. The channel code directly affects the performance of such an iterative IC process. A "too weak" channel code in conjunction with a highly loaded system might prevent successful detection while a "too strong" channel code might diminish the spectral efficiency unnecessarily. Hence, a good code "matches" the system at the working point, i.e. for a given load in terms of layers at a specific Signal-to-Noise Ratio (SNR). The combination of different working points and corresponding optimized codes again leads to a set of Modulation and Coding Schemes (MCS) that can be chosen from according to the current requirements on rate and error correction capabilities. A thorough code design is therefore substantial in order to optimize the system throughput.

The role of the channel code becomes more essential when combining non-orthogonal channel access with multi-flow TWR. In that case, the decodability at multiple receivers simultaneously has to be ensured. This makes the code design problem prohibitively complex as the dimensionality of the problem increases with the number of communicating nodes. One possibility to tackle the problem is to split the overall code design into two distinct code design problems, namely for the multiple access phase and for the broadcast phase separately. This eases the overall problem significantly. However, these two design problems are not completely independent of each other as they are coupled via the required individual data rates. This means that the designed MCSs for the multiple-access phase and for the broadcast phase have to support the same data rates for all nodes.

A convenient way to perform such a code design and the design of iterative turbo-like systems in general are EXIT-charts [23]. The principal idea for EXIT-charts is to describe the information exchange between information processing systems by so-called information transfer characteristics. Given some assumptions on statistical properties of the processed information, this allows predicting the performance of iterative information processing systems as well as designing them in a convenient way.

Besides the applied channel code, the number of parallel transmitted layers also directly influences the rate and thus the system performance. Since both the number of superimposed layers N_L as well as the rate of the channel code R_{C,N_L}, determine the spectral efficiency $\eta = N_L R_{C,N_L}$, the system design process does not only consist in finding a matching code but also consist in finding a good combination of number of layers and code, i.e. a good MCS. Principally, the number of layers is a design parameter and therefore free to choose. If there are no further constraints on the code or the number of layers, it is favorable to choose a very high number of layers and a correspondingly very low rate code. This has two advantages. Firstly, the system's granularity is determined by the code rate of each layer. This means that each node can only change its rate in increments of the code rate by increasing or decreasing the number of transmitted layers. Secondly, the lower the code rate, the more degrees of freedom in the code design. This generally means a better possible matching of the code to the given number of layers. However, practicality prohibits a very large number of layers, as the overall decoding complexity increases linearly with the number of layers in the system. For practical systems, usually a trade-off between performance and complexity has to be found. Note that, in principle, for IDMA a common base rate is not required as IDMA offers to choose the applied channel code on each layer completely independently from all other layers. From a system design perspective, however, the restriction to a common base code is reasonable as it eases the code design process significantly.

A good choice for such a code design are Irregular Repeat Accumulate (IRA) codes [24] as they offer good error correction capabilities at a low encoding and a moderate decoding complexity making them well-suited for application in iterative detection schemes [25]. Another even simpler choice is irregular repetition codes. While those codes themselves have no coding gain, they are also very easy to design and perform well in iterative detection scenarios. Which of those codes to choose depends on the desired working point. The design of IRA codes for iterative detection schemes has been presented in [25]. In [22] this design method was applied to design multi-flow IDMA-based TWR systems using IRA codes. In [26] an alternate design method was presented.

10.3.3   Performance evaluation

The potential gains offered by TWR based on non-orthogonal channel access over conventional orthogonal channel access schemes are shown through numerical simulation results in terms of average throughput. Specifically, an IDMA-based four nodes TWR scheme is compared with an OFDMA-based legacy solution applying LTE parameters. The four nodes are distributed randomly and circularly uniformly with a maximum radius R_{max} around a common supporting RS, see Figure 10.11. The channels between all communicating nodes are determined employing a common distance-based path loss model. Additionally, small-scale fading is introduced, which is calculated based on the ITU WINNER models [27].

In Figure 10.13 the average throughput for the optimized IDMA-based system as well as the OFDMA-based legacy solution employing three different MCSs are shown. IDMA system optimization was performed for a working point of $R_{max} = 100$. As it

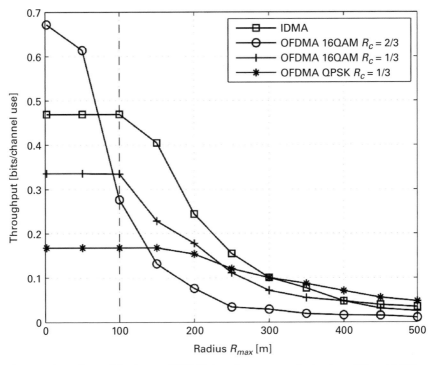

Figure 10.13   Average throughput of the legacy OFDMA-based system employing two different MCSs and the optimized IDMA-bases TWR system.

can be seen, the optimized IDMA system clearly outperforms the legacy solution for all three MCSs at the working point. The main reason for this is the spectral overloading that IDMA offers (see Section 7.3 in Chapter 7) combined with a thorough code design. Note that this scenario highly favors the OFDMA-based solutions as no impairments as CFOs, Doppler shift, etc. are considered. Under more practical assumptions, the gain of the IDMA system will be significantly larger.

10.4   Buffer-aided relaying

As mentioned earlier, one main problem with all practical relaying implementations is the half-duplex constraint. Especially, when using relays for range extension in a wireless backhaul, high throughput is essential.

This section provides an overview of different strategies for exploiting buffering in the relays that have been proposed over the last few years. Most of these schemes relate to scenarios where a single source wants to communicate with a single destination, using one or more relays. Only relaying for range extension will be covered here, i.e. it is assumed that the direct link between the source and destination nodes is too weak to be useful.

10.4.1   Why buffers?

Most relaying schemes are based on a fixed two-phase schedule, where a packet is transmitted from the source to a relay in one time slot and then forwarded from the relay to the destination in the next time slot. If packets can be stored at the relay, this strict two-phase schedule is no longer necessary, which provides additional degrees of freedom. As a first simple example, consider a scenario with a single relay. As long as the channel from the source to the relay is better than the channel from the relay to the destination, let the source keep transmitting data packets, which are accumulated in the relay buffer. In the opposite situation, the relay can forward previously buffered data to the destination. As shown in [28], such a channel-dependent opportunistic scheduling scheme can provide significantly improved spectral efficiency compared to a fixed schedule.

If several relays are within the communication range for both the source and the destination, then channel dependent opportunistic scheduling can be done by independently selecting between the different relays in the first and the second hop. If the number of relays is N, such a scheme provides diversity gains of order N, if one sticks to a regular two-phase schedule, and diversity gains of order $2N$, when the scheduling also selects if the source or a relay should transmit in each time slot.

Obviously, some care is needed to ensure that the queues are stable in the long run and to handle situations when a buffer is empty or full. See the next subsection for some details.

In the schemes mentioned so far, the source is idle in 50% of the time. However, this half-duplex limitation can be removed by letting the source node transmit data to one relay and simultaneously let another relay transmit data to the destination, see Figure 10.14. In this scheme, called 'space full duplex' [29], the data is effectively multiplexed into several parallel flows. The key enabling element of relay space Full-Duplex (FD) is the data buffers at the relays.

The throughput gains of these schemes come at the expense of an increased (and random) delay, because of the buffering. Therefore, buffer-aided relaying cannot be used in time-critical applications. On the other hand, a large number of applications, such as web browsing and software download, can tolerate some additional delay and at the same time benefit significantly from the increased average rate. Another drawback is the increased overhead for channel sounding and signaling.

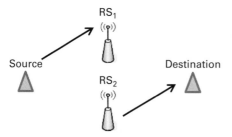

Figure 10.14   Illustration of 'space full duplex' transmission using buffer-aided relays.

10.4.2   Relay selection

An important aspect is how to select which relay should be transmitting and which should be receiving. In the following, different options for relay selection and rate adaption are discussed. The total throughput depends on a combination of the achievable rates between the source and the selected receiving relay and between the selected transmitting relay and the destination. This coupling makes the problem non-trivial.

As it is done in most of the literature on buffer-aided relaying, it is assumed that the data is decoded at the relays before being stored in the buffer, i.e. basically a DF operation. The data rates can either be fixed or adaptive (using adaptive modulation and coding). In addition, an ACK/NACK protocol can be used to handle channel outages.

As a benchmark, consider first a classical DF relaying with rate adaption but without buffering, where in each pair of time slots the rate is limited by the worst of the source-to-relay and the relay-to-destination hops. In case multiple relays are available to choose from, the relay maximizing the worst hop rate should be selected, so-called 'best relay selection'.

When buffering is added, but keeping the two-phase scheduling, the choice of relay can be done separately for the two links. Therefore, the relay with maximum channel gain from the source should be used for the source-relay link and the relay with maximum channel gain to the destination for the relay-destination link, resulting in the so-called max-max relay selection scheme [30]. In time slots where the corresponding buffer is full/empty, it is proposed to revert to a best relay strategy. By relaxing the scheduling order between the two links, the max-link scheme was proposed in [31], where in each time slot a selection is done between all the $2N$ available links.

For the case of space FD, i.e. allowing simultaneous transmission from the source and one of the relays, the max-max relay selection was proposed in [29]. A minor modification is needed since the transmitting relay has to be different from the receiving relay. As will be discussed further below, a complication in the space full-duplex schemes is the interference at the receiving relay, caused by the transmitting relay. Unless this inter-relay interference is negligible, the max-max relay selection is not the optimal choice. Instead, it was shown in [32], using a Lagrange relaxation, that even better performance can be obtained by choosing the pair of source-relay and relay-destination links that maximizes a linearly weighted combination of the two link rates, where the weighting only depends on the long-term channel statistics.

In practice, the buffer sizes have to be limited. Moreover, limiting the buffer sizes is also a way to bound the delays in the system, if the previously described relay selection schemes are properly modified to prioritize relays with full buffers.

A numerical comparison of the different basic schemes is provided in Figures 10.15 and 10.16 for a scenario with 5 relays, where all channels are Rayleigh fading with Additive White Gaussian Noise (AWGN), the average channel gain between relay and destination

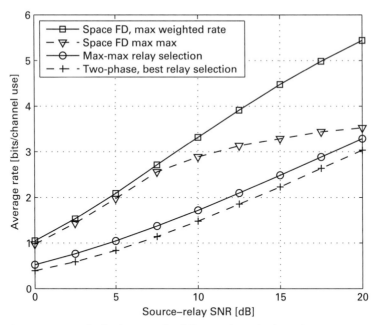

Figure 10.15   Average source-destination rate, for different relay selection schemes.

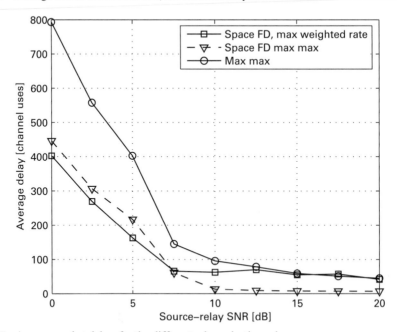

Figure 10.16   Average packet delay, for the different relay selection schemes.

was 3 dB weaker than the average source-relay gain, and the average relay-relay channel gain was 5 dB weaker the source-relay gain. Ideal rate adaption according to Shannon's rate expression is used. Clearly, the space FD schemes provide significantly higher average rates than relaying without buffers, but at the expense of a significant delay (which could be shortened by limiting the buffer sizes at each relay, still maintaining a large part of the rate improvements).

10.4.3   Handling inter-relay interference

In the space FD scheme, the signal from the transmitting relay, intended for the destination, will inevitably cause interference at the receiving relay. The main options to handle this inter-relay interference are:

- Physical separation between the relays. If the relays are sufficiently separated in distance and possibly by walls, the inter-relay channel will be weak enough. However, this is hard to obtain, given that both the source and the destination should be within hearing distance of all the relays.
- Interference cancellation. Multi-user detection techniques may be used at the relays to cancel the interference, see [33].
- Array antennas or directive antennas at the relays. MIMO techniques can be used for interference suppression or cancellation at both relays. A simple alternative if both source, destination and relays are mounted at fixed positions, for example in a wireless backhauling scenario, may be to use highly directive antennas with low side-lobes.

When using antenna arrays, the beamformers at the two selected relays should ideally be designed jointly to optimize the performance. Furthermore, the beamformer design and relay selection is coupled, since the beamformers depend on which two relays are selected, and since the spatial processing will influence the effective channel gains, which in turn determine the optimal relay selection. Such a joint optimization is complex both in terms of computations and signaling, therefore a number of simpler but suboptimal design strategies are outlined in [32].

10.4.4   Extensions

The basic idea of buffer-aided relaying, outlined above, can be extended to the two-way relaying as described in [34]. Further, the throughput gain provided by the buffer-aided relaying can be traded with reduction in power consumption. Furthermore, relay selection combined with buffering can be used to obtain physical layer security [35].

10.5   Conclusions

The benefits of multi-way traffic combined with **wireless network coding** are key for a spectrally efficient usage of in-band wireless backhauling in small-cell deployments. By considering multiple simultaneous transmissions jointly, **multi-flow relaying** can improve the spectral efficiency even further. Additionally, **buffer-aided relaying** is a mechanism that trades off delay for throughput while offering a degree of freedom in system design that is independent of wireless network coding and multi-flow relaying.

Therefore, it becomes natural to ask how all presented mechanisms can be combined in order to build transmission protocols that address more complex network scenarios. A rather simple scenario that shows how multi-flow transmission and buffer-aided relays work together has been presented in [34]. For more complex scenarios, the following questions remain unaddressed. For example, how to decide which flows should be coordinated and which ones should be treated simply as interference is still an open question. The next important question is how to gather the necessary CSI to operate the protocols. Note that in many cases with coordinated flows, such as CoMP, the cost of gathering CSI is substantial and undermines the performance gains. It is therefore required to take a proper architectural approach in order to integrate the presented ideas in a system-level design and achieve scalability.

References

[1] A. Osseiran, J. F. Monserrat, and W. Mohr, *Mobile and Wireless Communications for IMT-Advanced and Beyond.* Chichester: Wiley, 2011.

[2] J. N. Laneman, D. N. C. Tse, and G. W. Wornell, "Cooperative diversity in wireless networks: Efficient protocols and outage behavior," *IEEE Transactions on Information Theory*, vol. 50, no. 12, pp. 3062–3080, December 2004.

[3] B. Rankov and A. Wittneben, "Spectral efficient signaling for half-duplex relay channels," in Asilomar Conference on Signals, Systems and Computers, Pacific Grove, November 2005.

[4] P. Popovski and H. Yomo, "Physical network coding in two-way wireless relay channels," in IEEE International Conference on Communications, Glasgow, June 2007.

[5] E. Dahlman, S. Parkvall, and J. Sköld, *4G LTE/LTE-Advanced for Mobile Broadband*, 2nd edn. New York: Academic Press, 2013.

[6] J. I. Choi, M. Jain, K. Srinivasan, P. Levis, and S. Katti, "Achieving single channel, full duplex wireless communication," in ACM International Conference on Mobile Computing and Networking, Illinois, September 2010.

[7] M. Duarte and A. Sabharwal, "Full-duplex wireless communications using off-the-shelf radios: Feasibility and first results," in Asilomar Conference on Signals, Systems and Computers, Pacific Grove, November 2010.

[8] A. Sabharwal, P. Schniter, D. Guo, D. Bliss, S. Rangarajan, and R. Wichman, "In-band full-duplex wireless: Challenges and opportunities," *IEEE Journal on Selected Areas in Communication*, vol. 32, no. 9, pp. 1637–1652, September 2014.

[9] S. Goyal, Pei Liu, S.S. Panwar, R.A. Difazio, Rui Yang, and E. Bala, "Full duplex cellular systems: will doubling interference prevent doubling capacity?," *IEEE Communications Magazine*, vol. 53, no. 5, pp. 121, 127, May 2015.

[10] J. G. Andrews, H. Claussen, M. Dohler, S. Rangan, and M. C. Reed, "Femtocells: Past, present, and future," *IEEE Journal on Selected Areas in Communications*, vol. 30, no. 3, pp. 497–508, April 2012.

[11] J. Andrews, "Seven ways that hetnets are a cellular paradigm shift," *IEEE Communications Magazine*, vol. 51, no. 3, pp. 136–144, March 2013.

[12] Z. Ren, S. Stanczak, P. Fertl, and F. Penna, "Energy-aware activation of nomadic relays for performance enhancement in cellular networks," in IEEE International Conference on Communications, Sydney, June 2014.

[13] ICT-317669 METIS project, "Final report on the METIS 5G system concept and technology roadmap," Deliverable D6.6, April 2015, www.metis2020.com/docu ments/deliverables/

[14] C. D. T. Thai, P. Popovski, M. Kaneko, and E. de Carvalho, "Multi-flow scheduling for coordinated direct and relayed users in cellular systems," *IEEE Transactions on Communications*, vol. 61, no. 2, pp. 669–678, February 2013.

[15] H. Liu, P. Popovski, E. de Carvalho, Y. Zhao, and F. Sun, "Four-way relaying in wireless cellular systems," *IEEE Wireless Communications Letters*, vol. 2, no. 4, pp. 403–406, August 2013.

[16] H. Thomsen, E. De Carvalho, and P. Popovski, "Using Wireless Network Coding to Replace a Wired with Wireless Backhaul," *IEEE Wireless Communications Letters*, vol. 4, no. 2, pp. 141–144, April 2015.

[17] T.S. Han and K. Kobayashi, "A new achievable rate region for the interference channel," *IEEE Transactions on Information Theory*, vol. 27, no. 1, pp. 49–60, January 1981.

[18] M. Chen and A. Yener, "Multiuser two-way relaying: Detection and interference management strategies," *IEEE Transactions on Wireless Communications*, vol. 8, no. 8, pp. 4296–4305, August 2009.

[19] K. Kusume, G. Bauch, and W. Utschick, "IDMA vs. CDMA: Analysis and comparison of two multiple access schemes," *IEEE Transactions on Wireless Communications*, vol. 11, no. 1, pp.78–87, January 2012.

[20] L. Ping, L. Liu, K. Wu, and W. K. Leung, "Interleave-division multiple-access," *IEEE Transactions on Wireless Communications*, vol. 5, no. 4, pp. 938–947, 2006.

[21] P. A. Höher, H. Schöneich, and J. C. Fricke, "Multi-layer interleave-division multiple-access: Theory and practise." *European Transactions on Telecommunications*, vol. 19, no. 5, pp. 523–536, January 2008.

[22] F. Lenkeit, C. Bockelmann, D. Wübben, and A. Dekorsy, "IRA code design for IDMA-based multi-pair bidirectional relaying systems," in International Workshop on Broadband Wireless Access, Atlanta, December 2013.

[23] S. ten Brink, "Convergence of iterative decoding," *IEEE Electronic Letters*, vol. 35, no. 13, pp. 1117–1119, May 1999.

[24] H. Jin, A. Khandekar, and R. McEliece, "Irregular repeat accumulate codes," in International Symposium on Turbo Codes, Brest, September 2000.

[25] S. ten Brink and G. Kramer, "Design of repeat-accumulate codes for iterative detection and decoding," *IEEE Transactions on Signal Processing*, vol. 52, no. 11, pp. 2764–2772, November 2003.

[26] F. Lenkeit, C. Bockelmann, D. Wübben, and A. Dekorsy, "IRA code design for iterative detection and decoding: A setpoint-based approach," in IEEE Vehicular Technology Conference, Seoul, pp. 1–5, May 2014.

[27] International Telecommunications Union Radio (ITU-R), Guidelines for evaluation of radio interface technologies for IMT-Advanced, Report ITU-R M.2135, November 2008, www.itu.int/pub/R-REP-M.2135-2008

[28] N. Zlatanov, R. Schober, and P. Popovski, "Buffer-aided relaying with adaptive link selection," *IEEE Journal on Selected Areas in Communications*, vol. 31, no. 8, pp. 1530–1542, August 2013.

[29] A. Ikhlef, J. Kim, and R. Schober, "Mimicking full-duplex relaying using half-duplex relays with buffers," *IEEE Transactions on Vehicular Technology*, vol. 61, no. 7, pp. 3025–3037, September 2012.

[30] A. Ikhlef, D. S. Michalopoulos, and R. Schober. "Max-max relay selection for relays with buffers," *IEEE Transactions on Wireless Communications*, vol. 11, no. 3, pp. 1124–1135, March 2012.

[31] I. Krikidis, T. Charalambous, and J. S. Thompson. "Buffer-aided relay selection for cooperative diversity systems without delay constraints," *IEEE Transactions on Wireless Communications*, vol. 11, no. 5, pp. 1957–1967, May 2012.

[32] S. M. Kim and M. Bengtsson. "Virtual full-duplex buffer-aided relaying: Relay selection and beamforming," in IEEE International Symposium on Personal Indoor and Mobile Radio Communications, London, pp. 1748–1752, September 2013.

[33] N. Nomikos V. Demosthenis, T. Charalambous, I Krikidis, P. Makris, D. N. Skoutas, M. Johansson, and C. Skianis. "Joint relay-pair selection for buffer-aided successive opportunistic relaying," *Transactions on Emerging Telecommunications Technologies*, vol. 25, no. 8, pp. 823–834, August 2014.

[34] H. Liu, P. Popovski, E. De Carvalho, and Y. Zhao, "Sum-rate optimization in a two-way relay network with buffering," *IEEE Communications Letters*, vol. 17, no. 1, pp. 95–98, January 2013.

[35] J. Huang and A. L. Swindlehurst, "Buffer-aided relaying for two-hop secure communication," *IEEE Transactions on Wireless Communications*, vol. 14, no. 1, pp. 1536–1276, January 2015.

11 Interference management, mobility management, and dynamic reconfiguration

Michał Maternia, Ömer Bulakci, Emmanuel Ternon, Andreas Klein, and Tommy Svensson

This chapter covers network-level solutions aiming at enhancing end-user experience and bringing down the Operational Expenditures (OPEX) of the 5G deployments.

After analyzing contemporary trends and predictions for 5G, it becomes evident that there are certain aspects of 5G deployments that have to be taken into consideration when designing future network-level solutions. These aspects are predominantly related to the network densification and increasing heterogeneity. While network densification will result in much smaller distances between the Base Stations (BSs) [1], the heterogeneity of 5G will manifest itself in multiple dimensions e.g. cell types or operating frequency. [2]. This complicated heterogeneous environment, apart from dimensioning and planning problems, brings also opportunities for an efficient mapping of users or services to optimal (from the overall system perspective) access technologies or BS types. Aspects related to efficient signaling exchange in this heterogeneous environment (also in scenarios with control/user plane decoupling) are related to a Lean System Control Plane design (cf. Chapter 2).

Based on the previously mentioned factors, it can be easily inferred that one of the most desired 5G features is broadly understood flexibility. The flexibility can be achieved in different ways, starting from a flexible allocation of radio resources in Time Division Duplexing (TDD) mode according to instantaneous traffic conditions, through an efficient exploitation of different Radio Access Technologies (RATs) or network layers, and finally as a freedom to dynamically reconfigure the network using BSs that are automatically configured and self-backhauled. Besides, the kind and extent of Moving Networks (MNs) expected to be introduced in 5G can be perceived as a dynamic reconfiguration enabler and an important part of the Dynamic Radio Access Network (RAN) described in Chapter 2.

The performance of the 5G system can be further enhanced by exploiting context information understood as "(. . .) any information that can be used to characterize the situation of an entity. An entity is a person, place, or object that is considered relevant to the interaction between a user and an application, including the user and application themselves" [3]. Today, there are already many application domains and envisioned use cases where devices (such as sensors) or services (such as cloud services) communicate for unconscious support of people in their everyday life tasks. Meanwhile, context awareness is regarded as a key enabler for group management and content selection [3], situation management [4], group communications [5], improved system, transport, and service adaptations [6]–[8], as well as enhanced mobility support and resource management [9]–[12]. Context awareness is a vital part of the Localized Content and Traffic Flow concept described in Chapter 2.

The chapter begins with an overview of new forms of 5G deployments that may have potential impact on future network-level technology components in Section 11.1. Afterward, Section 11.2 depicts solutions and studies on interference and resource management. Section 11.3 focuses on mobility management and technical concepts aiming at an efficient detection of 5G BSs by users, for instance making use of context awareness. Finally, Section 11.4 discusses dynamic reconfiguration enablers that allow a flexible adaptation to instantaneous traffic conditions by means of a dynamic activation and deactivation of fixed BSs and MNs.

11.1   Network deployment types

Telecommunication cellular networks have been used extensively for both data and voice transmission over a long time and evolved into a complicated ecosystem following more and more diverse and challenging expectations of end users. This network evolution is predominantly related to the RAN part and the dominant trends are summarized in sections below and depicted in Figure 11.1.

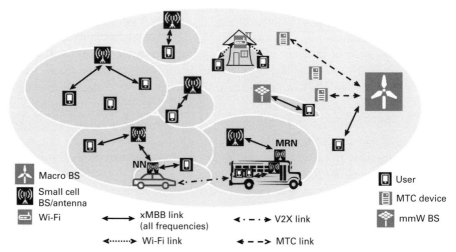

Figure 11.1   5G network deployment types including heterogeneous networks. NN stands for Nomadic Node, and MRN for Moving Relay Node.

11.1.1   Ultra-dense network or densification

Network densification is an inevitable process that is necessary to cater for the uptake of traffic expected toward 2020 and beyond. In fact, other major technologies that may solve the challenge of the so-called x1000 traffic growth, such as utilization of massive Multiple Input Multiple Output (MIMO) or bandwidth extension, may be difficult to pursue in some scenarios. In the context of network densification in 5G, the term Ultra-Dense Network (UDN) is often used. UDN denotes a large number of small cells (i.e. low-power nodes with limited range, such as micro pico and femto cells), that are deployed very close to each other, with inter-site distances of 10 m or even lower. In this group, femto nodes, i.e. BSs designed for residential or small office usage that are connected to Mobile Network Operator (MNO) networks using third party broadband connections, require special attention due to their potentially unco-ordinated deployment.

11.1.2   Moving networks

Due to the increasing numbers of mobile users demanding high-speed Internet access, public transportation vehicles like buses, trams and trains, as well as private cars, are becoming natural hotspots of mobile data communication. Therefore, it is expected that 5G will witness the introduction of MNs, which consist of:

- Moving Relay Nodes (MRNs) having the purpose of better serving in-vehicle users (see the concept of relaying and its main challenges for fixed deployments captured in Chapter 10). If an MRN controls its own resources, it can create a moving cell for the in-vehicle users.
- Nomadic Nodes (NNs) having the purpose to serve out-of-vehicle users, as a comple-ment to fixed network nodes and in a best effort manner.

Both solutions are integrated into vehicles. In the case of MRNs, two sets of antennas implemented in the interior and exterior of the vehicles can be used to circumvent the Vehicle Penetration Loss (VPL). Measurements have shown that the VPL can be as high as 25 dB in a minivan at the frequency of 2.4 GHz [13]. Even higher VPLs are foresee-able in the well-isolated vehicles of our interest, and in higher frequency bands. MRNs show good potential to improve the network performance, in terms of improved spectral efficiency and lowered outage probability, experienced by Vehicular Users (VUs) in noise-limited scenarios [14]. MRNs also show good performance gains in limited co-channel interference scenarios [15]. In case of NNs, an antenna set on top of vehicles is used to provide a backhaul connection and access link. This setup allows providing efficient broadband connectivity exactly where it is needed, i.e. for users in the proxi-mity of vehicles. A key advantage of NNs is the provisioning of relaying functionality for performance enhancement for the needed service time without a priori site leasing or site search. Furthermore, there are larger spaces available for the antenna and transceiver designs for vehicle-mounted NNs compared to conventional small cell BSs, allowing potential backhaul link enhancements and advanced relaying implementation. NNs are associated with some uncertainty about their availability, for instance, caused by human behavior (drivers), i.e. an NN may or may not be available in the target service region. Nevertheless, despite such uncertainty, a large number of NNs can be expected particularly in urban areas. During low-mobility (e.g. traffic jam) or stationary operation (e.g. a parked car), physical MRNs may be configured to serve also as NNs.

11.1.3   Heterogeneous networks

As mentioned, contemporary cellular networks have already evolved into a complicated ecosystem consisting of

- BSs operating with different output power and antenna location, hence of different cell size (i.e. macro, micro, pico and femto).
- A variety of different access technologies e.g. standardized by the 3rd Generation Partnership Project (3GPP) and Institute of Electrical and Electronics Engineers (IEEE), including e.g. Wi-Fi.

This non-homogeneous environment is often referred to as a Heterogeneous Network (HetNet) and is expected to further diversify in 5G, as one will potentially require different air-interface access solutions for handling extreme mobile broadband (xMBB) or massive or ultra-reliable Machine-Type Communications (mMTC or uMTC, cf. Chapter 4) [2]. This setup will be complemented by MNs, direct communication between users including Vehicle-to-anything Communication (V2X), cf. Chapters 4 and 5, and BSs operating in different frequency regimes as for instance millimeter waves (mmW), cf. Chapter 6.

11.2   Interference management in 5G

Proactive interference management is a necessary enabler for an efficient wireless cellular system. While Chapters 7 and 9 described already physical layer aspects of interference management, this section captures the network-level repercussions of proactive interference management. Since the capacity of modern cellular networks is interference-limited, there are already numerous publications and concepts that try to enhance the performance of Long-Term Evolution – Advanced (LTE-A) based solutions (see [16]–[18]). However, there are specific aspects foreseen for 5G, which are not present in legacy systems, which motivate new solutions for interference and radio resource management. These aspects, depicted in Figure 11.2, are related to

- Massive deployment of small cells. Since cell sizes in UDN will be smaller, the number of bandwidth hungry users per cell will be lower comparing to LTE-A deployments. This will lead to a more "bursty" activity pattern, i.e. transitions between active and inactive states will happen more often than in larger cells, where multiple users average this effect. An additional factor making the interference management in 5G even more challenging is a stronger role of direct D2D

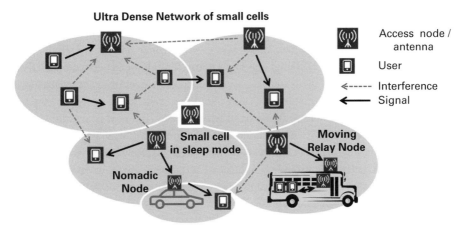

Figure 11.2 Interference management in a 5G UDN dynamic deployment.

transmission (cf. Chapter 5) and an increased number of user-deployed and potentially self-backhauled BSs.

- Utilization of a dynamic allocation of slots for uplink or downlink transmission in TDD mode. Such solutions are already discussed in 3GPP [19], but the specific air interface foreseen for dense deployments in 5G and described in Section 7.4 allows for an even more dynamic and instantaneous adaptation to traffic needs in each cell. This approach leads to a very efficient utilization of the available spectrum, but on the other hand increases the impact of cross-interferences (i.e. interferences caused by simultaneous UL and DL transmissions in adjacent cells), a side effect whose detrimental impact on the overall system performance is further boosted by reduced inter-site distance.
- Introduction of NNs and MRNs, which may refashion the network deployment in 5G. Such deployments add another degree of freedom for RRM in 5G, but, on the other hand, they also challenge static interference management schemes.

These observations imply that future access networks may experience more severe interference constellations than legacy systems, especially dynamic inter-cell interferences. This aspect calls for reconsidering the level of centralization of RRM functionality in 5G, especially in a UDN deployment. While Chapter 9 introduced already some basics of interference suppression and cancellation methods, the techniques described herein focus on network-level implications and how these techniques and novel MAC layer solutions could be exploited in 5G resource management. The remainder of this section then covers the implications of the use of moving networks with a particular focus on moving relay nodes, showing their potential to serve the needs of in-vehicle users in an emerging 5G system.

11.2.1 Interference management in UDNs

The architectural approaches to enable RRM mechanisms in 3GPP standards have changed visibly in the last generations of wireless cellular technologies. Early 3G releases relied on a Radio Network Controller (RNC) operating in a centralized manner, meaning that the RNC was taking decisions on radio resource allocation for BSs that were under its control. Later 3G releases, which introduced High Speed Downlink Packet Access (HSDPA) and High Speed Uplink Packet Access (HSUPA), moved the functionality of radio resource allocation to individual BSs. These BSs had no possibility for explicit information exchange about their scheduling decisions, and therefore operated in a fully decentralized/standalone manner. The next generation of 3GPP standards (i.e. LTE and LTE-A) introduced the X2 interface, which facilitated the exchange of scheduling related information between BSs (DL Relative Narrowband Transmit Power and UL High Interference Indicator [20]), hence enabling distributed RRM. In 5G, a question mark is now once again put on the level of coordination of RRM, for instance in the context of UDN deployments.

First cellular networks were used mostly to carry voice, for which a symmetric Frequency Division Duplexing (FDD) mode was suitable. However, modern cellular networks handle predominantly data traffic and, therefore, a TDD mode is desired for efficient exploitation of available spectrum in ultra-dense deployments, which target mainly broadband data transmission. In order to provide short latencies and to minimize the size of data buffers, a shortening of subframe length is also necessary. Based on the harmonized OFDM concept [21], the frame structure for UDNs can be shortened down to e.g., 0.25 ms for centimeter wave frequencies. Since each subframe contains demodulation reference signals (symbols allowing for estimation of the channel and its covariance matrix), a native support for cross-link interference mitigation is provided that can be exploited by e.g. Interference Rejection Combining (IRC) receivers.

11.2.1.1 Performance of UDNs using dynamic TDD

The performance of an uncoordinated indoor UDN deployment using a flexible UL/DL frame structure with an average inter-site distance of 5 m, is compared with the baseline LTE-A TDD frame based on a fixed allocation of UL/DL slots. As depicted in Figure 11.3, the introduction of this new frame structure brings visible benefits in form of reduced packet transmission time for most challenging transmission conditions, even when the fixed UL/DL slot allocation is proportional to the volume of UL and DL traffic and when IRC receivers are used (more information on the evaluation settings can be found in [22]).

Further performance gains can be observed when one moves from a decentralized/standalone RRM scheme to a centralized one. In the first case, BSs take individual decisions on the allocation of radio resources having only their own radio measurements and reports from devices that they serve. In a completely centralized approach, a central entity gathers information and makes the scheduling decisions for multiple BSs. Already in contemporary LTE-A networks, there are technical solutions that facilitate this way of decision making, e.g. baseband hotels (centralized RAN or cloud RAN) where processing capabilities are located in a common pool, and the connection toward radio frequency units at the local site is realized using optical fibers (cf. Chapter 3). Such fronthaul connection is required, since radio samples need to be transmitted, and not data

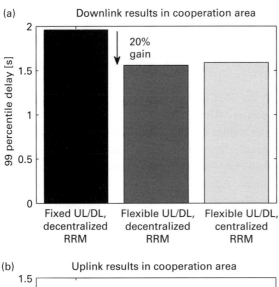

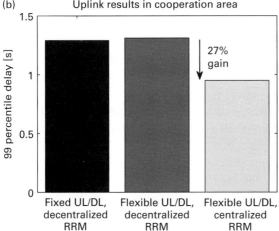

Figure 11.3 Performance gains of flexible UL/DL frame structure optimized for TDD operations in UDN and decentralized vs. centralized RRM [23], reprinted with permission (Lic. no. 3663650682565).

of the user. For a centralized RRM approach, an additional requirement on the latency comes from the timing of the radio measurements and the exchange of scheduling decisions that need to be adjusted to the ongoing radio and interference conditions. However, there are technical concepts that would allow operators to benefit from centralized RRM without the need of having a dense network of optical fiber, or other forms of reliable fixed backhaul, thus making it possible to utilize backhaul provided by a 3rd party with limited reliability. The proposal described in [22], for instance, exploits over-the-air signaling and control/user plane decoupling to provide coordination of resource scheduling among a dense network of 5G small cells. In such deployments, a macro cell could play the role of a centralized coordinator, receiving channel state and buffer reports from the small cells within its coverage, using similar technology as for regular radio access (but e.g. operating on a different frequency). The centralized scheduler would use available knowledge from controlled cells to provide optimal scheduling decisions. In one control/user plane decoupling variant, these scheduling decisions could be forwarded directly to the users. Another variant assumes forwarding scheduling decisions to small cells, which then apply these to their assigned users. In both cases, by utilizing over-the-air signaling a tight time alignment could be guaranteed for delay-sensitive RRM signaling.

Assuming a 5G UDN using a TDD frame structure [24], a fully centralized versus a standalone/decentralized resource scheduling schemes are compared in Figure 11.3, showing noticeable improvements when moving toward more coordinated scheduling schemes. Apart from the fully centralized and decentralized/ standalone RRM approaches, there are a large number of distributed RRM methods that proactively manage radio interference in cellular networks and could be potentially interesting for UDNs. One of the approaches evaluated for 5G and described in [22] relies on the exchange of scheduling coordination information among cells grouped in clusters. Here, the information exchange is limited to cells within the cluster, therefore the related signaling overhead can be significantly reduced. Another approach [25] utilizes game theory, where the BSs use different long-term strategies for resource allocation and exchange these long-term decisions (along with information on the channel gains or experienced interferences) in order to minimize the overall losses (i.e. regrets).

Finally, when discussing resource management, it should be mentioned that context awareness can be also used to improve the end-user experience and overall system performance (cf. Section 11.3.3 for other applications of context awareness in 5G). In fact, the context information related to the prediction of user positions can be used for an efficient distribution of radio resources [22]. For example, the transmission of delay-sensitive traffic can be prioritized when a user is predicted to move into unfavorable transmit conditions due to e.g. a coverage hole, congested cell, tunnel [26][27].

11.2.2   Interference management for moving relay nodes

In order to mitigate the high VPL when serving in-vehicle users, it is sufficient to configure MRNs as relay nodes. However, to handle the advanced interference scenarios in a UDN, it is beneficial that the MRNs constitute moving cells, i.e. cells controlled by a moving BS. However, as compared to traditional pico- or femto-cell deployments, new interference management schemes taking into account the special characteristics of moving cells are particularly needed to efficiently use the resources in ultra-dense deployment scenarios:

- An in-band MRN has no dedicated backhaul link connection, and thereby, it needs to share time and frequency resources with the users connected to the cell providing the wireless backhaul (e.g. a macro cell).

- Due to the mobility of moving cells, the interference situation is changing rapidly, which makes it more complicated to manage compared to stationary pico- or femtocells.

In particular, under the assumption of half-duplex MRNs, more resources (with respect to full-duplex operations) are needed for the backhaul links in order to compensate for the half-duplex loss. In addition, the MRN access links may interfere with the macro users in proximity of the vehicle. Thus, there is a need to design interference management and resource allocation schemes such that the improvement of the throughput of the users inside vehicles does not significantly degrade the performance of the macro users.

11.2.2.1 Performance of moving relay nodes

Interference management for MRNs in a UDN has been investigated in [22] and [28]. The simulation setup assumes a realistic urban environment and separate frequency bands for the macro and micro users, i.e. full-duplex but not full in-band MRN, as illustrated in Figure 11.4. Macro users and backhaul links of MRNs are subject to intercell interference from neighboring macro sectors. The micro users are subject to intercell interference from neighboring micro sectors, and potentially the interference from the access links of nearby MRNs. The access links of MRNs suffer from the interference from nearby micro sectors.

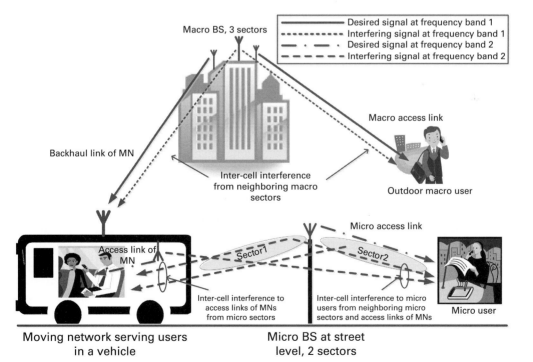

Figure 11.4   Illustration of desired and interfering signals in the considered MN scenario using MRNs in a UDN [29], Creative Commons Attribution (CC BY).

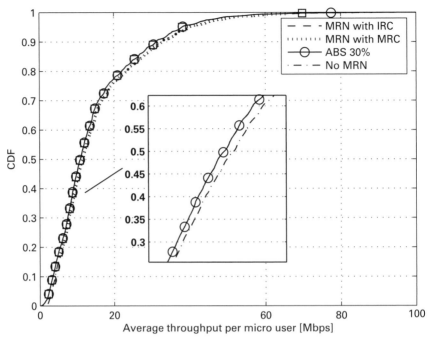

Figure 11.5   The cumulative distribution function (CDF) of micro user throughput when the VPL is 30 dB, [29], Creative Commons Attribution (CC BY). All the curves except "No MRN" are overlapping.

Figure 11.5 shows the CDF of the throughput of micro users in the presence of MRNs in the system. Outdoor users and VUs were equipped with single antennas, while multi-antenna receivers were used for MRN backhaul links. In order to ensure fairness, modified proportional fairness scheduling [30] was employed along with Almost Blank Subframes (ABSs) [31]. As seen, the performance of the micro users is not significantly impacted by MRNs if the VPL is high. In fact, the bandwidth of the macro cell is relatively large, and therefore not so many ABSs are needed at the macro cells. Further, high VPL can also attenuate the interference from the access links of the MRNs to the nearby micro users.

Figure 11.6 shows the 5th-percentile versus 90th-percentile throughput of VU and macro users when using different interference coordination or interference cancellation schemes. As seen, using IRC at the backhaul links of the MNs ("VU, MRN with IRC") significantly improves the performance at the VUs compared to direct BS transmission ("VU served directly by macro BS"). Other schemes can improve the 5th-percentile VU throughput, either by improving the desired signal strength (e.g. by using Maximum Ratio Combining (MRC) at the backhaul of the MRNs), or reducing the interference (e. g. through configuring ABSs at the macro cells). Furthermore, from these results one can see that for the macro users, there is no significant performance change if IRC or MRC is used by the MRNs. However, as the use of ABSs consumes resources at the macro cells, this scheme will have some impact on the macro user performance.

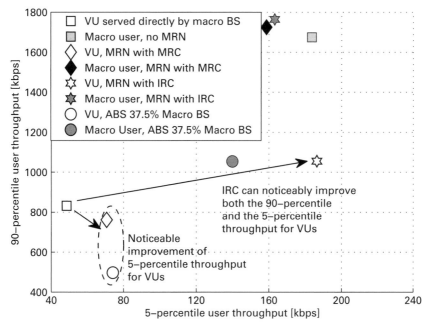

Figure 11.6   The 5th-percentile vs. 90th-percentile throughput of VU and macro user when the VPL is 30 dB, [29], Creative Commons Attribution (CC BY).

The above results show that MRNs can be used also in dense urban scenarios if appropriate interference mitigation techniques are used. However, the backhaul link of the MRN is the bottleneck. In order to boost its capacity, one can exploit the Channel State Information at the Transmitter side (CSIT) even at a high speed, through the so-called "predictor antenna" concept [32][33]. In the predictor antenna concept, dedicated antennas are deployed in the front row of an antenna array and used to look ahead at the channel that will be experienced in the near future by the transmit/receive antennas of the array. When the CSIT is available, well-known advanced transmitter based MIMO techniques can be used for high speed vehicles. When the CSIT is not available, MIMO schemes based on a closed loop would fail due to the inaccuracy of feedback information in a fast-changing wireless channel. The predictor antenna concept has also been used in combination with Massive MIMO [34][35], where substantial energy savings can be obtained in the MRN backhaul link.

When CSIT is available, a tight integration of the MNs in HetNets might also be possible, i.e. moving cells taking part in Coordinated Multi-point (CoMP)-like interference coordination and joint transmission/reception schemes, with soft-handover of the moving cells between macro BSs.

Finally, the backhauling of MRNs can potentially allow in-band full duplex transmission, since the size of the vehicle might enable sufficient spatial separation of transmit and receive chains [36], avoiding or sufficiently attenuating self-interference.

11.2.3   Interference cancelation

Previously described proactive interference management solutions rely on providing favorable scheduling conditions for the active users or BSs. It is achieved by allocation of specific resources to the appropriate users or BSs, based on the channel conditions, transmit power and the interference coupling between the transmitters and the receivers. The goal of this allocation is, typically, to minimize the level of interference experienced by the receivers with ongoing data transmission, in order to reach assumed performance metrics such as throughput maximization, fairness, packet delay minimization, etc.

Interference Cancellation (IC) methods, however, may in fact benefit from an intentional introduction of strong interference, which can be removed at the receiver side. In particular, if the level of individual interferers experienced at the receiver side is higher than the useful signal, these interferers can potentially be reliably subtracted before decoding the transmission of interest. Depending on the implementation, several IC methods are possible. In Parallel Interference Cancellation (PIC), the interfering signals from several sources are jointly removed, while in Successive Interference Cancellation (SIC) the intruding signals are cancelled iteratively (where the strongest interferer is removed first). IC methods can be also divided, depending on the receiver's capability to decode the interfering signals. If the receiver can perform only symbol-level detection, it is called soft IC. In case the receiver is decoding the spurious transmission and reconstructing the interference signal, it is called hard interference cancellation, which requires the knowledge of, e.g., modulation and coding scheme or resource blocks for the cancellation process. Network Assisted Interference Cancellation and Suppression (NAICS) is a good example of a promising IC method already introduced in LTE, but with a strong potential for extension and more pronounced usage in 5G (cf. Chapter 7).

There are several reasons why IC schemes will be beneficial in 5G. In UDN deployments, the interference constellations will by very dynamic, especially when employing a flexible UL/DL TDD air interface. Additionally, the close proximity of the BSs will lead to a situation where signals from BSs can interfere with the uplink reception in the neighboring sites. Furthermore, underlay direct D2D transmission reusing resources used for a regular cellular transmission can benefit from the intentional introduction of strong interference in conjunction with reliable IC.

Apart from that, UDN deployments will likely be time synchronized, which further reduces the complexity of the IC approaches. In case of wide area deployments, time synchronization is also likely as it is a prerequisite for techniques such as CoMP.

11.3   Mobility management in 5G

In the early phases of 5G deployments, legacy technologies will carry a vast majority of the overall cellular traffic. Therefore, one of the desired functionalities of 5G should be the ability to map a specific traffic or service to the most appropriate layer or RAT. In order to exemplify this functionality, non-delay critical traffic coming from stationary users could be carried using Wi-Fi solutions, broadband connectivity for video services for users on the move could be provided via LTE-A and its evolution, while the traffic that requires extremely short latencies or ultra high reliability would be handled using 5G radio access.

Apart from improving mobility issues known from contemporary networks [37]–[40], 5G deployments will bring several new aspects that call for novel mobility related solutions:

- Network densification may also challenge contemporary handover schemes. A particularly interesting research question is whether it is still more efficient to keep the paradigm of network-driven handover, or if a User Equipment (UE)-based approach could work better in UDN deployments[1]. The network densification process imposes also the need for new ways of small cell detection that are signaling and energy efficient.
- Operations in the mmW regime are very attractive for 5G due to the availability of large contiguous bandwidth chunks (cf. Chapter 12), but from a propagation point of view, shadowing effects in mmW are much more severe comparing to the ones experienced when using lower carrier frequencies. As a consequence, handovers may occur more often and the cell borders may be much sharper. An additional difficulty (from the mobility point of view) comes from the combination of mmW and techniques like massive MIMO, which is challenging for cell detection procedures as the search for narrow-beamed signals may be particularly cumbersome for UEs.
- Direct D2D transmission provides numerous benefits that exploit proximity, hop and reuse gains. Still, from the mobility perspective, moving D2D users with an ongoing data transmission require novel criteria for handover decisions in order to avoid excessive signaling exchange e.g. in inter-cell D2D operations (cf. Chapter 5).
- With the rise of MNs, the mobility support will be essential not only for users but also for wireless BSs on the move. Such moving wireless BSs will enable efficient group mobility with reduced signaling overhead and more stable communication links even at very high speeds.

11.3.1   User equipment-controlled versus network-controlled handover

Depending on where the handover is triggered and where the target cell is determined, three different approaches for handover mechanisms could be considered:

- **UE-controlled handover**: The UE monitors the quality and/or strength of the signals from cells in proximity and triggers the handover process based on pre-established criteria. In this case, the UE selects the target cell. This approach is used e.g. in DECT or Wi-Fi (limited mobility support).

[1] In Sections 11.3.1 and 11.3.2.1 the term User Equipment (UE) is used to indicate the difference between hardware (e.g. hand-held device, MTC device) and human. In other parts of Chapter 11 the term "user" is used jointly for both.

Table 11.1 Benefits and drawbacks of different handover types.

Handover type	Benefits	Drawbacks
UE controlled	Direct measurement of channel status and interference situation Handover decision based on UE status Handovers only if necessary Small signaling overhead	No network control HO failures if target cell is not prepared
Network controlled	Easy control of UE behavior Small signaling overhead No need to go through initial access procedure	Missing knowledge on the interference situation on the UE side
UE assisted and network controlled	Direct measurement of channel status and interference situation Easy control of UE behavior No need to go through initial access procedure	Measurement reports from UE required

Table 11.2 Handover interruption time assuming 0.25 ms slot duration.

Performance indicator	Network controlled inter-site [ms]	Network controlled intra-site [ms]	UE controlled inter-site [ms]	UE controlled intra-site [ms]
Interruption time UL	1.25	0.5	6.5	6.5
Interruption time DL	4.5	1.5	6.0	6.0

- **Network-controlled handover**: Contrary to the previous method, the network infra-structure (e.g. BS) monitors the received signals and determines if the UE should be associated with another cell by means of handover. The network also determines the target cell.
- **UE-assisted and network-controlled handover**: This approach was adopted by modern cellular 3GPP systems (e.g. WCDMA or LTE). Its underlying principle is that the UE monitors the received signal strengths from cells in proximity and reports these back to the network, which takes the decision on the handover and its target cell.

All approaches have their benefits and drawbacks that are captured in Table 11.1.

Aside from the signaling overhead and success ratio, the most important performance indicator for handover procedures is the interruption time. In a network-controlled approach, it is determined as the time between a UE receiving a HO command and the UE successfully transmitting a packet to the target cell (uplink) or receiving data packets from the target cell (downlink). In a UE-autonomous approach, it is defined as the time between the UE successfully sending the "BYE" message and the UE successfully transmitting a packet to the target cell (uplink) or receiving data packets from the target cell (downlink). Approximate values for handover interruption time are captured in Table 11.2. For the calculation of these values, the UE synchronization in the context of network-controlled handover is assumed to take 1 ms (and presupposes that the target BS identifier is known to the UE). For the UE-controlled handover, the UE synchronization and BS indicator reading time is assumed to take 5 ms.

For active UEs (having ongoing data transmission), a network based handover leads to smaller interruption times. It also provides a higher success rate due to the network-based selection of the target cell, where it is possible to reserve resources in advance. UE assistance is needed for a correct estimation of the interference situation at the UE side. However, for inactive UEs (without data in the buffer), a more slim UE-based procedure should be defined (comparable to idle mode re-selection). Additionally, this approach could be beneficial for high-mobility UEs as the handover reaction time (i.e. time between detecting a channel degradation that could trigger a HO decision and the UE receiving/sending a "BYE" message) may be shorter than for network-based handover. In the latter case, the main factor contributing to this delay is the network handover decision making time and latency of the interface between source and target cell (in the inter-site case). A potential way to decrease the UE-based handover-interruption time is e.g. a one-step-RACH based on sending in one RACH message both preamble and data.

11.3.2   Mobility management in heterogeneous 5G networks

The aforementioned heterogeneity of access possibilities in future 5G systems imposes challenges for efficient mobility support and resource management. In this regard, herein, the mobility management is tackled from three perspectives: Mobility support for users in a multi-RAT and multi-layer environment, mobility support for users connected via D2D communications, and mobility support for wireless BSs on the move, e.g. MRNs.

With the increased heterogeneity in wireless networks, an appropriate RAT and layer mapping is of paramount importance for the end users (cf. Chapter 2). The shadowing effects in the mmW range are much more severe compared to lower frequency bands, resulting in much sharper cell borders relative to today's wireless networks. In particular, because of the propagation characteristics of higher frequency bands above 6 GHz implying heavier signal deterioration, mmW communication is likely mostly available when there is a Line-of-Sight (LOS) link between the users and BSs. The autonomous mmW cell search by users necessitates frequent measurements on multiple carriers, which can imply increased power consumption as well as increased delay in finding the cells.

As BSs may serve both mmW bands and frequency bands below 6 GHz, measurements on lower-frequency bands can be utilized to detect whether there is a LOS link between the users and BSs. The detection can rely on the received signal characteristics, such as channel impulse response, root mean square delay spread, or received signal strength indicators. For instance, in case of a strong LOS detection in below 6 GHz range, there is a high probability of having the mmW coverage as well, especially when below 6 GHz and mmW antennas are at the same location. When mmW communication is not available, the mmW transceiver can be switched off. In addition, unnecessary measurements of the mmW range can be eluded for the users. Accordingly, energy savings can be attained both from the user and network perspectives.

11.3.2.1   Fingerprints coverage for multi-RAT and multi-layer environments

Network-assisted small cell discovery can substantially enhance the cell search in multi-RAT and multi-layer environments envisioned for 5G. The assistance information can consist of radio fingerprints of the coverage carrier, e.g. pilot signal powers of neighboring macro cells, where the radio fingerprints can correspond to a small cell location (belonging to a different frequency band, i.e. layer, or other RAT) or a handover region. While being connected to a macro cell carrier, the UE performs neighbor cell measurements and compares the obtained measurements with the fingerprints received from the network. In case of a fingerprint match, the UE reports the matched fingerprint to the network. Consequently, the network can configure targeted measurements on a specific carrier for finding the small cell in proximity.

In the exemplary setup outlined in [22], the network provides UEs with a set of macro cell radio fingerprints (consisting of macro cell ID and pilot power for the 3 strongest macro cells). Each of these sets corresponds to a small cell location, where it is assumed that small cells operate on a different frequency layer than the macro cells. Moreover, it is assumed that the UE performs inter-frequency measurements periodically as in today's networks; hence, the energy consumption due to the inter-frequency cell search is proportional to the area where the UE searches for inter-frequency small cells. With an increasing number of fingerprints per small cell, a more accurate mapping of small cell coverage can be obtained, which allows an efficient network-enhanced discovery of BSs operating on different frequencies. Yet, the larger number of fingerprints also implies increased signaling overhead. However, results from [22] indicate that already 1 to 3 fingerprints provide good performance improvement. For example, with one fingerprint sample per small cell, up to 65% energy saving with 95% accuracy of small cell coverage can be obtained, when compared to a UE searching for inter-frequency small cells in its entire environment.

11.3.2.2   D2D-aware handover

D2D communication is seen as an integral component of a 5G system, which will enable complementary transmission methods compared to legacy solutions. Nevertheless, despite the promising benefits of D2D communications (cf. Chapter 5), these can become challenging in the context of mobility. Hence, D2D-aware mobility management is crucial. For instance, at the cell-edge, D2D user pairs may often be handed over to a neighboring cell at different times. Therefore, delay-sensitive D2D services may not be maintained if more than one BS is involved in the control of D2D communications, and in the presence of a non-ideal backhaul connection between the involved BSs imposing additional delay.

One approach within the D2D-aware mobility management is the synchronized D2D handover for the D2D pair provided that the link quality allows for it, i.e. the condition

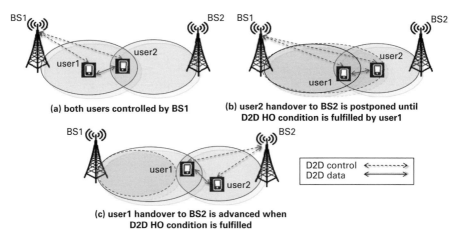

Figure 11.7   D2D-aware handover scheme.

that neither of the D2D users should experience a radio link failure is fulfilled. An illustration for the synchronized D2D handover and the associated gains in terms of increased continuous connected time of a D2D pair is depicted in Figure 11.7. In this scheme, in order to achieve synchronized handover for the D2D pair, the handover of user2 is delayed until user1 can be handed over to BS2. This approach can be extended to a group of D2D users under the same BS. Therefore, when there is a new user joining a D2D group, it is controlled by the same BS already controlling the D2D group. The simulation results outlined in [41] show that the continuous time of stay under control of the same cell can be increased significantly with D2D-aware handover compared to an independent handover for each user (handover is performed individually for each user based on an *A3* event [42], where an *A3* event defines the case where the received signal power from a neighbor cell becomes a certain offset better than the signal power from the serving cell [42]). Reported gains were 235% and 62% for 20 m and 100 m maximum distances between the D2D users, respectively. It is worth noting that a synchronized D2D handover can be performed without compromising the reliability of the handover.

11.3.2.3   Handover for moving relay nodes

One characteristic of 5G systems is the wireless BS on the move, as introduced in Section 11.1.2. Compared to VUs being directly served by static BSs, such moving BSs, e.g. MRNs, can significantly improve the performance of VUs by efficiently handling group mobility and avoiding penetration loss due to the vehicle shields and windows. A generalized framework to optimize the handover parameters of the system that minimize the End-to-End (E2E) Outage Probability (OP) of the VUs is proposed in [43]. Therein, the performance at the VU served by a half-duplex MRN is compared with that of the baseline, i.e. direct single-hop BS to VU transmission. The position and handover parameters are optimized to minimize the average E2E OP at the VU. As shown in [43], at low VPL around 10 dB,

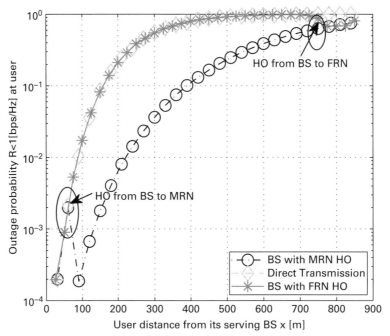

Figure 11.8   OP at the VU when VPL = 30 dB [43], FRN stands for Fixed Relay Node (deployed 805 m from the BS along the way of VU), reprinted with permission (Lic. no. 3666480737405).

the performance of MRN-assisted transmission is almost as good as the direct hop BS-to-VU transmission. However, as illustrated in Figure 11.8, at high VPL (e.g. 30 dB), there is a clear advantage to serve the VU by using the MRN assisted transmission. In particular when the VU is handed over to the MRN at the distance of ~100 m, the OP at the VU is significantly lowered thanks to circumvention of high VPLs.

11.3.3   Context awareness for mobility management

Network optimization is a closed-loop process [44]. In fact, mobile networks are considered as dynamic structures, where continuously new sites are deployed, capacity extensions are made and parameters are adapted to local conditions. In the past, MNOs have spent substantial efforts on manually tuning site-specific HO control parameters. However, in order to reduce the degree of human intervention in network optimization processes, optimal settings have to be identified and enforced autonomously. These settings yield an appropriate trade-off, e.g. with respect to connection drops, HO failures, and the occurrence of ping-pong HOs. In this scenario, the knowledge of the context is of paramount importance to improve the mobility management. In fact, as explained in Chapter 2, context awareness is not only an important component of Dynamic RAN and Localized Contents and Traffic Flows, but it can also improve the performance of mobility management schemes.

11.3.3.1   Exploitation of location information for mobility management

Recently, approaches such as [45]–[47] propose to adjust HO parameters according to user-specific behavior and mobility patterns. In particular, the approaches presented in [2] and [47] exploit additional vehicle context information, such as location and movement trajectory, to adaptively optimize street-specific HO parameters.

In order to enable reliable cooperative driver assistance services that improve traffic safety and efficiency in the future, a robust mobility support for vehicular users is required. As the traditional LTE HO optimization only aims at choosing a Cell Individual Offset (CIO) or Handover Margin (HOM) and exploiting the speed profile by adjusting the HO Time-To-Trigger (TTT), the driving direction and radio propagation properties are not considered. In contrast to, e.g., pedestrian users that move less predictably, vehicular users move along a defined path (i.e. streets) and within a certain speed range. Moreover, the position and trajectory of vehicles are usually well-known due to on-board Global Positioning System (GPS) and navigation systems. Thus, HO decisions can be improved based on this additional context information.

Two implementation variants can be envisioned, depending on the location of the HO decision unit. The first possibility is to employ a central unit that manages the HO parameters for each street. The vehicles send periodically the location information (e.g. the street ID) to the BS that is connected to the central unit. Then, based on an optimization algorithm, the corresponding HO parameters (HOM, TTT) will be sent to the vehicles. Another variant is to maintain a database of HO parameters at every vehicle. Based on the location information, the vehicles select the parameter for optimization. However, the database needs to be updated by the central unit in case a different optimization objective is targeted or a different traffic profile is observed.

In a heterogeneous radio access environment, the task of enabling robust and optimized mobility support across different RATs is quite challenging. Usually, neither user movements nor traffic demands are exactly known in advance and can only be estimated. Thus, continuous monitoring of RAN-specific and, in particular, mobility-related Key Performance Indicators (KPIs) is inevitable. In addition, measurement data and long-term KPI statistics can form a context history and be used for predicting trends with respect to important KPIs.

In many modern cellular systems, Self-Organizing Networks (SONs) solutions are used to exploit this measurement data and improve mobility management in an automated fashion. However, the configuration of HO control parameters in order to find an appropriate trade-off between a minimal number of connection drops, HO failures, and the occurrence of ping-pong HOs is one example where MNOs today still spend substantial effort on manually tuning these control parameters per service area. Reconfigurations are usually triggered upon detection of HO problems or upon integration of new BSs into the network.

In order to create awareness of the radio access situation of a considered service area, self-learning SON based approaches have been developed. For example, HO parameters are tuned between several cells in a cell pair-specific manner reflecting direction and impact of local user mobility as well as locally observed conditions, hence enabling context awareness.

Further, the developed approaches rely on a limited set of input parameters and do not require measurements performed and provided by users for optimizing network performance.

In particular, Fuzzy Logic (FL) based schemes have been found suitable for an autonomous adaptation of inter-RAT HO parameters [48][49]. Moreover, Fuzzy Q-Learning (FQL)-based approaches have already been successfully applied to the problem of dynamically and optimally tuning soft handover parameters [50][51], optimal resource sharing between real-time and non-real-time services in 3G networks [52], self-adaptation and optimization in HetNets [53], and self-tuning HO parameters for load balancing [54].

Further, recent approaches presented in [22] and [55], are able to self-optimize HO parameter settings according to locally observed conditions while enhancing multiple KPIs, such as connection dropping, HO failure, and ping-pong HO ratios. Furthermore, in the long-term, OPEX is significantly reduced when applying context awareness in SONs, while the users' quality of experience is improved due to more robust mobility support.

Figure 11.9 depicts the performance of various HO parameter optimization schemes with respect to an Overall Performance Indicator (OPI), which accounts for connection

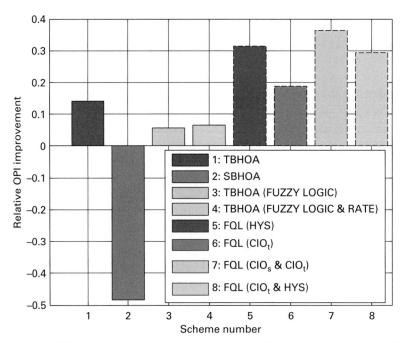

Figure 11.9   Relative OPI improvements per optimization scheme (fixed adaptation strategies #1-#4 are outperformed by FQL based solutions #5-#7[55]).

drops, HO failures, ping-pong HOs, and user satisfaction. In contrast to a 3GPP legacy system or optimization approaches relying on fixed adaptation strategies, such as Sum-Based Handover-Optimization Algorithm (SBHOA) and Trend-Based Handover-Optimization Algorithm (TBHOA), significant improvements of 32%, 19%, 36% and 29%, respectively, are achieved by different variants of the FQL-based self-tuning solution using different adaptation parameters such as Cell Individual Offset for source and target cells (CIO_s and CIO_t, respectively) or handover hysteresis (HYS). Here, scheme 7, FQL (CIO_s & CIO_t), is particularly suitable in case of direction-oriented user mobility, since it adapts HO parameters in a cell pair-specific manner. More details can be found in [22] and [55].

11.4   Dynamic network reconfiguration in 5G

Already nowadays, cellular networks consume a noticeable amount of global electrical energy [56] and without new improvements of the wireless network operations, this trend is expected to hold. A natural and straightforward solution is a dynamic activation and deactivation of BSs. Apart from operators' proprietary solutions, 3GPP have also made attempts to facilitate this [57], but dynamic activation and deactivation of BSs is hard to introduce in mature legacy systems due to backward compatibility issues. Therefore, it is of the uttermost importance that the 5G networks natively support this feature from the very first release of the standard.

One of the targets of the 5G system is also to handle the changing distribution of increasing traffic demand over time and space in an agile manner [58]. That is, the network needs to react quickly and dynamically to adapt to variations of service demand during a time period in a target region. One approach for providing required coverage and/or capacity is to deploy fixed small cells, such as pico-cells, femto-cells, and relay nodes overlaid by macro-cells. In today's mobile networks, small cells may be deployed by operators at certain locations with power supply facilities, and the locations can be determined, for example, via network planning. However, the full operation of such a dense fixed small cell deployment is not needed anytime and anywhere due to the changing distribution of traffic over time and space, and also not desired in order to achieve network energy savings. In this context, NNs can be put into operation complementing the existing network deployment to address the changing service requirements on demand.

11.4.1   Energy savings through control/user plane decoupling

An example of an architecture with decoupled control and user plane are HetNet deployments based on the Phantom Cell Concept (PCC) [59], as illustrated in Figure 11.10. It shall be noted that mobility handling using phantom cells for mmW communication systems is treated in Chapter 6. Systems deployed according to the PCC architecture are comprised of two overlaid networks:

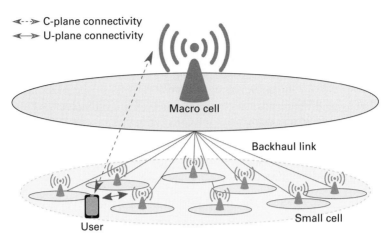

Figure 11.10   Representation of the PCC network architecture.

- A macro network, where macro cells operate using legacy standards (e.g. 3GPP LTE), guaranteeing backwards compatibility with legacy users, i.e. users which only support legacy standards.
- A small cell network, where each small cell is connected to a macro cell through a backhaul link.

PCC systems enable to separate the serving points of control plane (C-plane) and user data plane (U-plane), with e.g. macro cells responsible for C-plane connectivity and small cells handling U-plane connectivity.

In traditional HetNet small cell deployments, the connection establishment between a user and a small cell is independently managed by the user and the small cell. The user usually performs small cell channel estimation with the help of pilot symbols sent by small cells, and orders the measured small cell candidates by decreasing channel quality metrics (e.g. SINR). The user subsequently tries to connect to the best small cell candidate by attempting a Random Access (RA) procedure. Based on admission control parameters, the small cell can either accept or reject the connection request [60]. If the connection request has been rejected, the user can then try to connect to the subsequent small cell candidates, until a connection is successful or the connection procedure is abandoned. Figure 11.11 illustrates an example of this admission control-based trial and error connection paradigm, for which the user connects to the third-best small cell candidate after the connection requests to the first- and second-best small cell candidates have been rejected.

In architectures such as the PCC, it is assumed that a user always keeps C-plane connectivity to a macro cell, and in which the connection procedures are performed assuming the small cell user is assisted by the macro cell. In particular, the macro cell decides which small cell the user is connected to, based on the information received from different sources. A typical connection procedure for a macro-assisted small cell user is depicted in Figure 11.12.

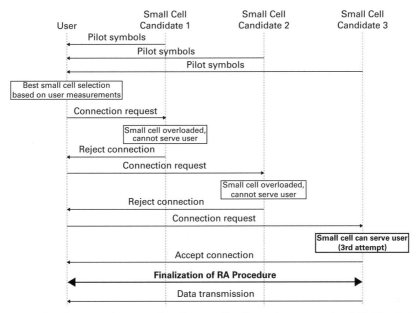

Figure 11.11 Example of a connection procedure for a small cell user in a conventional HetNet deployment.

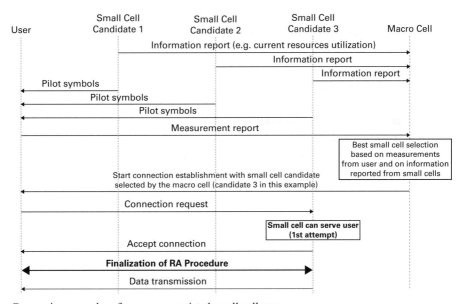

Figure 11.12 Connection procedure for a macro-assisted small cell user.

Unlike traditional admission control-based small cell connection procedures, the small cell determined as the best candidate by the macro cell in the macro-assisted schemes is guaranteed to be able to serve the connecting user. Consequently, the small cell user connection time and the amount of signaling required are significantly reduced, since only one RA procedure attempt is required.

Additionally, the connection procedure for a macro-assisted small cell user can be used to realize energy savings in PCC systems, in order to alleviate the energy consumption of densely deployed HetNets. The alleviation can be realized by putting small cells in a sleep mode, in which they consume a reduced (but still non-negligible) amount of energy with respect to the fully turned-on small cells. These cells in this sleep mode have the capability to be turned-on in a very short time. Hence, they can become operational whenever needed, e.g. in case of sudden traffic increases.

In traditional HetNet deployments, where the users autonomously connect to small cells, putting small cells to sleep could potentially impact the performance of the network, since the small cells in sleep mode typically do not transmit any pilot symbol or discovery signal and therefore cannot be detected by users autonomously [61][62]. Hence, the users connecting to the small cell network autonomously may end up with a suboptimal data rate.

The discovery issue can however be resolved when a C-plane/U-plane split system such as the PCC is used, in which the users are always connected to a central element (the macro cell). In such systems, the connection procedures of the users in a small cell can be completely handled by the macro cell, which is always aware of the status of small cells connected to it via the backhaul links. The macro cell can therefore wake up a given small cell by e.g. sending a wake-up signal through the backhaul link, in order to allow users to be served.

Energy savings in networks following the PCC architecture are realized by putting small cells into a sleep state (i.e. not serving any user) in which they consume a reduced amount of energy. In fact, in the sleep mode, small cells do not need to transmit discovery signals required for small cell discovery and initial channel estimation purposes. Thus, the level of interference in the small cell network will be reduced, leading to a potential increase in SINR of other small cell users, and therefore to an increased average throughput.

Simulation results indicate that up to 25% user throughput performance improvement can be observed when comparing the performance of the PCC system implementing energy savings to the PCC system not implementing any energy saving feature.

Several energy savings schemes for PCC systems based on the macro-assisted small cell user connection paradigm, are proposed in [22]. Some of these schemes can be based on downlink (e.g. small cells send discovery signals in a sporadic fashion) or uplink (e.g. small cells are woken up by users by receiving wake-up signals) signaling [60]. Other schemes rely on a database which stores the link quality estimates of all the potential users in a small cell. These estimates can be provided even when a small cell is in sleep mode [61][62].

Figure 11.13 shows the amount of achieved energy savings in the small cell network when a typical energy saving scheme is in use in a PCC system. The 0% energy savings case corresponds to a baseline system where all small cells are always on. As can be seen in Figure 11.13, up to 45% of energy savings in the small cell network can be achieved when a low number of users are present in the system. The amount of the achievable energy savings is network load dependent and decreases as the number of connected users increases.

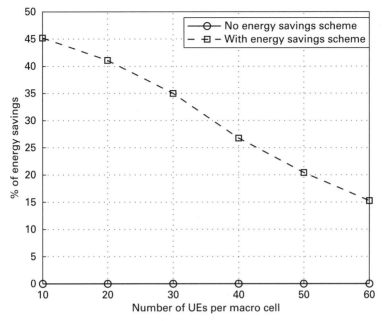

Figure 11.13   Achieved energy savings in the small cell network of a typically deployed PCC network.

11.4.2   Flexible network deployment based on moving networks

Section 11.2.2 introduced the concept of MNs, focusing on MRNs and interference management aspects. This section further extends the NNs concept depicted in Figure 11.14(a) and focuses on increasing the network capacity or extending the cell coverage area [63],[64] see Figure 11.14(b), as well as reducing the network energy consumption [64].

A nomadic network consists of unplanned small cells, which are not necessarily owned by the network operators, offering the possibility for relaying between users and BSs. While the location of operator-deployed relay nodes is optimized by means of network planning, the location of the NNs in a nomadic network is out of control of a network operator and, therefore, is considered to be random from the operator's perspective. Moreover, the availability and position of NNs may change over time (hence, the term "nomadic") due to battery state and vehicle movement. The NNs operate in a self-organized fashion and are in general activated and deactivated based on capacity, coverage, load-balancing or energy-efficiency demands. Hence, on one hand NNs provide an effective extension of the cellular infrastructure that allows for a dynamic network deployment and dynamic reconfiguration. On the other hand, due to their dynamic topology, nomadic networks also impose technical challenges. The appropriate management of such a large number of dynamic network nodes requires efficient RRM solutions to combat interference for performance enhancements,

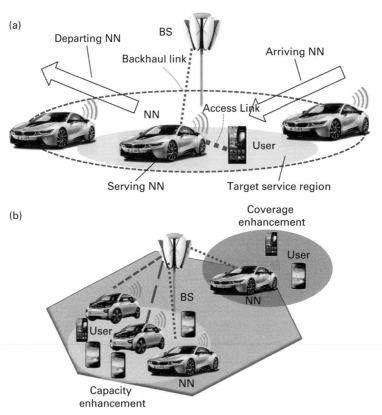

Figure 11.14 (a) Dynamic network deployment based on NN operation and (b) two key benefits of NN deployment.

alleviated security means in particular considering privately-owned nomadic networks, and an energy-efficient operation of NNs due to their limited battery lifetime.

To attain the aforementioned benefits of NNs, a flexible backhaul needs to be employed, where the capacity of the backhaul link between an NN and its serving BS plays a crucial role for the end-to-end user performance, especially when limited by severe fading characteristics. One possible cost-efficient realization for flexible backhaul is in-band relaying. Thanks to the availability of multiple NNs in a target service region, the flexible backhaul realization can be exploited by dynamic NN selection to overcome the limitations of the backhaul link and, thus, to further enhance the system performance [65]. Dynamic NN selection identifies the optimum serving NN based on the backhaul link quality, or more precisely, the SINR on the backhaul link, where due to low elevation and severe fading characteristics the backhaul can easily be the bottleneck on the end-to-end link. On this basis, the coarse NN selection takes into account long-term channel quality measurements based on shadowing, whereas the optimal NN selection relies on the short-term channel quality measurements based on both shadowing and multi-path fading.

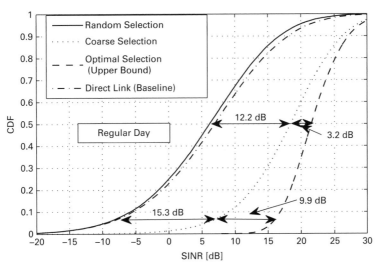

Figure 11.15   SINR gains with dynamic NN selection schemes [64], reprinted with permission.

Different NN selection approaches are compared in Figure 11.15 on the downlink assuming composite fading/shadowing conditions (Rayleigh-lognormal with 8 dB shadowing standard deviation). Direct link implies that there is no NN operation, and is considered as the baseline for comparison. Two cases can be distinguished: (1) the baseline case where the user is connected directly to the BS, and (2) the NN operation case where the user is indirectly connected to the BS via a NN. The NN availability is determined by the parking lot model [65]. Further, in the simulation set up, a maximum number of 25 NNs in the parking lot were considered. Relative to the random NN selection and direct link schemes, coarse NN selection provides significant SINR gains of about 15 dB and 12 dB at lower and median cumulative distribution function (CDF) percentiles, respectively. Optimal NN selection can further improve the performance at the cost of increased signaling overhead. The SINR gains can translate into end-to-end throughput gains of more than 100% relative to random NN selection depending on the access link quality as demonstrated in [65]. Therein, it is shown that the achievable gains are higher for an increased availability of NNs.

The NN availability can be further exploited for performing energy-aware network optimization. In the exemplary scenario, the energy consumption of an active BS is assumed to be 1 kWatt, whereas an NN consumes 10 Watt when it is actively transmitting, and the network consists of 7 BSs with an inter-site distance of 500 m, 50 randomly distributed users and 200 NNs from which the optimum ones are selected to minimize the total energy consumption for a given service requirement [64]. In Figure 11.16, system power consumption is shown over time where a strong inverse correlation is observed between the number of available NNs and the total energy consumption. This is attributed to the fact that a larger amount of NNs results in a higher probability of having suitable NNs to redirect data traffic and to shut down more BSs. Up to 20% energy savings are possible when many parked cars can serve as potential NNs.

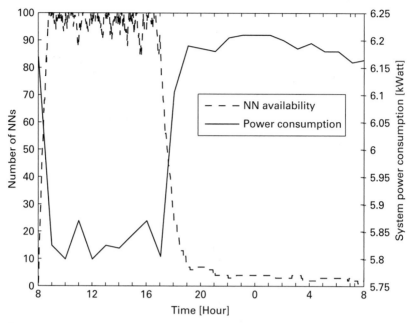

Figure 11.16   System power consumption with daily varying availability [64], reprinted with permission.

With CSIT available also for such NNs, using the predictor antenna concept introduced in Section 11.2.2, the moving cells can also play an integrated role in HetNets serving also outdoor users. Substantial energy savings have been reported using the approach in [66]–[68] by jointly optimizing the precoding, load balancing and HetNet BS operation mode (active or sleep) under per-user quality of service constraints and per-BS transmit power constraints.

11.5   Conclusions

In this chapter, layer 2 and 3 technical concepts related to interference and mobility management in 5G have been presented.

5G raises the need for novel interference management approaches, since in the expected very dense deployments the usage of flexible TDD, D2D and moving networks will lead to very different interference constellations than seen in legacy networks. It was shown that these novel interference constellations require more coordination between access nodes than in e.g. LTE-A, which is a fact that should be reflected in the design of 5G architecture and infrastructure interfaces such as the future X2 interface. As an example for novel interference management paradigms, a specific concept for moving networks was described.

The usage of very dense deployments, D2D, moving networks, mmWave communications and novel application requirements also suggest rethinking the design of mobility management concepts in 5G. For instance, it has been shown that for a high user activity level network-driven handover is clearly preferable, as used in legacy systems, whereas for connected yet inactive users, such as in the context of machine-type-communications, device-driven handover may be an interesting complementary option. Further, ultra-dense multi-layer deployments in 5G suggest exploiting radio fingerprints for mobility management. For D2D, it has been shown that it is beneficial to jointly handover pairs of devices that are currently using D2D transmission. In general, the exploitation of context information will play a key role in 5G, for instance for the optimization of handover parameters.

The chapter has been concluded with specific means for a fast activation and deactivation of cells for network energy efficiency reasons, or for the autonomous integration of nomadic network elements such as nomadic cells. Both schemes are crucial elements of the dynamic RAN, i.e. the capability of the system to adapt to rapid spatio-temporal changes in user needs and service mix.

References

[1] N. Bhushan, Li Junyi, D.Malladi et al., "Network densification: The dominant theme for wireless evolution into 5G," *IEEE Communication Magazine*, vol. 52, no. 2, February 2014.

[2] ICT-317669 METIS project, "Final report on the METIS system concept and technology roadmap," Deliverable D6.6, April 2015, www.metis2020.com/docu ments/deliverables/

[3] A. K. Dey, "Providing Architectural Support for Building Context-Aware Applications," PhD Thesis, College of Computing, Georgia Institute of Technology, December 2000.

[4] J. Holsopple, M. Sudit, M. Nusinov, D. F. Liu and H. Du. S. J. Yang, "Enhancing situation awareness via automated situation assessment," *IEEE Communications Magazine*, vol. 48, no. 3, pp. 146–152, March 2010.

[5] C. Janneteau, J. Simoes, J. Antoniou et al., "Context-aware multiparty networking," in ICT-Mobile Summit, Santander, April 2009, pp. 1–11.

[6] P. Bellavista, A. Corradi, and C. Giannelli, "Mobility-aware connectivity for seamless multimedia delivery in the heterogeneous wireless internet," in IEEE Symposium on Computers and Communications, Santiago, July 2007.

[7] J. Antoniou, F. Pinto, J. Simoes, and A. Pitsillides, "Supporting context-aware multiparty sessions in heterogeneous mobile networks," *Mobile Network and Applications*, vol. 15, no. 6, pp. 831–844, December 2010.

[8] P. Lungaro, Z. Segall, and J. Zander, "Predictive and context-aware multimedia content delivery for future cellular networks," in IEEE Vehicular Technology Conference, Taipei, May 2010, pp. 1–5.

[9] P. Bellavista, A. Corradi, and C. Giannelli, "Mobility-aware management of internet connectivity in always best served wireless scenarios," *Mobile Networks and Applications*, vol. 14, no. 1, pp. 18–34, February 2009.

[10] P. Makris, D. N. Skoutas, and C. Skianis, "A survey on context-aware mobile and wireless networking: On networking and computing environments' integration,"

IEEE Communications Surveys & Tutorials, vol. 15, no. 1, pp. 362–386, April 2012.

[11] F. Pantisano, M. Bennis, W. Saad, S. Valentin, and M. Debbah, "Matching with externalities for context-aware cell association in wireless small cell networks," in IEEE Global Communications Conference, Atlanta, December 2013, pp. 4483–4488.

[12] M. Proebster, M. Kaschub, T. Werthmann, and S. Valentin, "Context-aware resource allocation to improve the quality of service of heterogeneous traffic," in IEEE International Conference on Communications, Kyoto, June 2011, pp. 1–6.

[13] E. Tanghe, W. Joseph, L. Verloock, and L. Martens, "Evaluation of vehicle penetration loss at wireless communication frequencies," *IEEE Transactions on Vehicular Technology*, vol. 57, pp. 2036–2041, July 2008.

[14] Y. Sui, A. Papadogiannis, and T. Svensson, "The potential of moving relays: A performance analysis," in IEEE Vehicular Technology Conference, Yokohama, May 2012, pp. 1–5.

[15] Y. Sui, A. Papadogiannis, W. Yang, and T. Svensson, "Performance comparison of fixed and moving relays under co-channel interference," in IEEE Global Communication Conference Workshops, Anaheim, December 2012, pp. 574–579.

[16] J. Ellenbeck, C. Hartmann, and L. Berlemann, "Decentralized inter-cell interference coordination by autonomous spectral reuse decisions," in European Wireless Conference, Prague, June 2008, pp. 1–7.

[17] L. Garica, G. O Costa, A. Cattoni, K. Pedersen, and P. Mogensen, "Self-organising coalitions for conflict evaluation and resolution in femtocells," in IEEE Global Telecommunications Conference, Miami, December 2010, pp 1–6.

[18] L. Zhang, L. Yang, and T. Yang, "Cognitive interference management for LTE-A femtocells with distributed carrier selection," in IEEE Vehicular Technology Conference Fall, Ottawa, September 2010, pp. 1–5.

[19] 3GPP TR 36.828, "Further enhancements to LTE Time Division Duplex (TDD) for Downlink-Uplink (DL-UL) interference management and traffic adaptation (Release 11)," Technical Report, TR 36.828, V11.0.0, Technical Specification Group Radio Access Network, June 2012.

[20] 3GPP TS 36.423, "X2 application protocol (X2AP) (Release 12)," Technical Specification, TS 36.423, V12.5.0, Technical Specification Group Radio Access Network, March 2015.

[21] E. Lahetkanges, K. Pajukoski, J. Vihriala et al., "On the flexible 5G dense deployment air interface for mobility broadband," in International Conference on 5G for Ubiquitous Connectivity, Akaslompolo, November 2014, pp. 57–61.

[22] ICT-317669 METIS project, "Final report on the network-level solutions," Deliverable D4.3, March 2015, www.metis2020.com/documents/deliverables/

[23] V. Venkatasubramanian, M. Hesse, P. Marsch, and M. Maternia, "On the performance gain of flexible UL/DL TDD with centralized and decentralized resource allocation in dense 5G deployments," in IEEE International Symposium on Personal, Indoor, and Mobile Radio Communications, Washington, September 2014.

[24] ICT-317669 METIS project, "Proposed solutions for new radio access," Deliverable D2.4, March 2015, www.metis2020.com/documents/deliverables/

[25] P. Sroka and A. Kliks, "Distributed interference mitigation in two-tier wireless networks using correlated equilibrium and regret-matching learning," in European Conference on Networks and Communications, Bologna, June 2014, pp. 1–5.

[26] R. Holakouei and P. Marsch, "Proactive delay-minimizing scheduling for 5G Ultra Dense Deployments," in IEEE Vehicular Technology Conference, Boston, September 2015.

[27] H. Abou-zeid, H. S. Hassanein, and S. Valentin, "Optimal predictive resource allocation: Exploiting mobility patterns and radio maps," in IEEE Global Communications Conference, Atlanta, December 2013, pp. 4877–4882.

[28] Y. Sui, J. Vihriala, A. Papadogiannis, M. Sternad, Y. Wei, and T. Svensson, "Moving cells: A promising solution to boost performance for vehicular users," *IEEE Communications Magazine*, vol. 51, no. 6, pp. 62–68, June 2013.

[29] Y. Sui, I. Guvenc, and T. Svensson "Interference Management for Moving Networks in Ultra-Dense Urban Scenarios," *EURASIP Journal on Wireless Communications and Networking*, Special Issue on 5G Wireless Mobile Technologies, April 2015.

[30] M. R. Jeong and N. Miki, "A simple scheduling restriction scheme for interference coordinated networks," *IEICE Transactions on Communications*, vol. E96-B, no. 6, pp. 1306–1317, June 2013.

[31] S. Sesia, I. Toufik, and M. Baker, *LTE – The UMTS Long Term Evolution: From Theory to Practice*, 2nd ed. West Sussex: John Wiley & Sons Ltd., 2011.

[32] M. Sternad, M. Grieger, R. Apelfrojd, T. Svensson, D. Aronsson, and A. Belen Martinez, "Using 'predictor antennas' for long-range prediction of fast fading for moving relays," in IEEE Wireless Communications and Networking Conference, Paris, April 2012, pp. 253–257.

[33] N. Jamaly, R. Apelfrojd, A. Belen Martinez, M. Grieger, T. Svensson, M. Sternad, and G. Fettweis, "Analysis and measurement of multiple antenna systems for fading channel prediction in moving relays," in European Conference on Antennas and Propagation, The Hague, April 2014.

[34] D. Thuy, M. Sternad, and T. Svensson, "Adaptive large MISO downlink with predictor antenna array for very fast moving vehicles," in International Conference on Connected Vehicles and Exp, Las Vegas, December 2013, pp. 331–336.

[35] D. Thuy, M. Sternad, and T. Svensson, "Making 5G adaptive antennas work for very fast moving vehicles," *IEEE Intelligent Transportation Systems Magazine*, vol. 7, no. 2, 2015.

[36] V. V. Phan, K. Horneman, L. Yu, and J. Vihriala, "Providing enhanced cellular coverage in public transportation with smart relay systems," in IEEE Vehicular Network Conference, Jersey City, December 2010, pp. 301–308.

[37] S. Fernandes and A. Karmouch, "Vertical mobility management architectures in wireless networks: A comprehensive survey and future directions," *IEEE Communication Surveys Tutorials*, vol. 14, no. 1, pp. 45–63, September 2012.

[38] I. Cananea, D. Mariz, J. Kelner, D. Sadok, and G. Fodor, "An on-line access selection algorithm for ABC networks supporting elastic services," in IEEE Wireless Communications and Networking Conference, Las Vegas, April 2008, pp. 2033–2038.

[39] G. Fodor, A. Eriksson, and A. Tuoriniemi, "Providing QoS in always best connected networks," *IEEE Communications Magazine*, vol. 41, no. 7, pp. 154–163, July 2003.

[40] E. Gustafson and A. Jonsson, "Always best connected," *IEEE Wireless Communications Magazine*, vol. 10, no. 1, pp. 49–55, February 2003.

[41] O. N. C. Yilmaz, Z. Li, K. Valkealahti, M.A. Uusitalo, M. Moisio, P. Lundén, and C. Wijting, "Smart mobility management for D2D communications in 5G networks," in IEEE Wireless Communications and Networking Conference Workshops, Istanbul, April 2014, pp. 219–223.

[42] 3GPP TS 36.331, "Evolved Universal Terrestrial Radio Access (E-UTRA); Radio Resource Control (RRC); Protocol specification (Release 12)," Technical Specification, TS 36.331 V12.5.0, Technical Specification Group Radio Access Network, March 2015.

[43] Y. Sui, Z. Ren, W. Sun, T. Svensson, and P. Fertl, "Performance study of fixed and moving relays for vehicular users with multi-cell handover under co-channel interference," in International Conference on Connected Vehicles and Expo, Las Vegas, December 2013, pp. 514–520.

[44] Next Generation Mobile Networks (NGMN) Alliance, "NGMN Use Cases related to Self Organising Network, Overall Description," Technical Report, May 2007, www.ngmn.org/uploads/media/NGMN_Use_Cases_related_to_Self_Organisin g_Network__Overall_Description.pdf

[45] T. Jansen, I. Balan, J. Turk, I. Moerman, and T. Kürner, "Handover parameter optimization in LTE self-organizing networks," in IEEE Vehicular Technology Conference, Ottawa, September 2010, pp. 1–5.

[46] A. Awada, B. Wegmann, D. Rose, I. Viering, and A. Klein, "Towards self-organizing mobility robustness optimization in Inter-RAT scenario," in IEEE Vehicular Technology Conference, Budapest, May 2011, pp. 1–5.

[47] P. Legg, G. Hui, and J. Johansson, "A simulation study of LTE intra-frequency handover performance," in IEEE Vehicular Technology Conference, September 2010, Ottawa, pp. 1–5.

[48] S. Luna-Ramirez, F. Ruiz, M. Toril, and M. Fernandez-Navarro, "Inter-system handover parameter auto-tuning in a joint-RRM scenario," in IEEE Vehicular Technology Conference, Singapore, May 2008, pp. 2641–2645.

[49] S. Luna-Ramirez, M. Toril, F. Ruiz, and M. Fernandez-Navarro, "Adjustment of a Fuzzy Logic Controller for IS-HO parameters in a heterogeneous scenario," in IEEE Mediterranean Electrotechnical Conference, Ajaccio, 2008, pp. 29–34.

[50] R, Nasri, Z. Altman, and H. Dubreil, "Fuzzy-Q-learning-based autonomic management of macro-diversity algorithm in UMTS networks," *Annals of Telecommunications*, vol. 61, no. 9–10, 2006, pp. 1119–1135.

[51] M. J. Nawrocki, H. Aghvami, and M. Dohler, *Understanding UMTS Radio Network Modelling, Planning and Automated Optimisation: Theory and Practice*. Chichester: John Wiley & Sons, 2006.

[52] R. Nasri, Z. Altman, and H. Dubreil, "Optimal tradeoff between RT and NRT services in 3G-CDMA networks using dynamic fuzzy Q-learning," in IEEE International Symposium on Personal, Indoor and Mobile Radio Communications, Helsinki, 2006, pp. 1–5.

[53] Z. Feng, L. Liang, L. Tan, and P. Zhang, "Q-learning based heterogeneous network self-optimization for reconfigurable network with CPC assistance," *Science in China Series F: Information Sciences*, vol. 52, no. 12, December 2009, pp. 2360–2368.

[54] P. Munoz, R. Barco, I. de la Bandera, M. Toril, and S. Luna-Ramírez, "Optimization of a fuzzy logic controller for handover-based load balancing," in IEEE Vehicular Technology Conference, Budapest, May 2011, pp. 1–5.

[55] A. Klein, "Context Awareness for Enhancing Heterogeneous Access Management and Self-Optimizing Networks," PhD thesis at University of Kaiserslautern, ISBN 978-3-8439-2030-8, 2015.

[56] M. Webb et al., "Smart 2020: Enabling the low carbon economy in the information age," in IEEE/ACM International Symposium on Cluster, Cloud and Grid Computing, London, 2008.

[57] Evolved Universal Terrestrial Radio Access (E-UTRA); Study on energy saving enhancement for E-UTRAN, 3GPP TSG RAN, TR 36.887, V12.0.0, June 2014.

[58] NGMN-Alliance, "5G white paper – Executive version," NGMN 5G Initiative, Technical Report, December 2014, www.ngmn.org/uploads/media/141222

[59] H. Ishii, Y. Kishiyama, and H. Takahashi, "A novel architecture for LTE-B: C-plane/U-plane split and Phantom Cell concept," in IEEE Globecom Workshops, Anaheim 2012, pp. 624–630

[60] Views on Small Cell On/Off Mechanisms, 3GPP TSG RAN, R1-133456, August 2013.

[61] E. Ternon, P. Agyapong, L. Hu, and A. Dekorsy, "Database-aided energy savings in next generation dual connectivity heterogeneous networks," in IEEE Wireless Communications and Networking Conference, Istanbul, 2014, pp. 2811–2816.

[62] E. Ternon, P. Agyapong, L. Hu, and A. Dekorsy, "Energy savings in heterogeneous networks with clustered small cell deployments," in IEEE Wireless Communications Systems, Barcelona, 2014, pp. 126–130.

[63] A. Osseiran, F. Boccardi, V. Braun et al., "Scenarios for 5G mobile and wireless communications: The vision of the METIS project," *IEEE Communications Magazine*, vol. 52, no. 5, pp. 26–35, May 2014.

[64] O. Bulakci, Z. Ren, C. Zhou et al., "Towards flexible network deployment in 5G: Nomadic node enhancement to heterogeneous networks," in IEEE International Conference on Communications, London, June 2015.

[65] O. Bulakci, Z. Ren, C. Zhou et al., "Dynamic nomadic node selection for performance enhancement in composite fading/shadowing environments," in IEEE Vehicular Technology Conference, Seoul, 2014, pp. 1–5.

[66] Y. Sui, A. Papadogiannis, W. Yang, and T. Svensson, "The energy efficiency potential of moving and fixed relays for vehicular users," in IEEE Vehicular Technology Conference, Las Vegas, 2013, pp. 1–7.

[67] J. Li, E. Björnson, T. Svensson, T. Eriksson, and M. Debbah, "Optimal design of energy-efficient HetNets: Joint precoding and load balancing," in IEEE International Conference on Communications, London, June 2015.

[68] J. Li, E. Björnson, T. Svensson, T. Eriksson, and M. Debbah, "Joint precoding and load balancing optimization for energy-efficient heterogeneous networks," *IEEE Transactions on Wireless Communications*, vol. 14, no. 10, pp. 5810–5822, October 2015.

12 Spectrum

Mikko A. Uusitalo, Olav Tirkkonen, Luis Campoy, Ki Won Sung, Hans Schotten, and Kumar Balachandran

This chapter examines and investigates the available choices for spectrum in the 5G system. The relative attractiveness of specific choices of spectrum for various 5G scenarios is identified. The chapter is presented in two parts.

The first part provides an overview of the spectrum landscape leading up to the deployment of 5G. Several new frequency bands with differing regulatory restrictions are expected to be suitable as candidates. It is likely that some bands will require shared use of spectrum. Consequently, this chapter identifies and describes the relevant future modes of spectrum access. Further, this chapter investigates the resulting technical requirements pertaining to shared spectrum use for the 5G system.

The second part of this chapter describes the most important and promising technology components for spectrum access in more detail. A key concept discussed is a flexible spectrum management architecture accompanied by a spectrum-sharing toolbox composed of components that will aid and arbitrate access to spectrum assets, subject to various sharing criteria. Finally, techno-economic analysis of spectrum and related enablers are presented.

12.1 Introduction

Spectrum is a key resource for any radio access network. The availability of spectrum has consistently driven the mobile communication industry through four generations of cellular radio systems, providing telecommunications services with ever-increasing capacity. Early generations of systems were built to provide mobile telephone services, then expanded to handle information services, rich communication services, and media delivery.

The drivers for high network capacity are (1) availability of spectrum while accounting for abundance, cost of acquisition and operation (2) demand in terms of the traffic that is driven through the network, (3) diversity of services that can maintain load in a network across all hours of a day, (4) the multiplexing capability of the Internet Protocol (IP), and (5) the computational ability provided by the advances in semiconductor technology increasing processing power and lowering storage cost. The last four

5G Mobile and Wireless Communications Technology, ed. A. Osseiran, J. F. Monserrat, and P. Marsch. Published by Cambridge University Press. © Cambridge University Press 2016.

of these factors constitute a seemingly endless growth potential to the market, while the first, availability of spectrum, is currently under stress.

The availability of spectrum in suitable frequency ranges, and the efficiency with which it is used, do affect the achievable network capacity and performance. While appropriate market-based pricing of spectrum has consistently motivated the tremendous improvements in network capacity through increased spectral efficiency, scarcity of spectrum could lead to stagnation of the mobile telecommunication industry.

12.1.1   Spectrum for 4G

At the time of writing, 3GPP [1] had defined 47 frequency bands for LTE Release 12, all of which are pertinent to specific national or regional jurisdictions and some of which may overlap. Together, these frequency bands cover roughly 1.3 GHz of independently addressable spectrum.

A single LTE carrier can have a variety of carrier bandwidths ranging from 1.4 MHz to a maximum of 20 MHz. Release 10 of the LTE specification introduced Carrier Aggregation (CA) as a way to increase the system bandwidth for a single LTE deployment. The standard allows the aggregation of multiple carriers, either within the same band or across different bands [2] to increase bandwidth of LTE signals from a maximum of 20 MHz in Release 8 MHz to 100 MHz within Release 10, where 100 MHz corresponds to five LTE carriers. The aggregation of multiple carriers has been the primary method of increasing transmission capability. Indeed, CA has been instrumental in LTE Release 10 achieving the ITU-R qualification for IMT-Advanced [3]. In reality, CA has been useful in improving operator access to spectrum in more modest ways, e.g., in allowing an asymmetrical expansion of bandwidth for downlink, or in expanding a 2×5 MHz block to wider bandwidths. CA also improves capacity through improved multiplexing with cross-carrier scheduling.

The 3GPP standard has specified more than 100 possible CA combinations, with up to three downlink (DL) carriers and two uplink (UL) carriers, including aggregation between FDD and TDD frequency bands. In addition, Release 13 of the 3GPP standard allows LTE to operate in unlicensed spectrum at 5 GHz in a dynamic manner coordinated with licensed bands; this is known as Licensed Assisted Access (LAA). Release 13 allows the use of downlink operation in the unlicensed channel, while Release 14 will introduce bidirectional communication for LAA.

By mid 2015, the bandwidth assigned to Mobile Service [4] varied between 600 MHz and 700 MHz. The allocated spectrum in some regions is shown in more detail in Table 12.1 below.

It must be noted that the numbers in the table are aggregate totals and individual regulatory jurisdictions may differ in the availability of spectrum. It is worth quoting a 2013 report by Plum Consulting [5]:

Our findings from published information indicate that in the high income countries of Europe, North America, and the Asia Pacific region some 500–600 MHz of spectrum is currently assigned

Table 12.1 Spectrum allocated to Mobile Services.

	EU	US	Japan	Australia
700/800 MHz	60	70	72	90
800/900 MHz	70	72	115	95
1500 MHz	–	–	97	–
1800/1900 MHz	150	120	210	150
2 GHz	155	90	135	140
2.3 GHz	–	25	–	98
2.6 GHz	190	194	120	140
Others	–	23	–	–
Total	625	594	749	713

Table 12.2 Frequency ranges identified or allocated for 4G during WRC-15.

Frequency range (MHz)	Allocation (MHz)	Regions
470–694/698	Subject to auction	Some countries in the Americas, APAC[1]
694–790	96	Global band, now includes Region 1[2]
1427–1518	91	51 MHz global, and 91 MHz in Region 2[3]
3300–3400	100	Global band, except in Europe/North America
3400–3600	200	Global band, includes most countries
3600–3700	100	Global band, except Africa, parts of APAC
4800–4990	190	Few countries in APAC and one in America

to mobile services. However, in most middle and low income countries in ASEAN and South Asia only some 300–400 MHz of spectrum is assigned to mobile services. Hence there is a spectrum "divide" of around 200 MHz between these groups of countries.

The situation in 2015 is not very different and is similar to what is provided in the table. The conclusion of World Radiocommunication Conference (WRC) 2015 has effectively increased the amount of spectrum globally assigned for IMT and 4G (by approximately 400 MHz), but the allocations have many regional differences and variations across national boundaries as well. Thus the size of the frequency ranges in Table 12.2 cannot be reasonably summed up or averaged on a regional basis. The process of releasing the spectrum for primary mobile services will follow normal regulatory processes.

The 5G system will benefit from evolution of LTE to expanded bandwidth and enhanced transmission techniques. It is also clear that a new air interface will be introduced to cope with frequency bands capable of supporting new requirements and use cases for mobile broadband. In particular, the *amazingly fast* scenario, and the *super*

---

[1] Asia-Pacific Region.

[2] Region 1 covers Europe, the Middle-East and Africa: Regional arrangements differ and may be less than 96 MHz, e.g. 2×30 MHz in a paired arrangement.

[3] Region 2 covers the Americas, Greenland and some of the eastern Pacific Islands.

real-time and reliable communications scenario will benefit from the introduction of a new air interface.

12.1.2   Spectrum challenges in 5G

Mobile data volumes to be supported by 5G mobile access will dramatically increase, and requirements on coverage, reliability and low latency will be strengthened. Accordingly, the quantity, quality, and usage efficiency of spectrum will be a key success factor for 5G (see Chapters 1 and 2). Mobile communication networks of past generations have largely benefited from designation of dedicated licensed spectrum identified for IMT in the ITU-R Radio Regulations. The ITU-R estimates that the mobile broadband industry will need access to somewhere between 1340–1960 MHz of spectrum by 2020 [6], without considering significant changes in the ability of mobile networks to support use cases differing from those of today. With an ambition to deliver 1000 times higher traffic capacity and 10 to 100 times higher typical user data rate [7], 5G will require significantly more spectrum and wider contiguous bandwidths than what is currently available or anticipated for mobile and wireless communication systems until 2020. While current use cases for mobile broadband are expected to expand into improved support of video and massive machine connectivity, future plans for 5G can admit far greater demands on capacity, driven by extreme use cases for MBB such as VR, large scale surveillance, autonomous vehicles, next-generation backhaul, moving networks, haptic feedback for remote operation of machinery, etc. (see Chapter 2 for information on 5G use cases).

Figure 12.1 presents a possible classification of spectrum access options. The categorization is in four quadrants, with one axis separating horizontal and vertical spectrum sharing, and the other separating exclusive and common rights of use. Exclusive use implies that appropriate rights are granted for licensed access. Common use includes unlicensed use of spectrum. Horizontal spectrum sharing corresponds to situations in which similar users of spectrum have equal priority for spectrum access, whereas vertical sharing corresponds to sharing between disparate services with differing priorities.

Dedicated and licensed spectrum will continue to be preferred for 5G. Licensed spectrum ensures a stable framework for investment so that coverage and QoS can be guaranteed. However, novel regulatory approaches and tools such as Licensed Shared Access (LSA) are expected to address the demand for spectrum [9][10]. LSA is best defined as a way to enable binary sharing between a primary incumbent user and a secondary mobile user, typically separated geographically. The incumbent user is usually a satellite system, radar or a fixed service that has primary allocation to the spectrum. The authorization for secondary use is provided through an LSA repository in the regulatory domain and an LSA controller function in the operator's Operational and Support Subsystem (OSS). The LSA repository implements a geolocation database that supports the authorization mechanism. Innovative ways of spectrum sharing and new

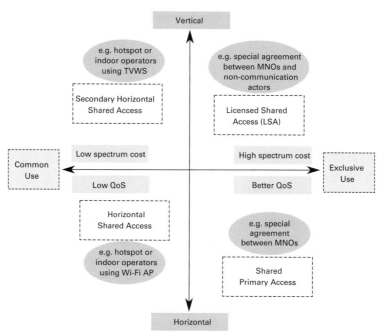

Figure 12.1   Spectrum Access Options [8] to address increasing demand for spectrum. TVWS stands for TV white space, and AP for access point.

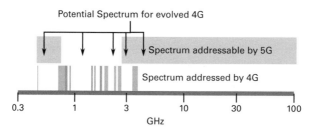

Figure 12.2   Spectrum addressable by 5G is shown in relation to current allocations to terrestrial mobile services.

regulatory approaches such as LSA will take a complementary role when coping with the demand for additional spectrum for mobile services.

Future access to spectrum for 5G in the 4–6 GHz range may have to depend extensively on sharing with incumbent users in many regions where repurposing of spectrum is not possible. Similarly, operation in mmW bands may initially be amenable to sharing arrangements between operators [11]. The experiences in the 60 GHz unlicensed band during the near term should serve to inform industry of future needs. Nevertheless, it will be important for the 5G system to have access to licensed spectrum in the cmW band. It is inevitable that the need for spectrum in the mmW band will eventually explode due to the abundant supply, but near-term preferences lie in expansion within cmW bands from 4 GHz to 30 GHz. Figure 12.2 illustrates the spectrum addressable by 5G systems. The sub-GHz spectrum is useful for some massive MTC scenarios for wide area coverage. The figure also shows detailed maps of 4G bands up to 3800 MHz defined up to 3GPP Release 12. Regions of spectrum that future releases of 4G can address are depicted as well.

Particular challenges arise when enabling complementary access to spectrum bands on a shared basis, especially at higher frequencies (e.g. mmW). Extensions of the LSA approach (e.g. limited spectrum pool and mutual renting) have been identified [11] to add flexibility. In mutual renting the spectrum resources in a band are subdivided into several blocks and each block is licensed to a single operator. Licensees then "rent" parts of their licensed resources to other operators. A spectrum pool allows an operator to obtain an authorization, usually a license, to use up to the whole band on a shared basis with a limited number of other known authorized users. It should be noted that high degrees of sharing do not typically lead to high spectrum efficiency. The 5G system will need high capacity spectrum capable of fielding systems with high spectral efficiency for new broadband services that demand very high throughput. When sharing is introduced, some capacity is lost in the limited ability to coordinate between unrelated networks, and different forms of sharing pose differing costs on the ability of service providers to maximize the capacity of their individual deployments. In this respect, the binary sharing proposed by LSA is less problematic than the introduction of other forms of tiered sharing that may admit horizontal usage. Any form of sharing that severely curtails uniform and ubiquitous deployment of services is suspect in its capability to maximize access to spectrum for a large pool of mobile users. Exclusively licensed allocations offer the best approach to spectrum policy for high capacity and coverage provisioning.

In general, 5G systems need to support different spectrum authorization modes and sharing scenarios in different frequency bands. To deal with this challenge, a Spectrum Sharing Architecture has been defined, together with a Spectrum Toolbox [11]. The toolbox consists of a set of Spectrum Technology Components (TeCs) which will enable the adoption of new mechanisms for spectrum sharing within the sharing architecture. These TeCs have been evaluated related to their capability in identifying new spectrum opportunities, enabling more flexible spectrum usage, as well as their mutual interdependencies. The evaluation shows that a Spectrum Controller entity is needed in the MNO networks, as well as an external Spectrum Coordinator entity.

In addition to IMT and its evolution to xMBB, spectrum needs arising from new use cases with new applications, such as mMTC and uMTC should be taken into account. The aim with 5G is to better address the needs of new vertical applications.

12.2   5G spectrum landscape and requirements

Future regulation and technology development will have to navigate a complex landscape of spectrum availability as illustrated in Figure 12.3. Multiple frequency bands, subject to different regulation, and including various forms of shared spectrum, are

Usage condition	License	Dedicated (Exclusive)	Dedicated	LSA	Unlicensed (Shared)
	Service	Primary	Primary		Secondary
Frequency band classification		Exclusive	Shared		License-free
Primary service allocation		Mobile	Mobile & other service(s)		Other service(s)
Band Examples		880–915/925–960 MHz Band	1452–1492 MHz Band	2300–2400 MHz Band	5150–5350 MHz Band

Figure 12.3 The future spectrum landscape for mobile communication systems consists of different bands made available under different regulatory approaches.

expected to be harnessed by wireless communication systems. To exploit these opportunities, spectrum use under these circumstances will have to develop certain agility and flexibility. Systems have to be able to interoperate across a variety of regulatory models and sharing arrangements.

Radio access networks have historically evolved by increasing the spectrum efficiency and spectrum availability in each new generation. In 5G radio access interfaces, a greater reliance on beamforming is expected to further improve spectral efficiency, especially in the higher cmW and mmW bands. Beamforming can improve the geometry of user plane and dedicated signaling links, i.e. by concurrently improving SNR along the best direction to reach a receiver or to listen to a transmitter, while reducing the amount of energy radiated along channel modes that are unlikely to reach a receiver with significant SNR (cf. Chapter 8 for more information on beamforming). The throughput of a 5G system is also expected to improve for xMBB traffic due to the ability to increase bandwidth by yet another order of magnitude over 4G.

Many desirable spectrum bands are between 3 GHz and 100 GHz, however, these bands are in regions of the electromagnetic spectrum that are assigned to other services, such as radiolocation, fixed services (point-to-point links), active and passive earth exploration, and satellite communication. The region from 3 GHz to 30 GHz is heavily used by many of these services. Therefore, mobile systems need to be able to coexist with incumbent services that may not move away. In some scenarios that involve indoor or short range access, new procedures for spectrum sharing will be needed that allow the benefit of greater coordination between service providers, without severely affecting the utilization efficiency and the achievable capacity. Spectrum sharing will therefore need to be addressed in both horizontal (co-primary), and vertical (with defined primary and secondary services) layers, enabling access to broader continuous spectrum bands for 5G systems. Wide area systems that aim to provide improved coverage as well as high capacity will need exclusivity of access to spectrum, and suitable licensing must be arranged for those.

There is in principle no limit to the degree to which carrier aggregation can be exploited. For example, the limitation on LTE to 100 MHz spectrum occupancy per system can be relaxed. However, it is significantly complex to aggregate over too many bands. As operation grows into higher frequency bands, the filter bandwidths in the front end can increase, and that affords an opportunity to aggregate many contiguous carriers.[4]

For mobile telecommunications, securing spectrum at lower frequencies is normally of high priority, as low-frequency spectrum provides the most economical way of rolling out network coverage, and guaranteeing the planned QoS. This is not surprising since system cost is reduced by achieving the greatest degree of connectivity with a minimum number of points of service. Thus, sub-1 GHz spectrum is very valuable for establishing effective mobile coverage in a cost-efficient way, albeit limited in its ability to serve high data rates [12].

As coverage layers mature in deployment, the availability of wider bandwidth spectrum becomes increasingly important for capacity. The final decision on the best matching spectrum band(s) and access regimes depends on the trade-off between needed bandwidth and the cost of network deployment.

12.2.1   Bandwidth requirements

Bandwidth requirements of key usage scenarios play a very important role in determining spectrum requirements for 5G. The following factors play a role:

1. System Availability requirements and QoS requirements including variability characteristics of diverse use cases,
2. Demands on the system influenced by extreme requirements such as bandwidth, reliability and latency, the density of users and infrastructure, and
3. Spectrum efficiency, e.g. in cases where techniques such as Coordinated Multi-Point Transmission and Reception (CoMP) are used across many antenna layers distributed spatially.

The ITU-R [6] has defined a process for estimating spectrum requirements for mobile communications, based on four essential issues:

- Definition of services,
- Market expectations,
- Technical and operational framework, and
- Spectrum calculation algorithm.

The process forecasts [13] the spectrum requirement for 2020 to be between 1340 MHz and 1960 MHz, depending on the expected variation in the density of users. Some

---

[4] A general rule of thumb is that passband widths for filters can at most be around 3%–4% of the carrier frequency. The state of the art here may improve over time in limited ways.

studies have however questioned the values produced for IMT spectrum demand generated by the ITU-R model [14], based on discrepancies on the traffic density or spectral efficiency figures actually used in the ITU-R model.

In [15], a new approach has been adopted for 5G bandwidth requirement calculation, where different foreseen 5G scenarios have been analyzed, using as main parameters:

- The individual user traffic models (driven by apps and services),
- The density of individual users,
- The QoS targets for each user,
- Frequency band reusability, and
- Spectrum efficiency.

Using the new approach, the most demanding use cases related to xMBB service, such as the ones associated with dense urban information society, resulted in bandwidth requirements varying from 1 GHz to 3 GHz. Other xMBB services in which individual UEs are not that highly demanding (as in shopping malls or stadiums), the bandwidth requirements decreased to values ranging from 200 MHz to 1 GHz. Other services, such as the ones associated with mMTC, are currently expected not to result in significant additional challenges from a bandwidth requirement point of view. In what follows, a rationale is provided for justifying specific allocations in various bands based on anticipated availability.

The 2G/3G/4G systems in service were operated in 2015 on frequency bands below 6 GHz. Meanwhile, the fixed Internet is transitioning to even higher bandwidths and many end users in metropolitan areas can anticipate data rates in excess of 20 Mbps. In extreme cases, fiber deployments are bringing 1 Gbps well within reach of end users. It is natural that the current crop of wireless systems must evolve, at the very least to match, and at best, surpass the ability of the fixed Internet to drive mobile traffic. Many in the industry are convinced that the frontier for new spectrum lies in higher frequency bands than considered before. Currently both cmW bands and mmW bands are considered. Regulators such as the US Federal Communications Commission [16] and Ofcom [17] have issued inquiries on industry interest in access to such bands and a copious record of responses from industry participants exists in the public record. There is interest in gaining consensus across the world on suitable cmW and mmW band options for 5G; the conclusion of WRC-15 had yielded agreement on a number of bands for further study toward WRC-19, viz. 24.25 – 27.5 GHz, 31.8 – 33.4 GHz, 37.0 – 43.5 GHz, 45.5 – 50.2 GHz, 50.4 – 52.6 GHz, 66 – 76 GHz and 81 – 86 GHz.

There are clear advantages to extending the reach of cellular radio systems from 6 GHz to 30 GHz. The foremost of these is the ability of the state of the art in supporting dense macro coverage in these bands; the front-end RF technology capabilities and associated building practices make such bands very appealing, and well worth the challenges in securing bands here for the mobile service during WRC-19.

Millimeter wave bands extend from 30 GHz to 300 GHz by definition. It is generally accepted that the ability of the semiconductor industry to produce mass-market integrated radio chips probably extends reasonably to around 100 GHz. The anticipated access to spectrum allocations in the 30 – 100 GHz band is much greater than below 30 GHz, but the field of application of technologies in such bands is rather limited, as discussed in Chapter 6. It is therefore expected that mmW bands will be occupied initially by very localized dense deployments of relatively low-range radio nodes that offer an abundance of capacity, but limited coverage in a statistical sense. An enterprise or public hot zone deployment of a 5G mmW radio technology should target spectrum blocks of around 500 MHz with at least a total allocation of 2 GHz or more per band to allow aggregation of bandwidth. As the state of the art improves, the range and coverage of systems in such bands may improve. Chapter 6 has more information on the hardware aspects and proposed building practices in such bands.

12.3   Spectrum access modes and sharing scenarios

Current radio regulations provide four frequency band classifications: Exclusive bands, Shared bands, License-free bands, and Receive-only bands[5]. Two service categories, Primary service and Secondary service are defined on the basis of ITU-R allocations in the regions.

Multiple frequency bands, subject to different regulatory regimes including various forms of shared spectrum concepts as illustrated in Figure 12.4, are expected to be available for mobile communication systems. Thus, 5G system design requires a high degree of flexibility to be capable of operating under different regulatory models and usage scenarios.

In general the use of radio spectrum can be authorized in two ways: Individual Authorization (Licensed) and General Authorization (License Exempt or Unlicensed). Authorization modes recognized as relevant for wireless communications are Primary user mode, LSA mode and Unlicensed mode.

Five basic spectrum usage scenarios are identified in Figure 12.4 for these authorization modes: dedicated licensed spectrum, limited spectrum pool, mutual renting, vertical

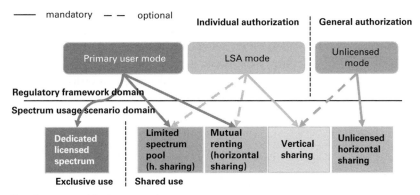

Figure 12.4   Spectrum usage/sharing scenarios.

---

[5] These are typically not important for mobile communications, which we restrict to be two-way.

sharing, and unlicensed horizontal sharing. The figure depicts the relationship between parts of the domains that are either necessary or supplementary, using solid and dotted lines respectively. More complex spectrum usage scenarios can be supported by combinations of the basic scenarios and authorization modes.

12.4   5G spectrum technologies

The focus in this section is on 5G spectrum technologies, in particular on new spectrum access modes as well as improved usage of new frequency bands. Improvements in usage of spectrum arise from exploitation of local variations in availability, which can happen in time, frequency and space. The combined usage of old and new frequency bands should also be taken into consideration. Therefore the main technical challenges that are seen for 5G in the area of spectrum are:

- Extracting value from new spectrum opportunities,
- Implementing efficient spectrum sharing, and
- Combining different spectrum assets for comprehensive treatment of coverage, mobility and capacity.

Effective spectrum management will need an internal Spectrum Controller entity in the MNO network, and an external Spectrum Coordinator [11].

12.4.1   Spectrum toolbox

A 5G system has to support all authorization modes and spectrum usage/sharing scenarios described in the previous section. A set of enablers or "tools" may be defined that need to be added to the typical portfolio of technical capabilities of today's cellular systems; see Figure 12.5. These enablers (presented in details in Section 12.4.2) either directly relate to spectrum sharing operation in a specific frequency range, or generally aim at providing a frequency agile and coexistence/sharing friendly radio interface design. In a particular situation, a specific technology may not have to support all the identified scenarios and hence only needs to implement a subset of technical enablers.

Besides enablers for spectrum sharing in order to cope with developments in spectrum regulation, expansion into higher frequency ranges is an additional novelty that is expected for 5G. It is obvious that system design and network building practices have

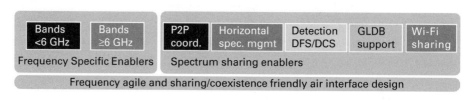

Figure 12.5   Spectrum enablers forming part of the Spectrum toolbox.

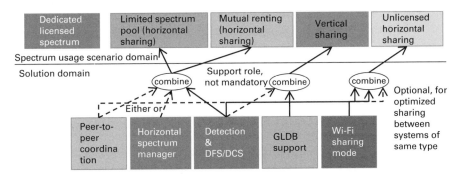

Figure 12.6   Technical enablers match to spectrum sharing scenarios.

to change due to the significantly different radio propagation conditions at higher frequencies. Hence, frequency-specific enablers for access to frequencies above 6 GHz are expected.

Figure 12.6 shows the different tools required to enable spectrum sharing, and how they relate to the scenarios described above. In some sharing scenarios, only one enabler is needed, but for others a set of enablers is needed. Note that some of the enablers and relations are optional (indicated by a dashed line connecting them to the respective scenario) meaning that they are not strictly required but may be helpful or desirable, or subject to design choices.

12.4.2   Main technology components

There is a rich variety of technologies related to spectrum sharing, cognitive radio and D2D that are addressed in literature. Many of these are pertinent to emerging 5G spectrum access. To fully populate the spectrum toolbox to be used in 5G cellular systems, some innovative new technologies are needed as well.

Wi-Fi sharing mode has special characteristics due to the unlicensed nature of spectrum. Ultimately a MNO does not have control of access of spectrum, and there is a requirement to leave transmission opportunities to other users. The CSMA/CA MAC of Wi-Fi [18] is in itself a prototypical spectrum sharing technology, addressing these issues. Listen-before talk and RTS/CTS techniques have been studied in the context of LTE frame structures [19]. LTE in unlicensed bands is being standardized in the form of Licensed Assisted Access (LAA) [20]. A precursor to the system will also be defined by the LTE-Unlicensed (LTE-U) Forum, an industry body headed by the operator Verizon [21]. In wider unlicensed bands, the fundamental problems of selecting carriers and allocating power on these have been identified [22]. These problems will remain unresolved in general when considering technologies designed by different standardization bodies employing different medium access protocols. A degree of cooperation between equipment of different MNOs is desirable, if multiple LTE-U networks are deployed in the same physical space. As depicted in Figure 12.6, a P2P coordination protocol may be used for this.

Detection and dynamic frequency selection (DFS)/dynamic channel selection (DCS) is a general tool for spectrum management. As observed in [23], cognitive spectrum sensing technologies developed in the cognitive radio context [24] are applicable to multi-operator sharing scenarios. In dynamical sharing, it is essential to identify the total interference level on a carrier, and potentially to be able to distinguish the source of interference. Information on the interference level from other MNO networks can be used as inputs for the Horizontal Sharing Manager (HSM), and P2P coordination protocols.

For dynamic vertical sharing, **Geolocation Database (GLDB) support** is essential. A GLDB is about storing information on the geographic locations where spectrum is available, or used by the incumbents. Other use of the spectrum is allowed at other locations, if it does not hurt the usage of the incumbent. For this, the interference caused to the incumbent is estimated. GLDB technologies are used for incumbent protection in TV white spaces [25] and similar solutions apply for secondary horizontal shared access, and for LSA repositories. Detailed maps of the radio environment need to be created [26] to be used as a basis for dynamic LSA negotiations. State of the art GLDB are created taking user density and terrain-based propagation into account (TeC06 in [11]). GLDB technologies show significant promise in providing support for Ultra-Reliable Communication in Vehicle-to-Vehicle (V2V) communication (TeC20 in [11]). By having access to GLDB, vehicles in a given region know the available spectrum for reliable V2V. When a MNO negotiates with a GLDB for access to vertically shared spectrum, clustering methods may be used to identify the groups of cells with similar requirements for spectrum and similar interference relations to LSA incumbents (TeC19 in [11]).

Spectrum sharing between MNOs can be realized either in a distributed or centralized fashion. In distributed operation, MNOs decide on the spectrum used by employing a **Peer-to-peer (P2P) coordination protocol** on top of sensing technologies. When operating on a specific carrier, which may be used by multiple MNOs, Physical/MAC layer coordination protocols may be used, in a similar fashion as in Wi-Fi sharing mode. Higher layer P2P coordination protocols would attempt to select suitable carriers for the MNOs. Ideally, this would be performed in a coordinated fashion. Full cooperation of MNO networks has been widely addressed in the literature. Cooperation based on exchange of full Channel State Information (CSI) was addressed in time domain [27], frequency domain [28] and with spatial [29][30] sharing. Cooperative game theory for spectrum sharing was discussed in [31], where full CSI may not be exchanged, but information about MNO utility, i.e. the gain an MNO may achieve by sharing, would have to be exchanged. Operator utility is related to MNO-specific RAN optimization objectives, and MNOs may not be willing to share neither utility nor channel state information with competitors. Approaches requiring limited information exchange have been developed [32][33]. In these, information such as the identity of certain interferers, or the relative prioritization of channels at certain locations is indicated to other MNOs. Further, a standardized protocol is agreed as a reaction to such information. This type of spectrum management is particularly well suited to multi-operator D2D scenarios (cf. Chapter 5), where devices served by multiple MNOs communicate directly, and MNOs need to agree about the spectrum resources used for this [34].

Centralized solutions for spectrum sharing between MNOs require a **Horizontal Spectrum Manager** (HSM). The HSM is operated by an external party, such as a spectrum broker, or a regulatory body. In some cases, some parts of MNO networks, such as their UDNs, may be geographically separated, and no coordination is needed between those networks. For this, dynamic frequency selection (DFS) or GLDB functionalities are needed, to assess the inter-operator separation, and a HSM may be used to decide, whether coordination is needed or not (cf. TeC03 in [11]). **Dynamic Frequency selection** is typically defined as the detection of energy or waveform signatures of primary users who are capable of pre-empting secondary spectrum use. If coordination of spectrum access is needed, centralized solutions to spectrum sharing are invariably related to value assessments of the spectrum made by MNOs. Ultimately, auction mechanisms would be used to resolve conflicts between the needs of different MNOs, based on these value assessments [35]. Ontological methods considering the specifics of MNO networks (cf. TeC17 in [11]), and fuzzy logic for finding MNO-specific strategies may be used [36]. In an MNO network, clustering methods may be used to find the geographical granularity of negotiations with the HSM (cf. TeC19 in [11]).

Various **technologies to enable flexible spectrum use with advanced air interfaces** may be relevant for 5G access. RF-coexistence is an important aspect of flexible spectrum use [37]. If there are significant amounts of fragmented spectrum available in narrow bandwidth pieces, there may be benefits of using multicarrier waveforms other than OFDM, which limit the out-of-band emissions [38]. Many DFS technologies and dynamic P2P coordination protocols would benefit from synchronization of radio frames between MNO networks. This can be done effectively in a distributed fashion over-the-air, based on listening to transmissions from all active base stations [39].

Some specific enablers are needed for **spectrum management of networks operating in high carrier frequencies**. An entity in the MNO network may, for example, estimate the fraction of Line-of-Sight (LOS) connections in the network. Depending on this, a carrier frequency is selected. Higher carrier frequencies are used in parts of the network with higher probability of LOS (cf. Section 3.5 in [40]).

12.5   Value of spectrum for 5G: a techno-economic perspective

Radio spectrum has been a valuable asset for MNOs, accounting for a significant part of capital expenditure. The valuation of spectrum directly relates to the decision of how much to spend on spectrum acquisition, and thus is essential for the investment strategies of the service providers. The importance of the spectrum valuation will be even more significant in 5G due to much higher bandwidth demand. Furthermore, it will be very difficult to assess the value of spectrum for 5G because of a much broader range of frequency bands to be exploited and more diverse ways of sharing and utilizing the spectrum. In this section, a techno-economic perspective of the spectrum valuation is briefly described. Here, the main focus is on xMBB services.

Spectrum auctions in the past years showed that differing prices were paid for the same chunk of frequency band depending on time and location. The varying economic conditions, the uptake of mobile technologies in the market and the interest of operators in the particular bands being offered tend to affect perception of value and pricing. This is a clear indicator that an absolute valuation for spectrum, i.e. a quantifiable €/Hz (Euro per Hertz) that may be inflation adjusted over time does not exist. A reasonable approach to address the changing dependency on business conditions is to measure the opportunity cost of alternative means of achieving a target system performance [41]. It is well known that there are three fundamental directions to the network capacity expansion: higher spectral efficiency, denser deployment of base stations (BSs), and more spectrum. Thus, to some extent, spectrum can be substituted by infrastructure investment, e.g. equipment upgrade and more BS sites. However, this picture is greatly complicated by the reality that large amounts of spectrum will be allocated at progressively higher frequency bands, and such bands may themselves be limited by poor propagation within the engineering constraints of system complexity and cost. The result of such analyses is calculated as the engineering value of spectrum in the literature.

Calculation of the engineering value of spectrum requires knowledge of the relationship between the cost of the infrastructure and spectrum. As the infrastructure cost is often modeled to be linearly related to the number of BSs [42], the problem can be simplified by obtaining the rate of substitution between the BS density and bandwidth to satisfy a certain performance target. A widely accepted assumption is that they are linearly exchangeable. This means that a combination of x times more bandwidth and y times denser BSs always yields xy times more capacity (see, e.g. [43], which summarizes the industry view on capacity expansion strategies). This assumption works well with traditional macro-cellular systems where BSs are fully loaded in busy hours. In an interference-limited system with saturated traffic, adding one more BS does not affect the SINR distribution of the system, whereas it increases the number of transmitters in the area. Thus, the area capacity increases linearly with the number of BSs [44].

Unfortunately, it seems that this linear relationship does not hold for 5G UDNs providing amazingly fast xMBB services [45]. Two different deployment regimes are considered: a sparse regime where the number of instantaneously active subscribers is larger than that of BSs, and an opposite dense regime. The sparse regime represents traditional macro-cellular networks, whereas the dense regime describes 5G UDN. Further, a cellular network is considered with homogeneous BS deployment and uniform user distribution. BS density varies from 0.01 to 100 times of the user density. System performance depends on cell utilization, which is a function of the ratio between BS and user densities. Figures 12.7 and 12.8 show the numerical results illustrating the relationship between spectrum and BS density in the sparse and dense regime, respectively [45]. While the same substitution rate is maintained throughout the sparse regime, further densification of BSs in the already dense environment turns out to be extremely ineffective. Even more than 20 times of densification is required for doubling the user data rate, which can be equivalently achieved by doubling the bandwidth. This is due to the diminishing spatial multiplexing

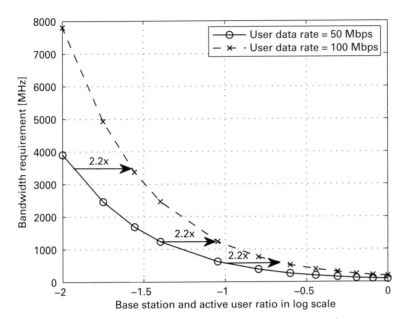

Figure 12.7   Bandwidth requirement as a function of BS density in the sparse regime.

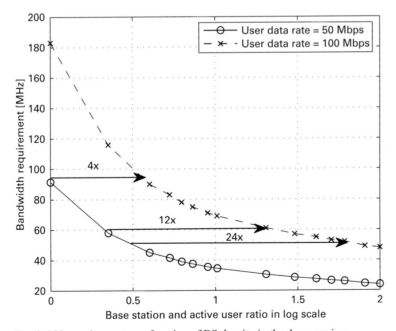

Figure 12.8   Bandwidth requirement as a function of BS density in the dense regime.

gain in the dense regime. In UDN, BSs remain idle during most of time in order to serve bursty traffic with very high data rate requirements. In this circumstance, the notion of area capacity is less meaningful, and adding more BSs does not immediately enhance system performance.

The analysis strongly motivates the need for more efficient use of spectrum (such as dynamic TDD), exploration of higher frequency bands (such as mmW), and flexible spectrum sharing for 5G systems, especially for amazingly fast services. Valuation of spectrum under various spectrum access modes and sharing options is a challenging task for 5G research, and may be impossible to perform in absolute terms for the time being. In general, the principle of the engineering value of spectrum is applicable to most scenarios. However, caution is needed in several aspects. For example,

- MNOs have limited ability to observe or control interference in shared spectrum environments, which may affect the QoS of the systems. An important question for MNOs is whether they need inter-operator interference coordination. The cost of cooperation, which is usually perceived very high in the business strategy domain, should be taken into account [46].
- Spectrum sharing can allow new entrants into the market, such as local MNOs providing indoor and hotspot capacity. For such newcomers, it may be difficult to measure their opportunity costs. An alternative would be to estimate the potential revenue, which can be made possible by the spectrum sharing [47].

12.6   Conclusions

Spectrum requirements

Sufficient amounts of spectrum need to be available in low spectrum bands in order to satisfy the requirement for seamless coverage of the 5G services xMBB and mMTC. Additional spectrum below 6 GHz is essential to cope with the mobile traffic in urban and suburban areas, and in medium density hotspots. Spectrum above 6 GHz is necessary for enabling wireless access in high density usage scenarios, i.e. to fulfill the high contiguous bandwidth demand for xMBB, and for wireless backhaul solutions for high capacity ultra-dense small cell networks.

The most demanding use cases related to xMBB services, such as the ones associated with dense urban information society, need between 1 GHz and 3 GHz of bandwidth. Other xMBB services in which individual users are not that highly demanding (as in shopping mall or stadium scenarios), the bandwidth requirements decrease to values ranging from 200 MHz to 1 GHz. Specific requirements that are pertinent to cmW or mmW bands are provided in the chapter. Other services, such as the ones associated with mMTC, do not present a challenge from bandwidth requirement point of view.

Types of spectrum

For the main 5G service of xMBB, a mixture of frequency spectrum comprising lower bands for coverage purposes and higher bands with large contiguous bandwidth for capacity is required. Exclusive licensed spectrum is essential to guarantee coverage obligations and QoS. This has to be supplemented by other licensing regimes, e.g. LSA or unlicensed access (e.g. LAA) to increase overall spectrum availability.

Frequency spectrum below 6 GHz is most suitable for mMTC applications. Spectrum below 1 GHz is needed to cope with severe coverage limitations. Exclusive licensed spectrum is the preferred option. However, other licensing regimes might be considered depending on specific application requirements.

Licensed spectrum is considered most appropriate for uMTC. For safety V2V and V2X communication the frequency band 5875 – 5925 MHz harmonized for Intelligent Transport Systems (ITS) [48] is an option.

There is evidence from studies that the 5G radio access technologies, capable of creating a new coverage system for extreme MBB services, are feasible within cmW bands. For mmW bands, short range coverage in localized and densely deployed environments may be more suitable, while these bands continue to be of use for longer range line-of-sight use for fixed links.

Licensing

Exclusive licensed spectrum is essential for the success of 5G to provide the expected QoS and to secure investments. Shared spectrum can be considered as well provided that predictable QoS conditions are maintained, e.g. by an LSA regime. License-exempt spectrum might be suitable as a supplementary option for certain applications.

References

[1] 3GPP TS 36.101, "User Equipment (UE) radio transmission and reception," Technical Specification TS 36.101 V12.9.0, Technical Specification Group Radio Access Network, October 2015.

[2] 3GPP, LTE-Advanced description, June 2013, www.3gpp.org/technologies/key words-acronyms/97-lte-advanced

[3] International Telecommunications Union Radio (ITU-R), "Requirements related to technical performance for IMT-Advanced radio interface(s)," Report ITU-R M2134, November 2008.

[4] J. S. Marcus, J. Burns, F. Pujol, and P. Marks, "Inventory and review of spectrum use: Assessment of the EU potential for improving spectrum efficiency," WIK-Consult report, September 2012.

[5] P. Marks, S. Wongsaroj, Y.S. Chan, and A. Srzich, "Harmonized Spectrum for mobile Service in ASEAN and South Asia: An international comparison," Plum Consulting Report for Axiata Berhad, August 2013.

[6] International Telecommunications Union Radio (ITU-R), "Methodology for calculation of spectrum requirements for the terrestrial component of International Mobile Telecommunications," Recommendation ITU-R M.1768-1, April 2013.

[7] A. Osseiran, F. Boccardi, V. Braun, K. Kusume, P. Marsch, M. Maternia, O. Queseth, M. Schellmann, H. Schotten, H. Taoka, H. Tullberg, M. A. Uusitalo, B. Timus, and M. Fallgren, "Scenarios for 5G mobile and wireless communications: The vision of the METIS project," *IEEE Com Mag*, vol. 52, no. 5, May 2014.

[8] A. A. W. Ahmed and J. Markendahl, "Impact of the flexible spectrum aggregation schemes on the cost of future mobile network," in International Conference on Telecommunications, Sydney, April 2015, pp. 96–101.

[9] CEPT ECC, "Licensed Shared Access (LSA)," ECC Report 205, February 2014, http://www.erodocdb.dk/Docs/doc98/official/pdf/ECCREP205.PDF

[10] Radio Spectrum Policy Group, "RSPG opinion in licensed shared access," RSPG Opinion, November 2013, http://rspg-spectrum.eu/rspg-opinions-main-deliverables/

[11] ICT-317669 METIS project, "Future spectrum system concept," Deliverable D5.4, April 2015, www.metis2020.com/documents/deliverables/

[12] International Telecommunications Union Radio (ITU-R), "Assessment of the global mobile broadband deployments and forecasts for international mobile telecommunications," Report ITU-R M.2243, November 2011.

[13] International Telecommunications Union Radio (ITU-R), "Future spectrum requirements estimate for terrestrial IMT," Report ITU-R M.2290-0, December 2013.

[14] LS Telecom AG., "Mobile spectrum requirement estimates: Getting the inputs right," September 2014.

[15] ICT-317669 METIS project, "Description of the spectrum needs and usage principles," Deliverable D5.3, April 2015, https://www.metis2020.com/documents/deliverables

[16] FCC, "In the matter of use of spectrum bands above 24 GHz," FCC 14–154, GN Docket No 14–177, October 2014.

[17] Ofcom, "Laying the foundations for next generation mobile services: Update on bands above 6 GHz," April 2015.

[18] K.-C. Chen, "Medium access control of wireless LANs for mobile computing," IEEE Networks, vol. 8, no. 5, pp. 50–63, September/October 1994.

[19] R. Ratasuk et al., "License-exempt LTE deployment in heterogeneous network," in International Symposium on Wireless Communications Systems, Paris, August 2012, pp. 246–250.

[20] 3GPP TSG-RAN, "Chairman summary," 3GPP workshop on LTE in unlicensed spectrum, RWS-140029, June 2013.

[21] Alcatel-Lucent, Ericsson, Qualcomm Technologies, Samsung and Verizon "Coexistence study for LTE-U SDL," LTE-U Technical report v. 1.0, February 2015, www.lteuforum.org/uploads/3/5/6/8/3568127/lte-u_forum_lte-u_technical_report_v1.0.pdf

[22] R. Etkin, A. Parekh, and D. Tse, "Spectrum sharing for unlicensed bands," IEEE Journal on Selected Areas in Communications, vol. 25, no. 3, pp. 517–528, April 2007.

[23] P. Karunakaran, T. Wagne, A. Scherb, and W. Gerstacker, "Sensing for spectrum sharing in cognitive LTE-A cellular networks," in IEEE Wireless Communications and Networking Conference, Istanbul, April 2014, pp. 565–570.

[24] T. Yucek and H. Arslan, "A survey of spectrum sensing algorithms for cognitive radio applications," IEEE Communications Surveys & Tutorials, vol. 11, no. 1, pp. 116–130, January 2009.

[25] D. Gurney et al., "Geo-location database techniques for incumbent protection in the TV white space," in IEEE International Dynamic Spectrum Access Networks Symposium, Chicago, October 2008, pp. 1–9.

[26] K. Ruttik, K. Koufos, and R. Jäntti, "Model for computing aggregate interference from secondary cellular network in presence of correlated shadow fading," in IEEE International Symposium on Personal, Indoor and Mobile Radio Communications, Toronto, September 2011, pp. 433–437.

[27] G. Middleton, K. Hooli, A. Tölli, and J. Lilleberg, "Inter-operator spectrum sharing in a broadband cellular network," in IEEE International Symposium on Spread Spectrum Techniques and Applications, Manaus, August 2006, pp. 376–380.

[28] L. Anchora, L. Badia, E. Karipidis, and M. Zorzi, "Capacity gains due to orthogonal spectrum sharing in multi-operator LTE cellular networks," in International Symposium on Wireless Communications Systems, Paris, August 2012, pp. 286–290.

[29] E. A. Jorswieck et al., "Spectrum sharing improves the network efficiency for cellular operators," *IEEE Communications Magazine*, vol. 52, no. 3, pp. 129–136, Mar. 2014.

[30] S. Hailu, A. Dowhuszko, and O. Tirkkonen, "Adaptive co-primary shared access between co-located radio access networks," in International Conference on Cognitive Radio Oriented Wireless Networks, Oulu, June 2014, pp. 131–135.

[31] J. E. Suris, L. A. DaSilva, Z. Han, and A. B. MacKenzie, "Cooperative game theory for distributed spectrum sharing," in IEEE International Conference on Communications, Glasgow, June 2007, pp. 5282–5287.

[32] G. Li, T. Irnich, and C. Shi, "Coordination context-based spectrum sharing for 5G millimeter-wave networks," in International Conference on Cognitive Radio Oriented Wireless Networks, Oulu, June 2014, pp. 32–38.

[33] B. Singh, K. Koufos, O. Tirkkonen, and R. Berry, "Co-primary inter-operator spectrum sharing over a limited spectrum pool using repeated games," in IEEE International Conference on Communications, London, June 2015, pp. 1494–1499.

[34] B. Cho et al., "Spectrum allocation for multi-operator device-to-device communication," in IEEE International Conference on Communications, London, June 2015, pp. 5454–5459.

[35] J. Huang, R. Berry, and M. Honig, "Auction-based spectrum sharing," *Mobile Networks and Applications*, vol. 11, no. 3, pp. 405–418, June 2006.

[36] K. Chatzikokolakis et al., "Spectrum sharing: A coordination framework enabled by fuzzy logic," in International Conference on Computer, Information, and Telecommunication Systems, Gijón, July 2015, pp. 1–5.

[37] S. Heinen et al., "Cellular cognitive radio: An RF point of view," in IEEE International Workshop on Cognitive Cellular Systems, Rhine river, September 2014.

[38] J. Luo, J. Eichinger, Z. Zhao, and E. Schulz, "Multi-carrier waveform based flexible inter-operator spectrum sharing for 5G systems," in IEEE International Dynamic Spectrum Access Networks Symposium, McLean, April 2014, pp. 449–457.

[39] P. Amin, V. P. K. Ganesan, and O. Tirkkonen, "Bridging interference barriers in self-organized synchronization," in IEEE International Conference on Self-Adaptive and Self-Organizing Systems, London, September 2012, pp. 109–118.

[40] ICT-317669 METIS project, "Final report on network-level solutions," Deliverable D4.3, April 2015, www.metis2020.com/documents/deliverables/

[41] B. G. Mölleryd and J. Markendahl, "Analysis of spectrum auctions in India: An application of the opportunity cost approach to explain large variations in spectrum prices," *Telecommunications Policy*, vol. 38, pp. 236–247, April 2014.

[42] J. Zander, "On the cost structure of future wireless networks," in IEEE Vehicular Technology Conference, Phoenix, May 1997, vol 3, pp. 1773–1776.

[43] J. Zander and P. Mähönen, "Riding the data tsunami in the cloud: Myths and challenges in future wireless access," *IEEE Communications Magazine*, vol. 51, no. 3, pp. 145–151, March 2013.

[44] A. Ghosh et al., "Heterogeneous cellular networks: From theory to practice," *IEEE Communications Magazine*, vol. 50, no. 6, pp. 54–64, June 2012.

[45] Y. Yang and K. W. Sung, "Tradeoff between spectrum and densification for achieving target user throughput," in IEEE Vehicular Technology Conference Spring, Glasgow, May 2015, pp. 1–6.

[46] D. H. Kang, K. W. Sung, and J. Zander, "High capacity indoor and hotspot wireless systems in shared spectrum: A techno-economic analysis," *IEEE Communications Magazine*, vol. 51, no. 12, pp. 102–109, December 2013.

[47] A. A. W. Ahmed, J. Markendahl, and A. Ghanbari, "Investment strategies for different actors in indoor mobile market in view of the emerging spectrum authorization schemes," in European Conference of the International Telecommunications Society, Florence, October 2013, pp. 1–19.

[48] CEPT ECC, "ECC Decision (08)01: The harmonised use of the 5875–5925 MHz frequency band for Intelligent Transport Systems (ITS)," ECC/DEC/(08)01, March 2008, www.erodocdb.dk/docs/doc98/official/pdf/ECCDec0801.pdf

13 The 5G wireless propagation channel models

Tommi Jämsä, Jonas Medbo, Pekka Kyösti, Katsuyuki Haneda, and Leszek Raschkowski

5G wireless propagation channel models are crucial for evaluation and comparison of the performance of different technology proposals, and for assessment of the overall performance of the foreseen 5G wireless system. This chapter elaborates on the main challenges of 5G channel modeling and describes the new proposed channel models.

Two different channel-modeling approaches, stochastic and map-based, are detailed. The purpose of the stochastic approach is to extend the traditional well-established WINNER [1] type of modeling for 5G. Some of the 5G requirements may however be hard to meet with stochastic modeling. For that reason, the map-based model, which is based on ray tracing, was also developed [2]. In order to parameterize and evaluate the models, extensive measurement campaigns have been conducted. The detailed description of the METIS channel models can be found in [2].

13.1 Introduction

The envisioned scenarios, use cases and concepts of 5G wireless communications, as described in Chapter 2, set new critical requirements for radio channel and propagation modeling. Some of the more important and fundamental requirements are the support of

- extremely wide frequency ranges from below 1 GHz up to 100 GHz,
- very wide bandwidths (> 500 MHz),
- full 3-dimensional and accurate polarization modeling,
- spatial consistency, i.e. the channel evolves smoothly without discontinuities when the transmitter and/or receiver moves or turns, for supporting highly dense scenarios,
- coexistence of different types of links in the same area such as cellular links with different cell sizes and Device-to-Device (D2D) connections,

5G Mobile and Wireless Communications Technology, ed. A. Osseiran, J. F. Monserrat, and P. Marsch. Published by Cambridge University Press. © Cambridge University Press 2016.

- dual-end mobility, i.e. both link-ends move simultaneously and independently, for supporting D2D and Vehicle-to-Vehicle (V2V) connections as well as moving base stations,
- high spatial resolution and spherical waves for supporting very large antenna arrays, massive MIMO and beamforming,
- elevation extension for supporting 3D models and
- specular scattering characteristics especially for high frequencies.

Moreover, the model should provide spatially consistent characteristics for different topologies and between different users. Realistic small-scale fading, for example, would require multiple users to share a common set of scattering clusters.

Currently recognized and widely used channel models, like the 3GPP/3GPP2 Spatial Channel Model (SCM) [3], WINNER [1][4], ITU-R IMT-Advanced [5], 3GPP 3D-UMi and 3D-UMa [6], and IEEE 802.11ad [7], are found to be inadequate for 5G in that they do not meet these requirements [8]. While common channel models such as SCM, WINNER and IMT-Advanced were designed for frequencies of up to 6 GHz, there are other models available, such as the IEEE 802.11ad, which focus on the 60 GHz band. Whereas those models are applicable only for a specific frequency range, the METIS channel models described in this chapter cover the full frequency range from cellular bands below 1 GHz up to 100 GHz.

The parameterization of the stochastic model is derived from the literature and from propagation measurements conducted for relevant scenarios, while the map-based model is justified by comparing its outputs with measurements. It should be noted that the stochastic model is a Geometry-based Stochastic Channel Model (GSCM) further developed from WINNER and IMT-Advanced. The map-based model uses ray tracing in a simplified 3D model of the environment. In addition, a hybrid model is also proposed in [2] to allow scalability of the model in which users of the model may combine elements from the map-based model and partly utilize the stochastic model.

The key objectives of this chapter are to

- identify the 5G propagation channel requirements,
- provide channel measurements at various bands between 2 GHz and 60 GHz, and to
- derive channel models fulfilling the requirements.

13.2   Modeling requirements and scenarios

Two factors determine requirements on the channel model. The first factor considers the scenarios from the environment and user perspective, while the second one concerns the technology components needed to provide the required end-user services. The requirements for 5G channel models are discussed in Section 13.2.1. The propagation scenarios are defined based on the usage scenarios and the technology components, and are described in Section 13.2.2.

Table 13.1 Requirements for 5G channel models.

Category	Requirement
Scenarios	Wide range of propagation environments
Spectrum	Frequency range from below 1 GHz up to 100 GHz
	Support of system bandwidths greater than 500 MHz (high delay resolution)
Antenna	Support of very large array antennas (non-planar (spherical) waves and high angular resolution)
	Modeling of large arrays beyond consistency interval
System	Spatial consistency of Large Scale (LaS) parameters for small cells, moving cells, D2D, M2M, V2V, MU-MIMO, etc.
	Dual-end mobility
	Moving environment
General	Physical realism (the model needs to be validated by the means of a sufficient amount of measurement data)
	Reasonable complexity for a given application

13.2.1 Channel model requirements

The main challenges of the modeling approach are to account for higher frequencies and wider bandwidths, together with much larger array antennas in terms of element numbers as well as of the physical size with respect to the wavelength. The wide bandwidth and large antenna array size brings about the need for a drastically better channel model resolution in both the delay and spatial domains. The main 5G channel model requirements are shown in Table 13.1 and discussed in detail in the following subsections. Here, the consistency interval means the maximum distance within the large-scale parameters can be approximated to be constant.

13.2.1.1 Spectrum

In addition to the frequencies below 6 GHz, the following bands above 6 GHz have been prioritized [9]: 10 GHz, 18–19 GHz, 28–29 GHz, 32–33 GHz, 36 GHz, 41–52 GHz, 56–76 GHz, and 81–86 GHz, from which the high priority bands are 32–33 GHz, 43 GHz, 46–50 GHz, 56–76 GHz and 81–86 GHz. The ultimate goal for the 5G channel model is to define continuous functions for all channel model parameters and propagation effects for the full frequency range from below 1 GHz up to 100 GHz. As there are greater bandwidths available in the high frequency regime, systems with bandwidths of 500 MHz and above are most likely.

13.2.1.2 Antenna

Current channel models [1][5] assume plane wave propagation (far field), which only holds for small array antenna sizes (i.e. only the phase difference between antenna elements is taken into account). An important technology component of 5G mobile communications is the use of very large array antennas for massive MIMO and pencil beamforming (cf. Chapter 8 for more information on massive MIMO). For these highly directive antennas or large array antennas, a substantially non-realistic

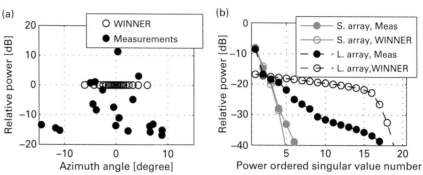

Figure 13.1   (a) Power angular distribution according to the WINNER model and measurements [10]. (b) Corresponding MIMO singular value distributions at 40 GHz carrier frequency for a Small (S) and a Large (L) array antenna, respectively of sizes (0.1 m × 0.1 m) and (1 m × 1 m).

performance will be experienced using current modeling approaches. These channel models need to be improved in the angular resolution. Furthermore, large arrays require non-planar (spherical) wave modeling replacing the commonly used plane wave approximation. Very large arrays may lead to a situation where propagation conditions, like e.g. shadowing, vary over the array, i.e. the array is larger than the consistency interval of the channel. For very large arrays and massive MIMO, the following parameters of the channel paths have to be modeled accurately: azimuth and elevation angles, amplitude and polarization, delay, and correlation distances of large-scale parameters.

The importance of an improved directional channel modeling is illustrated in Figure 13.1. In particular, Figure 13.1(a) shows the distributions of path power versus azimuth angle for a real measured channel (urban macro-cell) and the WINNER model, while Figure 13.1(b) shows the corresponding distributions of MIMO channel singular values.

For the small antenna array (0.1 m × 0.1 m) the singular value distributions of measurements and the WINNER model agree well. However, for the large array (1 m × 1 m) the distribution of the WINNER model is nearly optimal (uniform), which is clearly unrealistic comparing with the measured channel. It is clear that, in order to provide adequate modeling for 5G, the power angular distribution needs to be substantially revised for improved agreement with measurements.

13.2.1.3   System

The 5G communication system is going to consist of various link types. Important aspects are the expected decrease of cell sizes from traditional macro- and microcells to pico- and femto-cells and future moving base stations as well as direct D2D connections between user terminals. These various types of links will co-exist in the same area. Moreover, the link density is expected to grow tremendously. All these features set new requirements to channel modeling. The current most commonly used channel models [1][3][5] are drop-based, meaning that the scattering environment is randomly created for each link. The corresponding performance of spatial techniques like MU-MIMO is exaggerated, because the model assumes independent scatterers also in the case of nearby mobiles, which is not the case in reality. As mentioned earlier, spatial consistency means that the channel evolves smoothly without discontinuities when the transmitter and/or receiver moves or turns. It also means that channel characteristics are similar in closely located links, e.g. two nearby users seen by the same base station.

Dual-end mobility and moving environments require different Doppler models, different spatial correlation of LaS and Small-Scale (SS) parameters than in the conventional cellular case.

13.2.1.4   Additional requirements

The model should be realistic, i.e. based on real propagation measurements or physical propagation studies. Furthermore, it should be validated against measurement results. Furthermore, the model should be implementable based on the unambiguous model description.

Simulations of a wide range of propagation scenarios and network topologies set different requirements to model accuracy and complexity. For example, the link number in a massive sensor network is huge, but it may be based on very simple transceivers, each equipped with a single antenna, which allows simplifications in the angular domain. On the contrary, angular information is crucial in massive MIMO simulations. Therefore, other simplifications may be considered given the individual system requirements or use case.

13.2.1.5   Summary of channel model requirements

The 5G channel model requirements described above are based on generic 5G assumptions and scenarios. The requirements cover different aspects such as spectrum, antennas, and system. In addition, some generic aspects such as simulation complexity were briefly discussed. The 5G channel model requirements are both challenging and extensive. Further, none of the existing models fulfills all the 5G propagation requirements [2].

13.2.2   Propagation scenarios

The 5G vision, where access to information and sharing of data is available anywhere and anytime to anyone and anything, leads to a wide range of propagation scenarios and network topologies that have to be considered. Furthermore, the wireless network has to serve a wide range of users that may be either stationary or mobile, while they may be communicating directly using a D2D link. The wireless system should work reliably in any propagation scenario, including Outdoor-to-Outdoor (O2O), Indoor-to-Indoor (I2I), Outdoor-to-Indoor (O2I), dense urban, wide area, highway, shopping mall, stadium, etc.

Table 13.2 Propagation scenarios.

Propagation Scenario	Outdoor/ Indoor	Supported Link Types (stochastic model)
Urban Micro	O2O, O2I	BS-MS, D2D, V2V
Urban Macro	O2O, O2I	BS-MS, BH
Rural Macro	NA	BS-MS, D2D, V2V, BH
Office	I2I	BS-MS
Shopping Mall	I2I	BS-MS
Highway	O2O	BS-MS, V2V
Open Air Festival[1]	O2O	BS-MS, BH, D2D
Stadium	O2O	Not Applicable (NA)

The network topologies should support not only cellular, but also direct D2D, M2M, and V2V links as well as full mesh networks.

The considered propagation scenarios are defined by physical environments, link types (e.g. Base Station (BS) to Mobile Station (MS), or Backhaul (BH)), cell types, antenna locations, and supported frequency range. These propagation scenarios are given in Table 13.2 for the map-based and stochastic model. The map-based model supports all the link types.

13.3   The METIS channel models

It is a considerable challenge to provide adequate 5G channel modeling, regarding scenarios, propagation characteristics and complexity. The obvious approach would be to extend present stochastic modeling like the WINNER and ITU IMT-Advanced models. The problem though is that these models are empirical, requiring extensive measurements to assess a vast number of model parameters and their cross correlations. As the required number of degrees of freedom for 5G channel modeling is very high since highly resolved directional characteristics must be spatially consistent also for dual-end mobility, it is not feasible to determine accurately all of those parameters by measurements.

An obvious alternative for providing 5G channel modeling in an efficient way, and still meeting the requirements regarding propagation scenarios and radio channel characteristics, is to use ray-tracing. The main advantage is that the model is inherently spatially-consistent and that only a few of the model parameters need to be calibrated by measurements. The main drawbacks are the need for a geometrical model of the environment and the high computational complexity. In order to provide a scalable complexity depending on the needs of the model user, both the traditional stochastic approach and the ray tracing based approach are used. As the ray tracing

---

[1] The herein open-air-festival scenario corresponds to the large-outdoor-event use case described in Chapter 2.

based approach requires a 3D building geometry map, it is referred to as the "map-based model".

13.3.1 Map-based model

The map-based model provides accurate and realistic spatial channel properties suitable for applications like massive MIMO and advanced beamforming. It automatically provides spatially consistent modeling also for challenging cases such as the dual-end mobility of D2D and V2V links. The model is based on ray-tracing combined with a simplified 3D geometric description of the propagation environment. The significant propagation mechanisms, i.e. diffraction, specular reflection, diffuse scattering, and blocking, are accounted for. Building walls are modeled as rectangular surfaces with specific electromagnetic material properties. The map-based model does not contain any explicit path loss model. Instead, the path loss, shadowing as well as other propagation characteristics are determined based on the map layout and, optionally, on a random distribution of objects accounting for people, vehicles and trees, etc.

13.3.1.1 General description

A prerequisite for any ray-tracing-based model is a geometrical description of the environment – i.e. a map or a building layout defined in a three dimensional (3D) Cartesian coordinate system. The level of map details does not need to be high. Only building walls and possibly other fixed structures have to be defined. A city geometry – Madrid Grid – is illustrated in Figure 13.2. An important target, when developing the

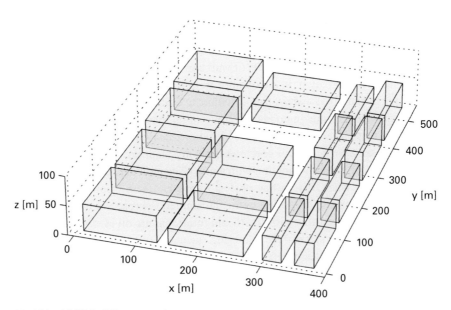

Figure 13.2   Madrid grid 3D building geometry.

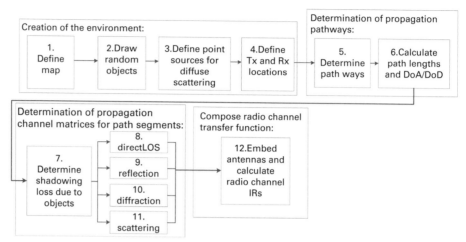

Figure 13.3   Block diagram of the map-based model.

model, has been to keep it as simple as possible while still fulfilling the model requirements.

A block diagram of this channel model is illustrated in Figure 13.3 with numbered steps of the procedure to generate radio channel realizations. On higher level, the procedure is divided into four main operations: creation of the environment, determination of propagation pathways, determination of propagation channel matrices for path segments, and composition of the radio channel transfer function. The main operations are briefly described hereafter, while the detailed procedure description is given in [2].

13.3.1.2   Creation of the environment

The first four steps in Figure 13.3 are for creating the environment. A 3D map containing coordinate points of wall corners, where walls are modeled as rectangular surfaces, is created. In the outdoor-to-indoor case, both outdoor and indoor maps and the location of indoor walls within a building block are defined. Then random scattering/shadowing objects, representing humans, vehicles, etc., are distributed on the map. Thereafter, the object locations can be either defined based on a known regular pattern, like the spectator seats in a stadium environment, or drawn randomly from a uniform distribution with a given scenario dependent density. Some care must be taken when distributing objects to prevent them from being too close, i.e. closer than half the object width, to the transmitter, the receiver, walls or to each other. All surfaces, like walls, with significant roughness are divided into smaller tiles, which is required to model diffuse scattering.

Step 4 specifies the transceiver locations or trajectories. For providing an accurate multi-antenna channel model output, the location of each element of a large array antenna is specified individually in a three dimensional space. However, for simplicity, when the array antennas are small and the radiation patterns are defined with

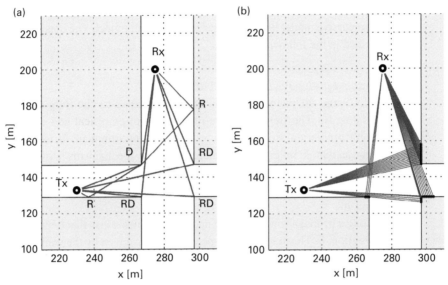

Figure 13.4   (a) Pathways due to flat exterior walls and corresponding interactions: diffraction (D), specular reflection (R), and specular diffraction (RD); and (b) pathways due to rough surfaces.

a common phase center (or a measurement center), it is adequate to specify only locations of transmitter and receiver antenna phase centers, not locations of individual elements. In this case, common propagation parameters are utilized for all array elements, and the spatial separation of antennas is taken into account in the final step (i.e. Step 12).

13.3.1.3   Determination of propagation pathways

In Steps 5 and 6 all significant pathways are determined. First, those pathways that are accounted for by the map geometry are determined. For this purpose, diffraction, specular reflection and diffuse scattering need to be modeled as shown in Figure 13.4. The output of Step 5 contains interaction types (direct, reflection, diffraction, object scattering, diffuse scattering) and the coordinates of interaction points for each path segment of each propagation pathway. As shown in Figure 13.4(b) the diffuse scattering due to rough surfaces is accounted for by distributing point scatterers on the surface of the exterior walls. For each pathway between two consecutive nodes that are in Line of Sight (LoS) with respect to such a point scatterer, there is a corresponding pathway via that point scatterer.

13.3.1.4   Determination of propagation channel matrices

The next steps of the procedure consist of shadowing loss due to blocking objects and determination of propagation transfer functions. Only a high-level approach is described here, but diffraction and scattering are discussed in more detail in the following.

Two diffraction models are proposed. The first and most accurate model is based on the Uniform Theory of Diffraction (UTD). A drawback of the UTD approach is its high complexity. For this reason, a substantially simpler modeling approach, based on the Berg recursive model [9] is proposed as the default model. The details of this model are described in the following section.

Diffracted pathways by Berg recursive model

The LOS and diffracted pathways can be described by the Berg recursive model in Steps 8 to 10. The Berg recursive model is semi-empirical and designed for signal strength prediction along streets in an urban environment. It is semi-empirical in the sense that it reflects physical propagation mechanisms without being strictly based on electromagnetics theory. It is based on the assumption that a street corner appears like a source of its own when a propagating radio wave turns around it.

Along a propagation path, each node contributes a loss that depends on the change in direction, θ. The total loss at a specific node j is given by the well-known expression for free-space path loss between isotropic antennas where a fictitious distance d_j is used, i.e.

$$L_{j|dB} = 20\log_{10}\left(\frac{4\pi d_j}{\lambda}\right),\tag{13.1}$$

where λ is the wave length. The fictitious distance corresponds to the real distance but is multiplied by a factor at each diffraction node. The resulting fictitious distance d_j becomes longer than the real distance meaning that it accounts for diffraction loss when used in the free space path loss formula. An example with four nodes is shown in Figure 13.5. At each node j, the fictitious distance is given by the following recursive expression

$$\begin{cases} d_j = k_j s_{j-1} + d_{j-1} \\ k_j = k_{j-1} + d_{j-1} q_{j-1} \end{cases},\tag{13.2}$$

where s_j is the real distance between node j and its following node $(j+1)$, q_j is the angle dependence and is a function of θ_j. The initial values are $d_0 = 0$ and $k_0 = 1$.

The angle dependence for the fictitious propagation distance extension is given by the following expression

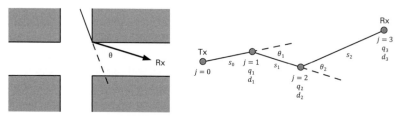

Figure 13.5   Example of a street corner acting as a node (left). Topological example with four nodes (right).

$$q_j = q_{90} \left(\frac{\theta_j}{90\deg} \right)^{\nu}, \tag{13.3}$$

where $\theta_j \in [0, 180]$ deg, q_{90} and ν are parameters determined by fitting the model to measurement data. It should be noted that diffraction nodes and corresponding angles are used only in the shadowed zone behind a corner. The parameter q_{90} accounts for the amount of diffraction loss caused by each node. The corresponding wavelength dependency is given by

$$q_{90} = \sqrt{\frac{q_\lambda}{\lambda}}, \tag{13.4}$$

where q_λ is a frequency dependent model parameter (the parameter range is given in [11]). The parameter ν (typically of a value around 1.5) accounts for how fast the loss changes in the transition zone between LOS and Non LOS (NLOS).

An example comparing measurements and the Berg recursive model in an indoor office environment is shown in Figure 13.6. It is evident that the corner diffraction is accurately modeled for both 60 GHz and 2.4 GHz. The receiver position is fixed in one corridor, while transmitter positions are placed in increasingly distant locations. The corridor corner is about 75 m away from the receiver. A striking result is the difference of about 15 dB between the two frequencies for distances about 80 m, i.e. with NLOS situation after the corridor corner. This difference is close to what is expected assuming that diffraction at the corridor corner is the dominating process.

Results from urban street microcell measurements, presented in Section 13.3.2.1, indicate however that the frequency dependency is minor suggesting that other scattering processes like diffuse scattering are dominant.

When calculating the diffraction loss due to the Berg recursive model, any specular interaction points are ignored. Furthermore, the segment length via any scatterer is modified to account for corresponding loss when using the recursive model. This is exemplified using Figure 13.7 where the corresponding distances s_j for the segments sg_i, are $s_0 = sg_1$, $s_1 = sg_2 + sg_3$, $s_2 = sg_4$ and $s_3 = 2 \, sg_5 sg_6 / R$. The former path is scattered by the object sc_1 using the scattering cross-section of a perfectly conducting sphere with radius R.

Scattering and blocking objects
Each path may be scattered, as well as shadowed, by objects like humans or vehicles. The effect of such scatterers is significant if they are located close to either end of the link (transmit or receive antennas).

Objects like vehicles and pedestrians will result in blocked as well as scattered paths. For the blocking model, each object is associated with a vertically oriented screen, which is used in combination with a simplified knife-edge diffraction model. In addition, the power of the scattered wave is modeled based on the scattering cross-section of a perfectly conducting sphere. For the details of the blocking and scattering model, the reader is referred to [11].

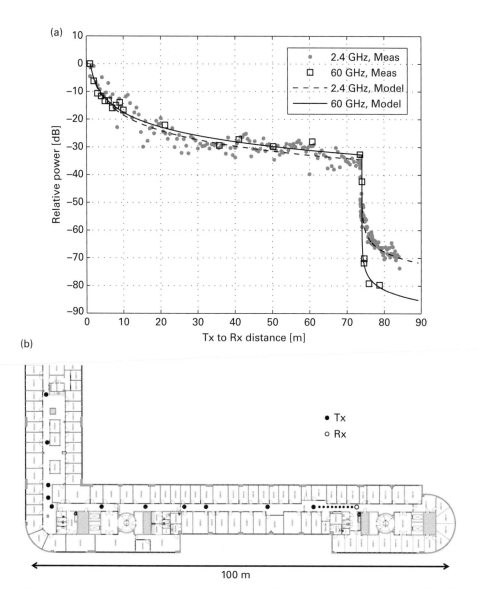

Figure 13.6 Measured signal strength relative to free space at 1 m distance for isotropic antennas at 2.4 GHz (grey dots) and 60 GHz (black squares) and Berg recursive fitted curves ($q_{90} = 2$ and 20 for 2.4 and 60 GHz respectively) (top) from a corridor office scenario (bottom).

Figure 13.8 illustrates the effect of randomly placed objects on the power azimuth spectrum along the Rx route. When an object is passed by the receiver, there is a corresponding received scattered wave which goes from the forward direction (0 degrees) to the backward direction (+/−180 degrees). This is in addition to the strong direct wave that is received from the backward direction (180 degrees).

In order to keep the complexity low, only significant scatterers are selected. Two types of such scatterers have been identified. The first type consists of scatterers that are in

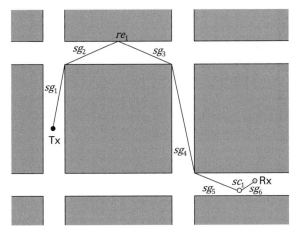

Figure 13.7 Example of a pathway subject to diffraction, specular reflection and object scattering.

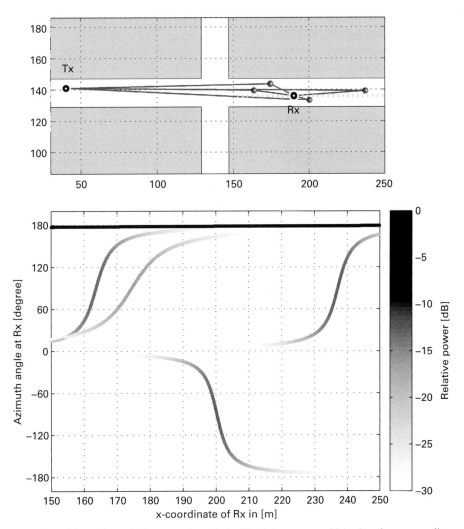

Figure 13.8 Scattering objects along the Rx route (top, dashed line) in a street grid (top) and corresponding receiver azimuth angle versus x location of the Rx route (bottom).

LOS to either the transmitter and/or the receiver. The second type consists of scatterers that are in LOS to two nodes, along a propagation pathway, which in turn are in NLOS to each other. One such example is a scatterer that is located in a street crossing and is in LOS to both Tx and Rx, which are located in NLOS with respect to each other along the two crossing streets. Moreover, the power of the scattered wave should be more than -40 dB relative to the strongest path.

13.3.1.5 Composing radio channel transfer function

The final step of the map-based model (Step 12 in Figure 13.3) generates a time variant radio channel transfer function by combining the known antenna radiation patterns with the propagation parameters determined in the previous steps. Complex polarimetric antenna radiation patterns are sampled with per-path directions of arrival and departure determined in Step 6. The propagation parameters required here are path delays, total attenuation by shadowing objects, 2x2 polarization matrices and divergence factors per path segment.

An example of the output of the map-based model for a moving Rx is shown in Figure 13.9. The center frequency is 2 GHz, the Tx (triangle) height is 15 m and the Rx (dots forming the vertical line in the upper left part) height 1.6 m. The Rx is

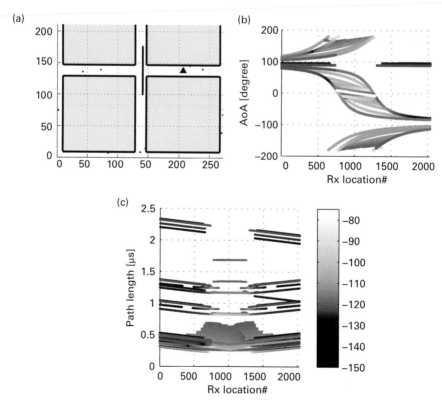

Figure 13.9   Propagation parameters of a transition scenario, AoA (left bottom) and path delay (right).

Table 13.3 Proposed extensions to GSCM.

Item	Extension	Comments
Path loss	New scenarios, D2D, mmWave	Scenario-specific path loss models, path loss for D2D cases, path loss models for 50–70 GHz.
Large-scale fading	Inter-site correlation	Use inter-parameter correlation to generate site-specific Large-Scale Parameter (LSP) maps
Shadowing	Spatial consistent shadowing	Using sum-of-sinusoids for low complexity.
Parameters	Updated parameterization	Channel model parameter tables updated based on literature review and METIS measurements.
Cluster locations	Define cluster coordinates	Spherical wave simulation by fixing the distances to First Bounce Cluster (FBC) and Last Bounce Cluster (LBC).
Time evolution	Smooth evolution of channel model parameters	First fixing the cluster locations and then calculating the Small-Scale Parameters (SSPs) from the geometry. Strongest clusters are selected all the time, which provides a smooth birth-death effect of clusters in case of moving devices.
Cluster power angular spectrum	Direct sampling of Laplacian shape	Direct sampling provides better correlation accuracy and a more realistic angular distribution for Massive MIMO

moving along a uniformly sampled straight line from south to north. The evolution of path delay and angular parameters along the course of Rx locations is depicted in the plots.

The number of paths increases in and close to the line-of-sight area, mostly because of diffuse scattering from the surrounding walls. Arrival angles evolve smoothly along the route, while departure angles are concentrated along the two street opening directions of the Tx location. The path gain is illustrated in decibels by grey scale bars. For visual clarity, the paths below -40 dB from the maximum path gain are not illustrated in the figure.

13.3.2   Stochastic model

The stochastic model is extended from the family of Geometry-based Stochastic Channel Models (GSCMs), mainly from 3GPP 3D Channel Models [6], which are an extension from WINNER+ models [4]. The proposed extensions are listed in Table 13.3 and discussed more in detail in the following subsections on path loss, sum-of-sinusoids calculation method for large-scale parameters, millimeter-wave parameterization, sampling of Laplacian shape, and dynamic model and spherical waves.

13.3.2.1   Path loss

The frequency range of the ITU-R UMi path loss model specified in M.2135 [5] was extended in the METIS project to cover the frequency range up to 60 GHz in three scenarios. Having chosen the models for the Manhattan-grid layout as a basis, they were tested against channel measurements performed in three different cities with eight radio frequencies [2]. The original models went through some modifications to improve agreement with the measurements. The resulting empirical METIS path loss models are summarized as follows.

Manhattan-layout LOS model: The model for the LOS scenarios yields

$$PL_{\text{LOS}}(d)_{|\text{dB}} = 10n_1 \log_{10}\left(\frac{d}{1\text{m}}\right) + 28.0 + 20\log_{10}\left(\frac{f_c}{1\text{GHz}}\right) + PL_{1|\text{dB}} \qquad (13.5)$$

for $10\text{ m} < d \leq d'_{\text{BP}}$ and

$$PL_{\text{LOS}}(d)_{|\text{dB}} = 10n_2 \log_{10}\left(\frac{d}{d'_{\text{BP}}}\right) + PL_{\text{LOS}}(d'_{\text{BP}})_{|\text{dB}} \qquad (13.6)$$

for $d'_{\text{BP}} \leq d < 500$ m, where f_c is the carrier frequency, d is the Tx-Rx distance and d'_{BP} is the effective Break Point (BP) distance, PL_1 is the path loss offset. The two symbols $n_1 = 2.2$ and $n_2 = 4.0$ represent the power decay constants before and after the BP, while the standard deviation of the shadow fading is $\sigma_S = 3.1$ dB. The differences of the METIS model compared to the original M.2135 model are summarized in two-folds. First, the BP distances were found to be much shorter than specified in M.2135 when estimating them from METIS path loss measurements and comparing them with the values of the original M.2135 model. Therefore, a frequency-dependent BP scaling factor was introduced to reflect the shorter BP distance observed in the measurements as

$$\alpha_{\text{BP}} = 0.87\exp\left(-\frac{\log_{10}\left(\frac{f_c}{1\text{GHz}}\right)}{0.65}\right), \qquad (13.7)$$

which leads to the effective BP distance given by

$$d'_{\text{BP}} = \alpha_{\text{BP}}\frac{4h'_{\text{BS}}h'_{\text{MS}}}{\lambda}, \qquad (13.8)$$

where λ is wavelength at the considered radio frequency, and h'_{BS} and h'_{MS} denote the effective BS and MS antenna heights. The effective antenna heights are given by the actual antenna heights h_{BS} and h_{MS} (greater than 1.5 meters each) by $h'_{\text{BS}} = h_{\text{BS}} - 1$ m and $h'_{\text{MS}} = h_{\text{MS}} - 1$ m to reflect the clutter effects on the ground such as cars.

It must be noted that the BP scaling factor in (13.7) is only valid for elevated base stations above 3 m. Two V2V measurements where the antenna height was 1.5 m above the ground did not follow this trend, where the BP scaling factor was 7.5 and 1.3 at 2.3 GHz and 5.25 GHz, respectively.

Another difference compared to the M.2135 model is the path loss offset PL_1; the offset allowed us to improve the overall agreement of the model with measurements.

Since the original M.2135 model is close to the free space path loss, the initial path loss reflects the effect of surrounding scattering environments. The offset is given by

$$PL_{1|\text{dB}} = -1.38 \log_{10}\left(\frac{f_c}{1\text{GHz}}\right) + 3.34. \tag{13.9}$$

Manhattan-layout NLOS model: The model for the NLOS scenarios is given by

$$PL_{\text{NLOS}} = PL_{\text{LOS}}(d_1)_{|\text{dB}} + 17.9 - 12.5n_j + 10n_j\log_{10}\left(\frac{d_2}{1\text{m}}\right)$$
$$+ 3\log_{10}\left(\frac{f_c}{1\text{GHz}}\right) + PL_{2|\text{dB}} \tag{13.10}$$

for 10 m $< d_2 <$ 1000 m and

$$n_j = \max\left(2.8 - 0.0024\left(\frac{d_1}{1\text{m}}\right), 1.84\right), \tag{13.11}$$

where d_1 and d_2 are the distances from the BS to a street crossing and from the MS to the cross section. BS and MS are located in the two crossing streets; $PL_{\text{LOS}}(d_1)$ is the path loss between the BS and the cross section derived from (13.5) and (13.6), and the last term PL_2 in (13.10) is the path loss offset for the NLOS model that depends on a street. Fitting of the model with measurements revealed that PL_2 and the shadow fading in NLOS scenarios are frequency independent, while shows notable dependence on streets, reflecting the fact that the coupling from the main to perpendicular streets depends highly on the building shape and vegetation at the street crossing. The street-dependent coupling level can be modeled as a random variable. The mean and standard deviation of PL_2 are −9.1 dB and 6.1 dB, while those of the shadow fading are 3.0 dB and 1.3 dB, respectively, over different streets. A normal distribution may work in reproducing the street-dependent parameter values, but more datasets are needed to ensure the statistical validity.

The NLOS model is a simplified version of the original M.2135 model in the sense, that it only considers the path loss from BS to MS, while the original model calculates the path loss from BS to MS and from MS to BS, and takes the smaller of them [12]. Although the path loss between BS and MS has to be the same regardless of the direction of signal transmission due to reciprocity, the path loss models do not necessarily hold this reciprocity. It was found that the simplified model works as good as the original model according to the available data sets.

13.3.2.2 Large-scale parameters based on sum-of-sinusoids

The creation of spatially consistent LSPs in the dual-end mobility case (in which both transmitter and receiver locations are defined by (x, y, z) coordinates) would lead to a six-dimensional (6D) LSP maps. Since the generation of such a 6D map would require a high computational complexity and extremely high memory consumption using traditional noise filtering methods, a different approach based on the sum-of-sinusoids is proposed. This method is described in [13] and has been further elaborated in [12].

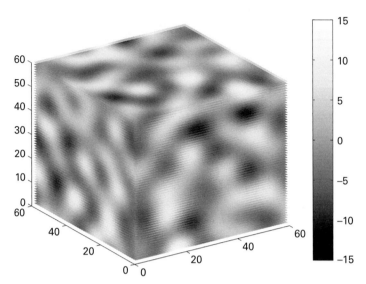

Figure 13.10 Example of a three-dimensional 60 × 60 × 60 meters shadowing map, grey scale indicates shadowing value in dB.

The method provides consistent joint correlation, and desired 1D correlation distances and standard deviation for the shadowing process.

The method is efficient in computational complexity and especially in memory consumption. Only $K(6 + 1)$ real numbers have to be stored in the memory (K is the number of waves). An example output in 3D space (instead of 6D for visualization purposes only) is illustrated in Figure 13.10. In principle, all the other spatial correlated large-scale parameters could be generated with this method as well.

13.3.2.3   mm-Wave parameterization

This subsection proposes measurement-based parameters for the stochastic model. A set of channel model parameters intended for the WINNER II [1] and the WINNER+ [4] models were produced for the 60 GHz band in an indoor shopping mall, an indoor cafeteria and on an outdoor open square. All the parameters are derived from short-range radio channel measurements, along with a deterministic field simulation tool with experimental proof [14] for the cafeteria and the square. The channel measurements were performed with 4 GHz bandwidth, resulting in a spatial resolution of 7.25 cm. The distance range between transmit and receive antennas was up to 36 m as shown in Table 13.4. The shopping mall and the square measurements include LOS as well as Obstructed-LOS (OLOS) scenarios, where the major blocking objects of the LOS are pillars in the shopping mall and lampposts and pedestrians in the open square.

The system and model parameters are given in Tables 13.4 and 13.5. The latter table covers large-scale parameters such as path loss, spread parameters in the delay and angular domains, shadow fading, and Rician K-factor; they are characterized by their mean μ and standard deviation σ in a log-scale of base 10. They are correlated random

Table 13.4 Settings for deriving the WINNER channel model parameters at 60 GHz for three short-range scenarios.

		Shopping mall		Cafeteria	Square	
Settings		LOS	OLOS	LOS	LOS	OLOS
BS-MS distance [m]	Min	1.5	4.0	1.0	6.4	6.4
	Max	13.4	16.1	13.1	36.3	36.3
Antenna height [m]	BS	2	2	2	6	6
	MS	2	2	1	1	1
Bandwidth [GHz]	Max	4		4	4	
Centre frequency [GHz]		63		63	63	
Dynamic range [dB]		20		20	20	

Table 13.5 Measurement-based channel model parameters at 60 GHz for three short-range scenarios.

		Shopping mall		Cafeteria	Square	
		LOS	OLOS	LOS	LOS	OLOS
Parameters/Models	Symbol	2D		3D	3D	
Path loss, dB	A	18.4	3.59	15.4	20.3	26.2
$PL = A\log_{10}(d/1\text{m}) + B$	B	68.8	94.3	767.1	67.5	70.5
Delay spread	$\mu_{\lg DS}$	-8.28	-7.78	-8.24	-8.82	-7.72
$\lg DS = \log_{10}(DS/1s)$	$\sigma_{\lg DS}$	0.32	0.10	0.18	0.37	0.32
Azimuth spread of departure (ASD)	$\mu_{\lg ASD}$	1.09	1.61	1.63	1.10	1.49
$\lg ASD = \log_{10}(ASD/1\text{deg})$	$\sigma_{\lg ASD}$	0.43	0.11	0.25	0.75	0.35
Azimuth spread of arrival (ASA)	$\mu_{\lg ASA}$	1.19	1.62	1.56	0.24	1.31
$\lg ASA = \log_{10}(ASA/1\text{deg})$	$\sigma_{\lg ASA}$	0.47	0.14	0.19	0.54	0.44
Elevation spread of departure (ESD)	$\mu_{\lg ESD}$	NA	NA	1.31	0.43	0.74
$\lg ESD = \log_{10}(ESD/1\text{deg})$	$\sigma_{\lg ESD}$	NA	NA	0.16	0.29	0.32
Elevation spread of arrival (ESA)	$\mu_{\lg ESA}$	NA	NA	1.28	0.77	0.95
$\lg ESA = \log_{10}(ESA/1\text{deg})$	$\sigma_{\lg ESA}$	NA	NA	0.23	0.96	1.19
Shadow fading, dB	σ_{SF}	1.2	2.1	0.9	0.3	3.5
K-Factor, dB	μ_{KF}	7.9	NA	-2.5	8.4	NA
	σ_{KF}	5.8	NA	2.4	2.2	NA
Delay distribution		Exponential				
AOD and AOA distribution		Wrapped Gaussian				
Cross-polarization ratio (XPR), dB	μ_{XPR}	29	29	29	29	29
	σ_{XPR}	6.5	6.5	6.5	6.5	6.5
Number of clusters		6	18	42	4	25
Number of rays per cluster		20	20	20	20	20
Cluster ASD, deg		0.5	0.5	0.5	0.5	0.5
Cluster ASA, deg		0.5	0.5	0.5	0.5	0.5
Cluster ESD, deg		NA	NA	0.5	0.5	0.5
Cluster ESA, deg		NA	NA	0.5	0.5	0.5
Per-cluster shadowing, std, dB		2.5	6.3	4.2	4.9	4.9

DS (Delay Spread), lg (log-scale), NA (Not Available), PL (Path Loss).

variables as detailed in [1]; the parameters governing the correlation are not shown here for the sake of conciseness, but can be found in [2]. Table 13.5 also contains selected parameters to define clusters and their sub-paths. It is found in the measurements that the propagation paths do not form clusters as apparent as those at lower frequencies (below 6 GHz). Therefore, it is assumed that each propagation path is considered as a single cluster. Nonetheless, it is still possible to use the WINNER channel model with the original definition of 20 sub-paths in a cluster, and reproduce channels that have consistent characteristics with the measurements in terms of delay spread and Rician K-factor [15]. The cluster angular spreads are set to a very small value of 0.5 degrees to make the 20 sub-path model a good approximate to measurements. The validity of the 20 sub-path model is subject to further tests based on measurements, since the model may result in too optimistic performance for very large array antennas as demonstrated in Figure 13.1. Furthermore, estimation of the cross-polarization ratios (XPRs) needs further measurements to improve their quality, since the values in Table 13.5 are taken from the IEEE 802.11ad channel model [7]. The detailed measurement description, the parameter extraction procedure, and the complete list of parameters are given in [2].

13.3.2.4   Direct sampling of Laplacian shape

Both, the original SCM model [3] and the latest 3D channel model specified by 3GPP in [6] (Step 7) propose an approximation of the Laplacian shaped Power Angular Spectrum (PAS) by 20 sub-paths with equal power. When the number of transmit and receive antennas is limited, the final simulation result is not damaged too much by this simplification. However, this approximation is not suitable for massive MIMO, because of the correlation error and spatially separable sub-paths. Whereas a small

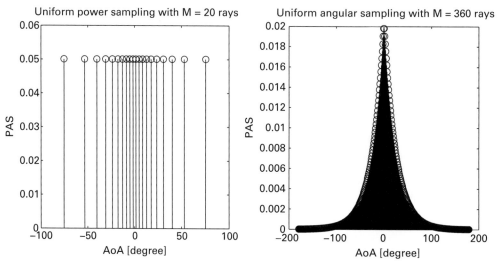

Figure 13.11   Equal amplitude (left) and direct sampling (right) of Laplacian-shaped power angular spectrum. x-axis is angle of arrival (or departure) in both cases and y-axis is the amplitude.

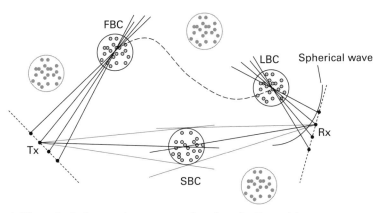

Figure 13.12   Adding spherical waves into geometry-based stochastic model.

array cannot detect individual sub-paths, a large array might due to its higher spatial resolution. A more accurate way to generate the ray offsets is to use direct sampling of the Laplacian shape as described in [11]. Therefore, it is proposed to apply this method in simulations with very large array antennas.

13.3.2.5   Dynamic modeling and spherical waves

For simulations with very large array antennas, a modeling of spherical waves is required. Therefore, it is necessary to define physical locations in terms of (x, y, z) coordinates of clusters (see Figure 13.12 where FBC stands for first bounce cluster, LBC for last bounce cluster, and SBC for single bounce cluster). Originally, SCM and WINNER models assume the AOD and AOA to be independent. This means that the resulting paths are generally multi-bounce paths in contrast to single bounce paths.

In case of multi-bounce, AOAs, AODs, and delays are defined based on the legacy GSCM principle. The maximum distance between Tx (or Rx) and the cluster location is determined from the geometry of Tx and Rx locations, AOA/AOD, and delay. This geometry is an ellipse in which the focal points are Tx and Rx and the length of the major axis equals the delay multiplied by the speed of light. The angles of departure and arrival define the directions from the Tx to the clusters and from the clusters to the Rx. The maximum cluster distance equals the distance between the Tx (or Rx) and the locus of the ellipse. The actual distance between the Tx (or Rx) and the cluster is randomly drawn between zero and the maximum distance (see Figure 13.13). When the location of the cluster is known, the phases per radiation element can be calculated explicitly from the geometry (see Figure 13.13).

In case of a Single Bounce Cluster (SBC) the modeling follows the same principles as the multi bounce case explained above. However, the SBC is located at the locus of the ellipse (see Figure 13.13). Because the AOA, AOD and delay are randomly drawn in the GSCM, most likely the geometry of these three parameters does not fit to the ellipse. Therefore, the SBCs are calculated directly based on two cluster parameters

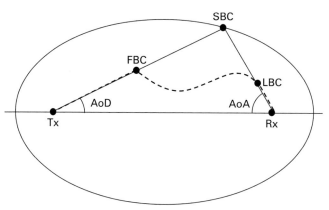

Figure 13.13　Determining the location of a cluster.

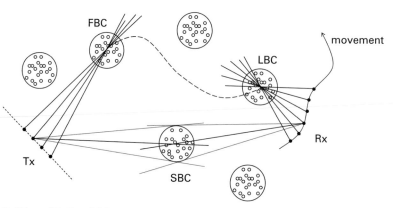

Figure 13.14　Drifting of LaS and SS parameters.

such as the delay and the AOD (or the delay and the AOA). Then, the direction at the other end is calculated by matching AOA (or AOD) with AOD (or AOA) and delay. In this case, 50% of the SBCs may be (randomly) chosen to be calculated based on the Tx-side cluster parameters and another 50% based on the Rx-side cluster parameters while the originally drawn cluster delays and also powers and cluster indexing are kept. However, the single bounce case has been so far excluded from the framework of the original SCM and WINNER models. In fact, the implementation of this approach would require further study.

After fixing the physical locations, drifting of LaS and SS parameters is enabled for a short distance movement as illustrated in Figure 13.14. Implementation of the drifting is straightforward and is fully based on the geometry. Delay, amplitude, phase, and angle of arrival are updated based on the Rx location. The sample density of Rx locations has to fulfill the Nyquist criterion, and the adequate filtering for denser time (location) samples has to be ensured. This feature is already part of the Quadriga channel model [17].

13.4 Conclusions

This chapter has introduced the requirements for 5G propagation and channel models, and has summarized the METIS channel models. It is found that many important requirements for 5G channel models are not fulfilled by the recently standardized MIMO channel models. For example, higher frequencies, massive-MIMO, and direct V2V links require features and parameters. Consequently, the METIS channel model addresses these much-needed features and parameters to model a 5G system. It is also noticed that different simulation scenarios and test cases may require different model features. Due to these diverse requirements, the METIS channel model is not a single model, it rather consists of both a map-based and a stochastic models.

References

[1] IST-4-027756 WINNER II project, "Channel models," Deliverable D1.1.2, version V1.2, February 2008.

[2] ICT-317669 METIS project, "METIS Channel Models," Deliverable D1.4, version v3, July 2015, www.metis2020.com/documents/deliverables/

[3] 3GPP TR 25.996, "Spatial channel model for multiple input multiple output (MIMO) simulations," Technical Report TR 25.996 V6.1.0, Technical Specification Group Radio Access Network, September 2003.

[4] CELTIC CP5-026 WINNER+ project, "Final channel models," Deliverable D5.3, V1.0, June 2010.

[5] International Telecommunications Union Radio (ITU-R), "Guidelines for evaluation of radio interface technologies for IMT-Advanced," Report ITU-R M.2135, December 2009, www.itu.int/pub/R-REP-M.2135-1-2009

[6] 3GPP TR 36.873, "Study on 3D channel model for LTE," Technical Report TR 36.873 V12.2.0, Technical Specification Group Radio Access Network, June 2015.

[7] A. Maltsev, V. Ergec and E. Perahia, "Channel models for 60 GHz WLAN systems," Document IEEE 802.11-09/0334r8, 2010.

[8] J. Medbo et al., "Channel modelling for the fifth generation mobile communications," in European Conference on Antennas and Propagation, The Hague, April 2014.

[9] ICT-317669 METIS project, "Description of the spectrum needs and usage principles,"Deliverable D5.3, September 2014. www.metis2020.com/documents/deliverables/

[10] J. Medbo et al., "Directional channel characteristics in elevation and azimuth at an urban macrocell base station," in European Conference on Antennas and Propagation, Prague, March 2012.

[11] ICT-317669 METIS project, "Initial channel models based on measurements," Deliverable D1.2, April 2014, www.metis2020.com/documents/deliverables/

[12] T. Jämsä and P. Kyösti, "Device-to-device extension to geometry-based stochastic channel models," in European Conference on Antennas and Propagation, Lisbon, April 2015.

[13] Z. Wang, E. K. Tameh, and A. R. Nix, "A sum-of-sinusoids based simulation model for the joint shadowing process in urban peer-to-peer radio channels," in IEEE Vehicular Technology Conference, Dallas, September 2005.

[14] J. Järveläinen and K. Haneda, "Sixty gigahertz indoor radio wave propagation prediction method based on full scattering model," *Radio Science*, vol. 49, no. 4, pp. 293–305, April 2014.

[15] A. Karttunen, J. Jarvelainen, A. Khatun, and K. Haneda, "Radio propagation measurements and WINNER II parametrization for a shopping mall at 61–65 GHz," in IEEE Vehicular Technology Conference, Glasgow, May 2015.

[16] W. Fan, T. Jämsä, J. Ø. Nielsen, and G. F. Pedersen, "On angular sampling methods for 3-D spatial channel models," *IEEE Antennas and Wireless Propagation Letters*, vol. 14, pp. 531–534, February 2015.

[17] S. Jaeckel, L. Raschkowski, K. Börner, L. Thiele, F. Burkhardt, and E. Eberlein, "QuaDRiGa: Quasi deterministic radio channel generator, user manual and documentation," Fraunhofer Heinrich Hertz Institute, Tech. Rep. v1.2.32-458, 2015.

14 Simulation methodology

Jose F. Monserrat, Mikael Fallgren, David Martín-Sacristán, and Ji Lianghai

A simulation methodology is needed in the 5G technical work in order to ensure consistency of results obtained, by means of a computer simulation. This methodology must comprise a procedure for calibrating the simulator, guidelines for evaluating, and a mechanism supporting and controlling the validity of the performed simulations. This chapter provides a methodology for simulation to align assumptions. The alignment allows for a direct comparison of different 5G technology components. The chapter is based on the experience of the authors in the simulation work performed in the framework of the International Mobile Telecommunications-Advanced (IMT-Advanced) definition and in METIS [1]. Finally, some relevant test cases and preferred models are introduced.

14.1 Evaluation methodology

In this section, methodology guidelines are given to enable consistent performance evaluations. The guidelines may serve as a framework with aligned assumptions, consistent choice of models and simulation reference metrics to ensure that the results can be compared. The results on different levels are not meant to be compared but to be used as possible input, e.g. link-level simulations can be used as input to system-level simulations but should not be compared to them. Below, the main performance indicators, as well as suitable channel and propagation models, are explained and defined. The main characteristics of the evaluation scenarios are out of the scope of this chapter, since they are thoroughly described in Chapter 2.

14.1.1 Performance indicators

The main performance indicators to be used in the evaluation of the 5G system are defined and explained hereafter. It should be noted that the material (of the performance indicators) is based on [1]–[4].

14.1.1.1 User throughput
The user throughput is defined as the total amount of received information bits at the receiver divided by the total active session time at the data link layer [2][3]. Active

5G Mobile and Wireless Communications Technology, ed. A. Osseiran, J. F. Monserrat, and P. Marsch. Published by Cambridge University Press. © Cambridge University Press 2016.

session time does not include the waiting time at the application layer, e.g. reading time for web-browsing, or back-off time introduced by TCP/IP's traffic control, and therefore it is, in general, different from the session length.

A second definition of the user throughput accounts for the whole session time, instead of only the active session time. Both definitions are equivalent for full buffer traffic model, which does not have neither reading nor back-off times.

A third definition considers that the user throughput is the average of the throughput experienced by all the packets received by the user [2]. Here the throughput of a packet is the packet size divided by the time used to transmit it from the data link layer at the transmitter to the same layer of the receiver. This definition is equivalent to the first one when the packet transmissions do not overlap in time.

14.1.1.2   Application data rate

The application data rate is defined as the data bit rate from the application layer of the user, i.e. data bits related to Transmission Control Protocol (TCP) and protocol overhead are excluded. This definition facilitates the comparison of technology components that can implement changes at any layer in the protocol stack.

14.1.1.3   Cell throughput

The cell throughput is defined as the total amount of received information bits in the cell under a pre-specified time interval [3]. The cell is defined as a single point of data aggregation for which the cell throughput is measured, e.g. a traditional Third Generation Partnership Project (3GPP) cell or a Wi-Fi access point.

14.1.1.4   Spectral efficiency

The spectral efficiency is defined as the aggregated user throughput divided by the aggregated spectrum used per measurement unit in the data link layer. Note that the aggregated spectrum includes the spectrum used for e.g. control and broadcast signaling. The measurement unit is a cell or an area unit, e.g. square kilometers.

The cell spectral efficiency is defined as the spectral efficiency where the aggregation is taking place per cell.

The normalized user throughput was defined in [3] as the user throughput divided by the channel bandwidth of the user's serving cell. This indicator is equivalent to a user spectral efficiency.

The cell edge user spectral efficiency is defined as the 5% point of the Cumulative Distribution Function (CDF) of the normalized user throughput [3].

14.1.1.5   Traffic volume

The traffic volume is defined at the application layer as the aggregated served traffic to all users, either in total for the setting or per area unit.

14.1.1.6   Error rate

The bit error rate is defined as the error rate of transmitted bits on the raw demodulation of the investigated technology.

The frame error rate is defined as the error rate of transmitted information blocks. For example, the information block can be a link-level codeword or a system-level transport block at the data link layer.

14.1.1.7 Delay

The application end-to-end delay is defined as the time elapsed from the application layer at the source to the application layer at the destination.

The Medium Access Control (MAC) layer delay is defined as the time elapsed from the MAC layer at the source to the MAC layer at the destination.

14.1.1.8 Network energy performance

The network energy performance is defined as the energy consumed to the number of served bits at the data link layer [4].

14.1.1.9 Cost

Cost is the amount of capital consumed to reach a certain solution. To enable an easy comparison, cost can be normalized by the system data rate, thus resulting in the metric of cost per served bit.

14.1.2 Channel simplifications

The choice of channel and propagation models for simulation evaluations should be made according to the required level of accuracy, but should take into account the computational complexity of the model. In fact, the channel propagation modeling heavily impacts the total computational burden of a simulator.

Stochastic and geometric models, as compared with ray-tracing option, are simpler to implement, but usually lack the required level of realism that 5G assessment requires. They use two different sets of channel parameters. The first one concerns small-scale parameters, including Angle-of-Arrival (AoA) and Angle-of-Departure (AoD) or delay of the rays. The second one is related to the large-scale parameters, such as shadow fading and path loss.

A reasonable alternative consists of a simplified ray-based approach for the characterization of the large-scale effects, followed by the use of a pure stochastic and geometric approach for the characterization of small-scale effects. This alternative, being much simpler than ray-tracing, still allows for a proper characterization of real environments.

14.1.2.1 Small-scale modeling

Concerning small-scale parameters characterization, International Telecommunications Union – Radiocommunication Sector (ITU-R) M.2135 models [5] are the ones most widely accepted by the research community. Although some propagation scenarios commonly considered in 5G studies, such as Device-to-Device (D2D) and Vehicle-to-Everything (V2X), are not covered by M.2135, a mapping could be defined between those propagation scenarios and the M.2135 channel models. A specific mapping is

Table 14.1 Small-scale models for the different propagation scenarios.

Propagation scenario	Model	Correlation length
Urban Micro O2O	ITU-R UMi	10
Urban Micro O2I	ITU-R UMi O2I	10
Urban Macro O2O	ITU-R UMa	50
Urban Macro O2I	ITU-R UMa	50
Indoor Office	ITU-R InH	10
D2D/V2X Urban O2O	ITU-R UMi with appropriate heights	10
D2D/V2X Urban O2I	ITU-R UMi O2I with appropriate heights	10
D2D Indoor Office	ITU-R InH with appropriate heights	10

summarized in Table 14.1. Urban, indoor, D2D and V2X scenarios are considered. O2O and O2I stand for Outdoor-to-Outdoor and Outdoor-to-Indoor, respectively. Concerning the channel models, the three ITU-R models are considered: the Urban Micro model (UMi) including its O2I variant, the Urban Macro model (UMa), and the Indoor Hotspot model (InH).

There are two issues to be solved concerning this small-scale characterization. The first issue is the validity of such models for dynamic simulations in which the position of users change over time. In this sense, it can be assumed that the conditions for rays and cluster generation remain static along a certain correlation length depending on the propagation scenario. After this distance, new cluster and rays must be generated according to the new geometry. The second issue is how to estimate the Line-of-Sight (LOS) or Non-Line-of-Sight (NLOS) conditions [5]. For synthetic simulations, these conditions are randomly selected. However, for realistic test cases, sight condition shall be re-evaluated for each correlation length based on the actual position of transmitter and receiver.

14.1.2.2   Large-scale modeling when base station is on the rooftop level

In this case, $PL(d)$, the total transmission path loss in decibels, is expressed as the sum of L_{fs}, the free space loss, L_{rts}, the diffraction loss from rooftop to the street, and L_{mds}, the reduction due to multiple screen diffraction past rows of buildings, that is,

$$PL(d) = \begin{cases} L_{fs} + L_{rts} + L_{mds} & \text{if} \quad L_{rts} + L_{mds} > 0 \\ L_{fs} & \text{if} \quad L_{rts} + L_{mds} \leq 0 \end{cases}. \tag{14.1}$$

Figure 14.1 illustrates the used geometry and the set of variables that impact the model response.

Let d be the mobile-to-base separation, the free space loss between them is given by

$$L_{fs} = -10 \log_{10} \left(\frac{\lambda}{4\pi d} \right)^2. \tag{14.2}$$

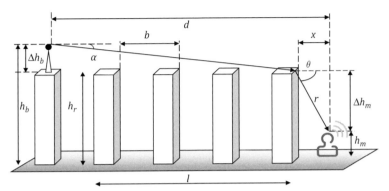

Figure 14.1   Geometry of the model.

The diffraction from the rooftop down to the street level gives the excess loss to the mobile station [6] is

$$L_{rts} = -20\log_{10}\left[\frac{1}{2} - \frac{1}{\pi}\arctan\left(sign(\theta)\sqrt{\frac{\pi^3}{4\lambda}r(1 - \cos\theta)}\right)\right], \quad (14.3)$$

where

$$\theta = \tan^{-1}\left(\frac{|\Delta h_m|}{x}\right), \quad (14.4)$$

$$r = \sqrt{(\Delta h_m)^2 + x^2}, \quad (14.5)$$

being Δh_m the difference between the last building height and the mobile antenna height, h_m, and x the horizontal distance between the mobile and the diffracting edges.

The multiple screen diffraction loss from the base antennas due to propagation past rows of buildings depends on the base antennas height relative to the building heights and on the incidence angle [7]. A criterion for grazing incidence is the settled field distance, d_s:

$$d_s = \frac{\lambda d^2}{\Delta h_b^2}, \quad (14.6)$$

where Δh_b is the difference between the base station antenna height, h_b, and the average rooftop h_r. Then, for the calculation of L_{msd}, d_s is compared to the length of the path, l (see Figure 14.1).

If $l > d_s$

$$L_{mds} = L_{bsh} + k_a + k_d\log_{10}(d/1000) + k_f\log_{10}(f) - 9\log_{10}(b), \quad (14.7)$$

where

$$L_{bsh} = \begin{cases} -18\log_{10}(1 + \Delta h_b) & \text{for} \quad h_b > h_r \\ 0 & \text{for} \quad h_b \leq h_r \end{cases} \tag{14.8}$$

is a loss term that depends on the base station height,

$$k_a = \begin{cases} 54 & \text{for } h_b > h_r \\ 54 - 0.8\Delta h_b & \text{for } h_b \leq h_r \text{ and } d \geq 500, \\ 54 - 1.6\Delta h_b d/1000 & \text{for } h_b \leq h_r \text{ and } d < 500 \end{cases} \tag{14.9}$$

$$k_d = \begin{cases} 18 & \text{for} \quad h_b > h_r \\ 18 - 15\dfrac{\Delta h_b}{h_r} & \text{for} \quad h_b \leq h_r, \end{cases} \tag{14.10}$$

and $k_f = 0.7(f/925 - 1)$ for medium sized cities and suburban centers with medium tree density, whereas $k_f = 15(f/925 - 1)$ for metropolitan centers. Note that frequency is expressed in MHz in these equations.

On the other hand, if $l \leq d_s$ then a further distinction has to be made according to the relative heights of the base station and the rooftops.

$$L_{msd} = -10 \log_{10}(Q_M^2), \tag{14.11}$$

where

$$Q_M = \begin{cases} 2.35\left(\dfrac{\Delta h_b}{d}\sqrt{\dfrac{b}{\lambda}}\right)^{0.9} & \text{for } h_b > h_r \\ \dfrac{b}{d} & \text{for } h_b \approx h_r, \\ \dfrac{b}{2\pi d}\sqrt{\dfrac{\lambda}{\rho}\left(\dfrac{1}{\vartheta} - \dfrac{1}{2\pi + \vartheta}\right)} & \text{for } h_b < h_r \end{cases} \tag{14.12}$$

$$\vartheta = \tan^{-1}\left(\dfrac{\Delta h_b}{b}\right), \tag{14.13}$$

and

$$\rho = \sqrt{\Delta h_b^2 + b^2}. \tag{14.14}$$

In this model, minimum coupling loss is set to 70 dB. Further, concerning outdoor-to-indoor characterization, the same approach as in the WINNER+ project [8] can be used.

14.1.2.3 Large-scale modeling when base station is much below the mean building height

The proposed model is based on the ITU-R UMi path loss model for Manhattan grid layout [5]. In general, this model distinguishes the main street, where the transmission point is located, perpendicular streets, and parallel streets. Figure 14.2 shows the geometry used.

If the receiver is in the main street, LOS path loss in decibels is calculated according to

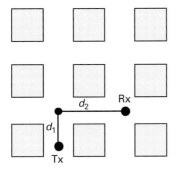

Figure 14.2   Upright projection of the geometry in this scenario.

$$PL_{LoS}(d_1) = 40\log_{10}(d_1) + 7.8 - 18\log_{10}(h'_b h'_m) + 2\log_{10}(f_c), \tag{14.15}$$

where d_1 is the distance in meters between transmitter and receiver, f_c is the frequency in GHz and h'_b and h'_m are the effective antenna heights in meters of transmitter and receiver, respectively. The effective antenna heights h'_b and h'_m are computed as follows

$$h'_b = h_b - 1, \ h'_m = h_m - 1. \tag{14.16}$$

Note that the 3D extension of the model depends only on varying h_m as desired. If the receiver is in a perpendicular street, then

$$PL = \min\Big(L(d_1, d_2), \ L(d_2, d_1)\Big), \tag{14.17}$$

where

$$L(d_k, d_l) = L_{LoS}(d_k) + 17.9 - 12.5n_j + 10n_j\log_{10}(d_l) + 3\log_{10}(f_c) \tag{14.18}$$

and

$$n_j = \max(2.8 - 0.0024d_k, 1.84). \tag{14.19}$$

For the sake of simplicity, the height used in the LOS formula will be the one of the receiver in Rx. It is worth noting that in case of being in a perpendicular street with distance less than 10 m between transmitter and receiver, then LOS conditions apply. Finally, for parallel streets, the path loss is assumed as infinite. Moreover, minimum coupling losses are set to 53 dB.

14.2   Calibration

This section describes in detail the steps to be followed in the creation and validation of a 5G simulator.

14.2.1   Link-level calibration

This section describes in detail a step-wise approach to execute the calibration process for a link-level simulator. Hitherto there is not a typical and detailed definition of 5G at link level. Therefore, as an intermediate step, the 5G link-level calibration can be done based on a Long-Term Evolution (LTE) link-level simulator. This intermediate step will make it easy to develop a similar methodology for the calibration of a fully detailed 5G technology. The methodology is based on breaking down the entire simulation chain of a link-level simulator into its single building blocks.

The next five subsections describe a set of five calibration steps defined for the downlink. For each calibration step, a proper reference for crosschecking is proposed. Section 14.2.1.6 deals with the specific calibration of the uplink, and Section 14.2.1.7 explains the end-to-end calibration process of LTE performance aligned with the 3GPP. Figure 14.3 summarizes the overall calibration process.

14.2.1.1   Calibration step 1 – OFDM modulation

The first step (step 1) of the calibration process consists in the validation of the OFDM Modulation/Demodulation (OMD) unit. It includes the validation of Fast Fourier Transform (FFT) and Inverse-FFT (IFFT). In order to do so, it is enough to focus on the inputs and outputs of this macro-block (i.e. no coding/decoding functionalities will be considered in this case).

The following assumptions are suggested for the calibration of the OMD unit. Both an Additive White Gaussian Noise (AWGN) channel and a Rayleigh fading channel should be considered in the assessment. For the latter channel, ideal channel estimation can be assumed. For both channels, it is suggested to use a simple receiver such as a Zero Forcing (ZF) receiver. The curves obtained by simulation in this first step should overlap the theoretical reference curves that can be found in literature (see, e.g. [9]).

14.2.1.2   Calibration step 2 – channel coding

In the next step (step 2) the coding functionalities of the LTE system must be included in the link-level simulator. The turbo coding performance depends highly on the turbo block size. Hence, the actual Physical Downlink Shared Channel (PDSCH) frame structure must be implemented and different Resource Block (RB) allocations have to be tested in all calibration steps after OMD unit calibration. Since the results obtained in this second step of calibration are LTE-specific, it makes sense to refer directly to the 3GPP documentation to perform the best-suited benchmark. In [10] turbo coding is evaluated assuming an AWGN channel, a ZF receiver and a maximum of only one Hybrid Automatic Repeat Request (HARQ) transmission, for different coding rates and number of RBs allocated to the transmission. Once this step is assessed, the Forward Error Correction (FEC) macro-block based on turbo coding is included in all the rest of calibration steps.

14.2.1.3   Calibration step 3 – SIMO configuration

In the next step (step 3) the entire transmission chain should be simulated with the channel model included. The following assumptions are suggested for this calibration

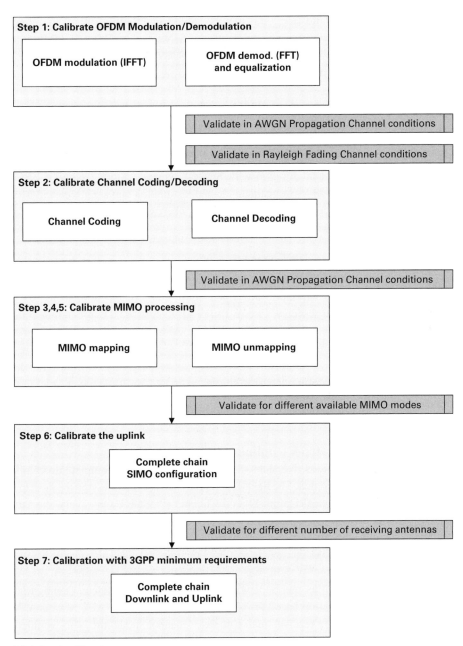

Figure 14.3   Link-level calibration process.

step: a fixed bandwidth of 10 MHz – 50 RBs – is simulated assuming the realistic channel estimation and different channels – Extended Pedestrian A (EPA), Extended Vehicular A (EVA) and Extended Typical Urban (ETU) [11] – with a Single Input Multiple Output (SIMO) 1x2 configuration [12]. The scenario could be validated through a direct comparison with the minimum requirements specified by the 3GPP in [13]. This validation refers to a threshold value for the system throughput for a given Signal to Interference plus Noise Ratio (SINR) value. Other 3GPP internal references are recommended in this validation step. Specifically, in [14] an extensive collection of results from different manufacturers is shown.

14.2.1.4   Calibration step 4 – MIMO configuration for transmit diversity

Similarly, the fourth step (step 4) validates the system by a direct comparison with the 3GPP minimum requirements [13]. The simulation assumptions for the calibration of the transmit diversity scheme proposed in LTE includes a Multiple Input Multiple Output (MIMO) 2x2 configuration with Space-Frequency Block Coding (SFBC), maximum bandwidth of 10 MHz, realistic channel estimation and a ZF receiver. More details on these assumptions can be found in [15], whereas the 3GPP results used for calibration are summarized in [16].

14.2.1.5   Calibration step 5 – MIMO configuration for spatial multiplexing

In the fifth calibration step (step 5) spatial multiplexing simulations are assessed. These can be divided into open loop and closed-loop scenarios. In case of open loop, large-delay Cyclic Delay Diversity (CDD) precoding must be implemented following the 3GPP standard. Although minimum requirements for two antenna configurations – 2x2 and 4x2 – are provided in [13], a more detailed set of results can be found in [17] and [18], respectively. On the other hand, in case of the closed-loop scenario, single layer and multiple layer spatial multiplexing configurations must be calibrated. Some specific results with 4x2 MIMO are summarized in [18].

14.2.1.6   Calibration step 6 – uplink

With the previous five steps, the macro-blocks of the complete downlink chain have been fully validated. The next step (step 6) consists in assessing the system in the uplink using the Physical Uplink Shared Channel (PUSCH). Obviously, the same five steps described for the downlink could also be valid for the reverse channel taking into account the particular characteristic of the Discrete Fourier Transform (DFT)-precoded OFDM that exists in this direction. Provided the macro-blocks testing, the calibration process will continue with the complete uplink transmission chain. The simulation assumptions will include a SIMO 1x2 antenna configuration, realistic channel estimation based on Maximum A Posteriori (MAP), Minimum Mean Square Error (MMSE) receiver and complete channel modeling. This simulation scenario corresponds to the one proposed in [19]. The minimum requirements for PUSCH are provided in the specification [20]. For the validation purpose, it is suggested to refer also to the 3GPP detailed results gathered in [21].

14.2.1.7   Calibration step 7 – 3GPP minimum requirements

Once the block-wise validation described above has been fulfilled, the entire simulation chain should be aligned to a valid reference for the particular system under consideration. In particular, the 3GPP LTE Release 9 framework can be used as a reference for the End-to-End (E2E) system performance comparison, by considering the minimum values provided in [13] and [20]. First, the simulator parameterization should be aligned with that of Common Test Parameters proposed in [13] and [20] and refer to the demodulation of PDSCH/PUSCH, in the following configurations: (1) single-antenna port, (2) transmit diversity, (3) open-loop spatial multiplexing and (4) closed-loop spatial multiplexing. For configuration, multiple references have been defined varying propagation conditions, Modulation and Coding Schemes (MCS) and number of used antennas. Furthermore, once the simulations have been performed, the obtained results should be compared with the minimum performance values reported in [13] and [19][20], for the PDSCH and PUSCH respectively.

14.2.1.8   Calibration step 8 – multi-link-level calibration

The calibration of the basic system blocks in a multi-link scenario is identical to the method for the single-link case described previously. This step focuses on the calibration of some of the most representative protocols for cooperative multi-hop transmission, which can be used as a starting point for further contributions on this topic. A comprehensive comparison of multi-hop protocols was carried out in [22]–[24], where the following protocols for two-hop relaying are considered: Amplify-and-Forward (AF) transmission, Decode-and-Forward (DF) transmission, and Decode-and-Reencode (DR) transmission. Analytical expressions to calibrate the outage probability of the different protocols can be found in [22] and [23]. Moreover, performance results for different end-to-end spectral efficiency values and relative positions of the relay are given. In addition, a calibration based on bit-error-rate performance can be done using the results in [24] as a reference.

14.2.2   System-level calibration

In the following, two calibration phases are presented. While the phase 1 intends to calibrate LTE, the phase 2 targets a more complex technology, LTE-Advanced.

14.2.2.1   Calibration phase 1 – LTE technology

The main assumptions are taken from 3GPP and ITU-R work [5][25], focusing on the urban micro-cell case, with carrier frequency 2.5 GHz, inter-site distance of 200 m, tilt 12 degrees and bandwidth 10 MHz for downlink and uplink, respectively. Table 14.2 summarizes the characteristics of the system.

For calibration, the following performance metrics need to be considered:

- Cell-spectral efficiency.
- Cell-edge user spectral efficiency.
- CDF of the normalized user throughput.
- CDF of the SINR. SINR will be collected after the MIMO decoder, thus resulting in a single SINR value per resource element allocated to the user. The final SINR per user is calculated as the linear average of all these values.

Table 14.2 Other simulation assumptions for calibration phase 1.

Issue	Assumption	Additional information
MIMO	1x2	Receiver diversity
Scheduling	Round Robin	
Cell selection	1 dB Handover margin	
Traffic model	Full Buffer	
Interference model	Explicit	
Channel state feedback	Realistic	5 ms / 5 RBs
SINR estimation	Perfect	
Feeder loss	2 dB	
Link-to-system model	Mutual Information Effective SINR Mapping	
Control overhead	3 OFDM symbols	
Receiver Type	MMSE	

Figure 14.4 shows the calibration for CDF of SINR and normalized user throughput. The cell spectral efficiency in this scenario is 1.2077 bps/Hz/cell, whereas the cell edge user spectral efficiency is 0.0267 bps/Hz.

14.2.2.2 Calibration phase 2 – LTE-Advanced with basic deployment

This calibration case is based on the assumptions made by 3GPP for the evaluation of IMT-Advanced [25]. The basic simulation assumptions for this calibration case can be found in [4][5], where the focus is on the urban micro-cell scenario with the same assumptions as above.

This scenario allows checking the validity of the implementation of spatial multiplexing using MIMO. In order to update this baseline system to the most updated version of 4G technologies, the assumptions summarized in Table 14.3 for the implementation of LTE-Advanced will be assumed.

For calibration purposes, Figure 14.5 shows the calibration material. The cell spectral efficiency in this scenario is 1.8458 bps/Hz/cell, whereas the cell edge user spectral efficiency is 0.0618 bps/Hz.

14.3   New challenges in the 5G modeling

This section describes a set of challenging characteristics in 5G modeling that require the development of new models for new features or a great increase in memory and/or computational complexity for an accurate evaluation. Some of these challenges have already been well addressed, like the use of real scenarios, the simulation of moving networks and the D2D link [1], while others are being tackled at this moment.

Table 14.3 Other simulation assumptions for calibration phase 2.

Issue	Assumption	Additional information
MIMO	4x2	Single User-MIMO scheme
Scheduling	Proportional Fair	Up to 5 users per subframe
		Priority to retransmissions
		Weight factor = 0.001
Cell selection	1 dB Handover Margin	
Traffic Model	Full Buffer	Other traffic models in a second round
Interference Model	Explicit	
Channel state feedback	Realistic	5 ms period / 5 RBs
SINR estimation	Perfect with synthetic error	error → lognormal 1 dB standard deviation
Feeder loss	2 dB	
Link-to-system model	Mutual Information Effective SINR Mapping	
Control overhead	3 OFDM symbols	
Receiver Type	MMSE	With inter-cell interference suppression capabilities

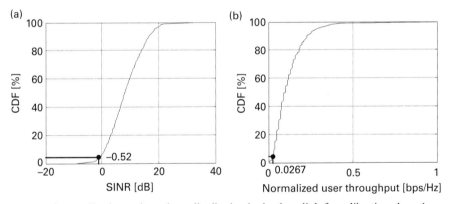

Figure 14.4 SINR and normalized user throughput distribution in the downlink for calibration phase 1.

14.3.1 Real scenarios

In order to obtain simulation results that reflect the system performance as precise as in reality, models exploited in simulation work should be able to reflect the characteristics of a real world scenario with a manageable computation complexity. Therefore, a simulation is only meaningful if it is established and aligned with its specific considered real scenario, in terms of environment model, deployment, propagation, traffic and mobility models. Invalidity of these models highly

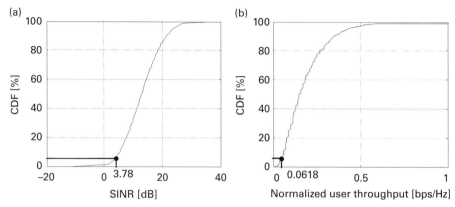

Figure 14.5   SINR and normalized user throughput distribution in the downlink for calibration phase 2.

contributes to lack of accuracy of simulation results, which motivates further strengthening those models that are not proper for 5G network evaluation. For instance, current channel models, such as ITU-R pedestrian and vehicular, lack the incorporation of elevation aspects with respect to transmit and receive antennas. For evaluating beyond 4G technologies, where massive and ultra-dense antenna deployments are to be expected, more precise and realistic channel models that take the elevation dimension and the resulting changes in radio propagation into account are required. It is demonstrated in literature [26] that very different results in terms of system performance gains can be obtained by different channel models. Hence, obtaining the appropriate modeling is crucial in order to get accurate results.

In order to check the capability to support new types of services, simulation and evaluation of different technology components will be conducted in 5G scenarios. Thus, current simulation assumptions need to be rephrased and aligned with the real scenarios, in order to provide a high validity of simulation results. Since strict communication reliability is required in some 5G scenarios, a high accuracy of simulation models is critical to obtain valid and meaningful simulation results. For instance, real scenarios should be captured in simulation as precisely as possible for emergency and traffic safety scenarios to avoid any misleading conclusion.

14.3.2   New waveforms

New waveforms are being proposed for 5G, such as Filter Bank Multi-Carrier (FBMC) and Universal Filtered Multi-Carrier (UFMC). Therefore, new physical layer abstraction must be provided to simulate the new characteristics of those waveforms in system-level simulations. In particular, the non-orthogonality and synchronism peculiarities that they have are quite different to previous physical layers. Research is being conducted on this topic and proposals are being published. For example, in [27] the focus in on FBMC although commonalities could be found with other waveforms.

14.3.3 Massive MIMO

Massive MIMO implies that the transmitter and/or the receiver has more than ten antennas. Typically, from tens to hundreds of antennas. In 4G, it is commonly assumed that only the base station has this high amount of antennas. However, in 5G both transmitter and receiver will be equipped with a high number of antennas.

In early phases of massive MIMO evaluations, the modeling can be simplified. For example, given a receiver with n_r antennas, the whole set of antennas could be replaced in simulation by a unique effective antenna with $10 \log_{10}(n_r)$ dB more gain than the antennas of the array. A more general approach is to consider that a whole antenna array could be represented in simulation by M effective antennas with $10 \log_{10}(n_r/M)$ dB more gain than the antennas of the original array. Each effective antenna would represent a set of n_r/M antennas of the original array. In this simple approach, the channel to each effective antenna would be equal to the channel of the central antenna of the set of replaced antennas. Another simplification that can be made is to assume that channels are uncorrelated, but in this case, it is suggested that the number of sets assumed is realistically chosen (based on literature reported values) in order to avoid overestimating the performance. If a more accurate analysis is required, channels between all the transmit antennas and all the receive antennas must be considered.

On the other hand, a realistic and accurate analysis of massive MIMO performance requires realistic channel models. Geometric-stochastic channel models are able to separate antenna and propagation effects. This fact simplifies evaluation of new transmission techniques based on different antenna configurations. The problem of many currently used models is that they lack a full-3D modeling and/or time-evolution. The former is needed to evaluate the gain of beamforming in elevation dimension for example, while the latter is needed for dynamic system-level simulations in which channel is tracked during a long period of time and not only over short snapshots. New models have been developed including these required characteristics. For example, 3GPP has extended the Spatial Channel Model Extended (SCME) to full dimension MIMO channel model in [28]. In addition, the QuaDRiGa channel model (see [29]) includes full-3D propagation modeling and time evolution. See Chapter 13 for a more detailed analysis of these challenges.

The huge size of the antenna arrays in massive MIMO has also important implications on the channel modeling. First, it may invalidate the assumption of plane wave propagation, i.e. considering only the phase difference of the signal between the antenna elements. Second, some propagation parameters that are usually considered fixed along the array may change in a non-negligible manner. Those problems have been studied and new channel models, like those presented in Chapter 13, solve these problems.

14.3.4   Higher frequency bands

The use of higher frequency bands, such as millimeter waves, opens a set of new challenges. Channel modeling is a first challenge for simulation at higher frequency bands. Ray tracing is a valid approach for real scenarios, or for synthetic scenarios with detailed description. Stochastic models are currently being developed as results from the measurements campaigns are being made available [30][31].

Frequency bandwidths available at higher frequency bands are considerably larger than those available at the frequencies used by 4G. If frequency sampling is kept equal in 5G simulations to the values used in 4G simulations, computational complexity could increase heavily. It can be noted that the delay spread found at higher frequencies is usually smaller than at lower frequencies, although the reported differences are not too big [32][33]. In scenarios where this frequency dependence is observed, larger frequency sampling than in 4G can be used without losing simulation accuracy. This is due to the higher coherence bandwidth that is inversely related to delay spread. The frequency sampling can be reduced in some evaluations without losing too much accuracy even if coherence bandwidth is not reduced. For example, consider an assessment with full buffer traffic model, a frequency-selective scheduler and 1 GHz bandwidth. It could be valid to divide the bandwidth in 100 portions of 10 MHz instead of considering portions of 180 kHz, as it is common in LTE. The rationale is that the consideration of 100 portions could be enough to get a multi-user diversity gain close to its maximum value. This option could not be valid with other kind of traffic models, especially with small packets, because in this case the use of big spectrum portions would be very inefficient. Obviously, results of a detailed evaluation should be compared to those of a less-detailed one in order to validate the proposed simplifications.

14.3.5   Device-to-device link

As a key enabler for 5G networks, D2D communication can be exploited, on one side, to offload traffic of cellular network, and on the other side, to decrease the latency. Depending on whether cellular resources are reused for D2D communication or not, mutual interference exists among cellular and D2D links or among D2D links, where the same physical resources are reused. In order to obtain reliable simulation results, the interference channel should be modeled properly. Since positions of D2D users are generated within runtime, path loss values for the above-mentioned interference links are hard to be pre-cashed due to the extreme long calculation time and large storage capacity.

Meanwhile, special consideration should also be taken into account regarding traffic models of D2D links, depending on the concrete environment, i.e. whether the D2D link is established between vehicles or pedestrians. The traffic model applied in simulations has critical impact on the evaluation of smart scheduling schemes for D2D communication.

Moreover, when V2X communication is evaluated under the scope of traffic safety and efficiency scenario, different simulation models compared with D2D communication should be defined. For instance, a different channel propagation model is foreseen due to the different operating frequency and lower antenna site as compared with cellular base station. Besides, the traffic model in this case specially focuses on a small package transmission with certain periodicity. It should be noticed, due to the special traffic model for V2X communication, that links generating interference will be changed very frequently. Therefore, due to the high reliability requirement of V2X communication, interference links should be properly generated for each package retransmission.

14.3.6   Moving networks

In legacy systems vehicles bring a penetration loss for their inside-vehicle users. However, due to a better computation capability, a larger antenna dimension and more storage space, communication facilities deployed in vehicles are more advanced than current mobile user devices. Therefore, vehicles play an important role for 5G networks to improve overall system performance. Advanced communication infrastructures enable every vehicle to have the flexibility to act as an access node for its carried users and users in the proximity. Vehicles can be considered in this case with antennas equipped on top. When simulation is carried out, the rule of vehicle infrastructure should be properly reflected. For instance, if information required by a user is already cashed by one vehicle located in proximity of this user, a V2X communication link is established to offload traffic from cellular network. Moreover, the vehicle acts as a transmitter and forwards the information directly to the user. In another case where required information by outdoor users is not cashed locally, both the wireless backhaul link between base station and vehicle and the link between vehicle and users should be simulated and aligned with legacy evaluation methodology. It should be noted that interference should be properly modeled if in-band wireless backhaul is exploited. Regarding inside-vehicle passengers, the vehicle penetration loss can be efficiently avoided by exploiting antennas deployed on top of the vehicle. The communication link between one vehicle and its carried passengers can be solved very well by current technologies and therefore does not need to be simulated.

14.4   Conclusions

This chapter has covered simulation methodology for 5G, listing the key performance indicators of interest, and proposing methodologies to simplify channel modeling and ensure the calibration of link and system-level simulators for a fair comparison of 5G concepts.

It has been pointed out that there are various challenges related to the simulation of a 5G system, such as the performance evaluation of novel waveforms or higher carrier frequencies, the modeling of massive MIMO, D2D communications or moving networks. However, various means have been presented to simplify the modeling of these aspects and keep the complexity of simulator development (and of simulation itself) tractable.

References

[1] ICT-317669 METIS project, "Simulation guidelines," Deliverable D6.1, November 2013, www.metis2020.com/documents/deliverables/

[2] ICT-317669 METIS project, "Scenarios, requirements and KPIs for 5G mobile and wireless system," Deliverable D1.1, May 2013, www.metis2020.com/documents/deliverables/

[3] International Telecommunications Union Radio (ITU-R), "Requirements related to technical performance for IMT-Advanced radio interface(s)," Report ITU-R M.2134, December 2008, www.itu.int/pub/R-REP-M.2134-2008

[4] INFSO-ICT-247733 EARTH project, "Most suitable efficiency metrics and utility functions," Deliverable D2.4, January 2012, www.ict-earth.eu/publications/deliverables/deliverables.html/

[5] International Telecommunications Union Radio (ITU-R), "Guidelines for evaluation of radio interface technologies for IMT-Advanced," Report ITU-R M.2135, December 2008, www.itu.int/pub/R-REP-M.2135-2008

[6] J. Medbo and F. Harrysson, "Channel modeling for the stationary UE scenario," in European Conference on Antennas and Propagation, Gothenburg, April 2013, pp. 2811–2815.

[7] International Telecommunications Union Radio (ITU-R), "Propagation data and prediction methods for the planning of short-range outdoor radiocommunication systems and radio local area networks in the frequency range 300 MHz to 100 GHz," Report ITU-R P.1411, October 1999, www.itu.int/rec/R-REC-P.1411-0-199910-S

[8] CELTIC / CP5-026 WINNER+ project, "Final Channel Models," Deliverable D5.3, June 2010, http://projects.celtic-initiative.org/winner+/deliverables_winnerplus.html/

[9] J. Proakis and M. Salehi, *Digital Communications*, 5th ed., New York: McGraw Hill, 2007.

[10] Nokia, "Ideal simulation results for PDSCH in AWGN," Work Item R4-071640, 3GPP TSG RAN WG4, Meeting #44bis, October 2007.

[11] 3GPP TS 36.521-1 V8.1.0, "User Equipment (UE) conformance specification Radio transmission and reception. Part 1: Conformance Testing," Technical Specification TS 36.521-1 V8.1.0, Technical Specification Group Radio Access Network, March 2009.

[12] Motorola, "UE demodulation simulation assumptions," Work Item R4-072182, 3GPP TSG RAN WG4, Meeting #45, November 2007.

[13] 3GPP TS 36.101 V9.22.0, "User Equipment (UE) radio transmission and reception," Technical Specification, TS 36.101 V9.22.0, Technical Specification Group Radio Access Network, March 2015.

[14] Ericsson, "Results collection UE demod: PDSCH with practical channel estimation," Work Item R4-080538, 3GPP TSG RAN WG4, Meeting #46, February
2008.

[15] Motorola, "Agreed UE demodulation simulation assumptions," Work Item R4-
071800, 3GPP TSG RAN WG4, Meeting #44bis, October 2007.

[16] Ericsson, "Collection of PDSCH results," Work Item R4-072218, 3GPP TSG
RAN WG4, Meeting #45, November 2007.

[17] Nokia, "Summary of the LTE UE alignment results," Work Item R4-082151,
3GPP TSG RAN WG4, Meeting #48, August 2008.

[18] Nokia, "Summary of the LTE UE alignment results," Work Item R4-082649,
3GPP TSG RAN WG4, Meeting #48bis, October 2008.

[19] Nokia Siemens Networks, "PUSCH simulation assumptions," Work Item R4-
080302, 3GPP TSG RAN WG4, Meeting #46, February 2008.

[20] 3GPP TS 36.104 V8.5.0, "Base Station (BS) radio transmission and reception,"
Technical Specification TS 36.104 V8.5.0, Technical Specification Group Radio
Access Network, December 2009.

[21] Ericsson, "Summary of Ideal PUSCH simulation results," Work Item R4-072117,
3GPP TSG RAN WG4, Meeting #45, November 2007.

[22] P. Herhold, E. Zimmermann, and G. Fettweis, "Cooperative multi-hop transmission
in wireless networks," *Computer Networks Journal*, vol. 3, no. 49, pp. 299–324,
October 2005.

[23] E. Zimmermann, P. Herhold, and G. Fettweis, "On the performance of cooperative
relaying protocols in wireless networks," *European Transactions on
Telecommunications*, vol. 1, no. 16, pp. 5–16, January 2005.

[24] E. Zimmermann, P. Herhold, and G. Fettweis, "On the performance of cooperative
diversity protocols in practical wireless systems," in IEEE Vehicular Technology
Conference, Orlando, October 2003.

[25] 3GPP TR 36.814 V2.0.1, "Further advancements for E-UTRA physical layer
aspects," Technical Report TR 36.814 V2.0.1, Technical Specification Group
Radio Access Network, March 2010.

[26] Ericsson, ST-Ericsson, "Elevation Angular Modelling and Impact on System
Performance," Work Item R1-130569, 3GPP TSG RAN WG1 Meeting #72,
February 2013.

[27] G. Wunder et al., "System-level interfaces and performance evaluation methodology for 5G physical layer based on non-orthogonal waveforms," in Asilomar
Conference on Signals, Systems and Computers, Pacific Grove, November 2013,
pp. 1659–1663.

[28] Timothy A. Thomas, Frederick W. Vook, Evangelos Mellios, Geoffrey S. Hilton,
and Andrew R. Nix, "3D Extension of the 3GPP/ITU Channel Model," in IEEE
Vehicular Technology Conference, Dresden, June 2013.

[29] S. Jaeckel, L. Raschkowski, K. Borner, and L. Thiele, "QuaDRiGa: A 3-D
multi-cell channel model with time evolution for enabling virtual field trials,"
IEEE Transactions on Antennas and Propagation, vol. 62, no. 6, pp. 3242–3256,
June 2014.

[30] M. R. Akdeniz, L. Yuanpeng, M. K. Samimi, S. Shu, S. Rangan, T. S. Rappaport,
and E. Erkip, "Millimeter wave channel modeling and cellular capacity

evaluation," *IEEE Journal on Selected Areas in Communications*, vol. 32, no. 6, pp. 1164–1179, June 2014.

[31] T. A. Thomas, H. C. Nguyenm, G. R. MacCartney, and T. S. Rappaport, "3D mmWave channel model proposal," in IEEE Vehicular Technology Conference, Vancouver, September 2014, pp. 1–6.

[32] T. Rappaport et al., "Millimeter wave mobile communications for 5G cellular: It will work!," *IEEE Access*, vol. 1, pp. 335–349, May 2013.

[33] T. Rappaport et al., "38 GHz and 60 GHz angle-dependent propagation for cellular and peer-to-peer wireless communications," in IEEE International Conference on Communications, Ottawa, June 2012, pp. 4568–4573.

Index